THE MICROFLOW CYTOMETER

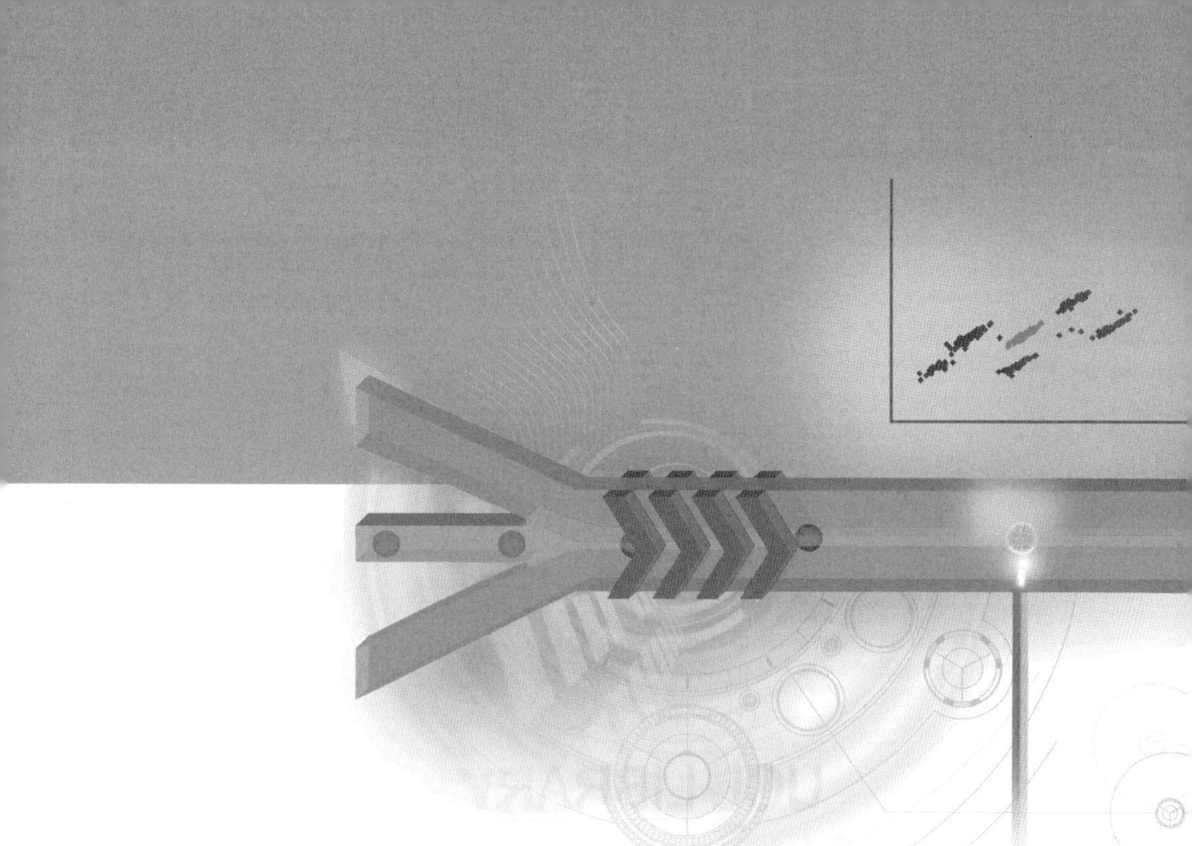

THE MICROFLOW CYTOMETER

Frances S. Ligler
Jason S. Kim

Naval Research Laboratory, USA

PAN STANFORD PUBLISHING

Published by

Pan Stanford Publishing Pte. Ltd.
Penthouse Level, Suntec Tower 3
8 Temasek Boulevard
Singapore 038988

Email: editorial@panstanford.com
Web: www.panstanford.com

British Library Cataloguing-in-Publication Data
A catalogue record for this book is available from the British Library.

THE MICROFLOW CYTOMETER

ISBN-13 978-981-4267-41-0
ISBN-10 981-4267-41-4

Printed in Singapore.

Preface

While there are numerous current volumes and journal articles on applications of flow cytometry, we could find few recent compendia focused on advances in flow cytometers. Clearly, flow cytometers are becoming smaller and more geared toward special-purpose applications and less sophisticated operators. Yet as potential system developers, we had to scan the literature in microfluidics, optics, electronics, and nanotechnology to assemble information on the state-of-the-art. The dissatisfaction and frustration resulting from our search for a digest of progress in flow cytometry produced the concept for this book. Our search for the leaders in each of the relevant sub-areas produced the selection of chapter authors who have kindly contributed their perspectives on the future challenges and opportunities for realization of microflow cytometers.

For the scientists and engineers interested in the future of flow cytometers, the following chapters describe the continuing development of inexpensive, portable flow cytometers through incorporation of microfluidic technologies and small optical components. The underlying microscale theories essential for microflow cytometry are discussed, as well as advances that are representative of the current state-of-the-art. Innovative component technologies and integration of the components into functional prototype devices are reviewed with a goal of automated analysis and manipulation of particles and cells. Currently available commercial "personal cytometers" are examined to highlight both strengths and areas for necessary improvement.

Chapters included are from prominent scientists and engineers, including Howard Shapiro — a keystone in flow cytometry from the start of the technology, Michael Ladisch — past chair of the Bioengineering Section of the US National Academy of Engineering, Wayne Roth and Colin Rich — corporate leaders in industrial development and manufacture of benchtop flow cytometers, and John Dzenitis — project leader for the BioWatch version 2 biosurveillance system. Other chapters by leading scientists focus on technical breakthroughs critical for next-generation systems.

We hope you enjoy the compilation of the technologies that we think will spur future development, as well as the lessons learned from current developers of flow cytometry instrumentation. Perhaps you will discover a "missing link" after reading this book that will revolutionize future microflow cytometers. If that is the case, we wish you a satisfying and fruitful future in flow cytometry.

With best regards to our readers.

Jason Kim and Fran Ligler
Naval Research Laboratory
October 2009

Contents

Chapter One

A History of Flow Cytometry and Sorting

Howard Shapiro
The Center for Microbial Cytometry and Howard M. Shapiro, MD, PC
283 Highland Avenue, West Newton, MA 02465-2513, USA
hms@shapirolab.com

Both the microfluidic and metrologic aspects of flow cytometry can be traced directly to Leeuwenhoek and Hooke's work in the late 1600s. Staining and improvements in microscopy made in the mid-1800s led to better understanding of what cells were and how they worked, but it was not until the mid-20th century that it became possible to perform accurate and precise quantitative analyses of cells in flow systems. Modern improvements in flow cytometry still owe more to progress in electronics and electro-optics than to advances in microfluidics.

1.1 INTRODUCTION

A flow cytometer examines a small volume of fluid and makes physical measurements to detect and characterize even smaller particles, which may or may not be cells, contained therein. A flow sorter has the added capability of separating particles with characteristics preselected by the user. The instruments themselves are, for the most part, designed and built by physical scientists and engineers to meet the needs of biomedical scientists and clinicians, and the transmission of information between the former and latter groups is not always as good as it might be. When I write about the history of the technology, which I have done several times in the past few years,[1–3] I try to consider what was done at each step in its evolution in the context of both what motivated the developers and what options were available to them at the time. Since the technology of cytometry, like so many other technologies, is now evolving far more rapidly than it was even a few decades ago, I find that my views of both the past and the future change with each passing year. This has thus far kept me from becoming bored and, I hope, from boring my readers. One can get new ideas from reading old stories; this may

The Microflow Cytometer by J S Kim & F S Ligler

www.panstanford.com
978-981-4267-41-0

be yet another reason to heed the admonition of the philosopher and poet George Santayana that "Those who cannot remember the past are condemned to repeat it."

1.2 MICROSCOPY, CELLS, AND CYTOMETRY — IN THE 1600s!

Although the use of lenses to correct vision came into practice in Italy around the Twelfth Century, it was not until the turn of the Seventeenth Century that spectacle makers there and in Holland used paired lenses to bring faraway objects closer, thereby inventing the telescope, and to bring nearby objects otherwise too small to see into view, inventing the microscope. By 1614, Galileo had described the appearance of a fly under the microscope;[4] it was earlier work with his telescope, rather than this foray into biology, that got him into trouble with the Establishment.

The term "cell" was appropriated from the work of Robert Hooke, whose *Micrographia*,[5] originally published in 1665, used the term to describe empty spaces visible in a thin slice of cork. The spaces themselves were bounded by what would now readily be recognized as the remnants of cell walls; Hooke's book did not describe or illustrate single cells, and there is no evidence that he saw them until the 1670s. At that time, he was asked to verify reports made in letters communicated to the Royal Society in London by Antonie van Leeuwenhoek, an industrious autodidact from Delft, Holland, who had observed blood cells, sperm, protozoa, and bacteria using small microscopes containing only a single lens but providing several times higher magnification than was available from compound microscopes used at the time by Hooke and others.[4]

To examine cells in liquid media, van Leeuwenhoek would draw up a small amount of specimen in a glass tube he had drawn "fine as a hair"; Hooke followed his lead, and thus we can attribute the first use of microfluidics in cytometry to the two men acknowledged as the fathers of the field (of the two, van Leeuwenhoek was the only acknowledged sperm donor). Both men took pains to report the size of objects they observed, and the number contained in a given volume of specimen, establishing them as true cytometrists; however, although Hooke would report his size measurements in inches, which were at least theoretically standardized in England, van Leeuwenhoek's Holland had no established standards of length, and he would instead typically compare the size of objects he observed to the size of a grain of coarse sand or the eye of a large louse.

When *Micrographia*[5] was published, very little was known of chemistry, but the atomistic theories of the Greeks had begun to come back into favor, and Hooke and his contemporaries, although aware of the limitations of their own microscopes, optimistically envisioned improvements in optics that would let them view objects far smaller than they could then see. Henry Power, referred to by Hooke as "the ingenious physitian", had published his own account of *Microscopicall Obseruations* in 1661, and, anticipating my own cytometric writing style by several hundred years, produced a lengthy verse *In Commendation of ye Microscope*,[6] asserting:

"Of all th' Inuentions none there is Surpasses
the Noble Florentine [presumably Galileo]'s Dioptrick glasses.
For what a better, fitter, guift Could bee in this world's Aged Luciosity.
To Helpe our Blindnesse so as to deuize
a paire of new & Artificiall eyes.
By whose augmenting power wee now see more
then all the world Has euer donn Before.
'Thy Atomes (Braue Democritus) are now
made to appeare in bulk & figure too.
When Archimide by his Arithmatick,
numbred the sands, had hee But knowne this trick.
Wee might haue seene each corn a massy stone,
& counted them distinctly one by one."

1.3 THE 1800s — CELL THEORIES, STAINING, AND BETTER MICROSCOPY

For almost 200 years from the time of Hooke and Leeuwenhoek, microscopists remained motivated primarily by intellectual curiosity rather than by a need or desire to identify causes and mechanisms of, or cures for, human diseases. Today's students of biology may learn the names of Matthias Schleiden and Theodor Schwann as the principal proponents of the theory that all living things are composed of cells, and medical students may hear of Rudolf Virchow as having set pathology firmly on a cellular foundation. Henry Harris's *The Birth of the Cell*[4] paints a more accurate picture of the larger cast of characters involved and the conflicts and controversies that arose as they sought to understand cell structure and function.

Until the mid-1800s, the optical quality of microscopes was relatively poor, making it difficult to distinguish structures in cells and tissues from artifacts, especially when the material being observed did not contain either pigments or constituents which differed significantly from one another in refractive index. Virchow defined leukemia as an excess of white blood cells, or leukocytes, in the 1850s, but it was not until the late 1870s that staining of cells, advanced considerably by Paul Ehrlich's experiments as a medical student with newly synthesized aniline dyes, made it much easier to identify and distinguish different white cell types. The task was further facilitated by the improved microscope optics developed by Zeiss and other manufacturers. Nonetheless, it was not until the 1880s that it had become accepted that cells gave rise to new cells only by mitotic division, and the role of the chromosomes in heredity was not elucidated until the next century.

Louis Pasteur and Robert Koch used microscopy to establish specific microbial causes for diseases such as anthrax, tuberculosis, and cholera, all of which, for one reason or another, remain of concern even in the modern world (the decidedly unmodern *Microbe Hunters*[7] is still worth reading for this story), and Ehrlich and Koch, working together, developed practical stains to distinguish mycobacteria, which cause tuberculosis and leprosy, from other bacterial species. Inspired by them, Christian Gram developed the staining technique that, although now considerably modified, still bears his name and is universally used to classify bacteria.

Until the mid-1800s, there were no objective means of recording the results of observations made using microscopy; even drawings made with the aid of a "camera lucida" that projected the image onto the drawing surface were subject to influence by an observer's perceptions. Daguerrotype photography was combined with microscopy in the 1840s, and, a few decades later, Koch championed the use of photography, which had by then seen considerable technical improvement, but there were no practical ways to quantify light intensity for some time thereafter.

Hooke and the other early users of compound microscopes more often than not illuminated specimens obliquely, visualizing them by reflected rather than by transmitted light. Simple microscopes such as those van Leeuwenhoek used were typically held up to the light, permitting transmitted light observation, although some of van Leeuwenhoek's descriptions of objects as light against a dark background suggest that he also used oblique illumination. The microscopes Pasteur, Koch, and Ehrlich used were fitted with substage condenser lenses, providing sufficient illumination for transmitted light microscopy. The Sun remained the most effective light source; one could arrange lenses as Hooke had to provide illumination from a lamp, but it was not until the late 1800s that bright artificial sources such as gas lamps with mantles, limelight, and carbon arcs became usable.

In the 1880s, Alphonse Laveran discovered the protozoan parasites that cause malaria, overcoming the considerable difficulty of detecting them in unstained specimens. Shortly thereafter, darkfield condensers became available, allowing carefully controlled oblique illumination to be delivered to a slide in such a way that only light scattered by material in the sample reached the eyepiece. This facilitated visualization of unstained material, and, because substantial amounts of light may be scattered from particles well below the resolution limit of transmitted light microscopy, also allowed small, formerly submicroscopic objects to be detected, although it was not possible to resolve their internal structure.

Ehrlich had established the principle of mixing acidic and basic dyes of different colors to stain blood leukocytes; this allowed him to distinguish three types of cells bearing cytoplasmic granules (which were therefore called granulocytes and also described as polymorphonuclear because their nuclei typically had multiple lobes) from mononuclear cells, later subclassified as lymphocytes (because they were the predominant cells in lymph) and monocytes. Granulocytes were described as basophilic, if they stained predominantly with the basic dye in the mixture, as acidophilic, if they stained predominantly with the acidic dye, and, because Ehrlich held the opinion that molecules of acidic and basic dyes combined to form a "neutral stain", as neutrophilic, if they stained with both dyes.

In the 1890s, a number of people discovered that the combination of the acid dye eosin and a chemically modified basic dye, methylene blue, greatly facilitated visualization of malaria parasites in red blood cells. Among such dye mixtures, that developed in 1904 by Gustav Giemsa became most popular. It is still the standard stain for identification of malaria parasites by microscopy and remains widely used for morphologic hematology, although routine clinical measurement of different blood cell types is now done by flow cytometry anywhere the technology is affordable and the infrastructure is adequate to support it. The popularity of Giemsa and related stains led to a change in granulocyte nomenclature; acidophil granulocytes are now called eosinophils.

By the mid-1800s, physicians had become interested in enumerating the numbers of cells contained in a given volume of blood. In order to keep cells well separated enough for counting and to keep the number actually counted low enough for an observer to accomplish the task in a reasonable time, it was necessary to dilute the blood. The earliest attempts at hemocytometry were made by depositing a known volume of diluted blood in a small area on a slide and counting all the cells in that volume; later, cells were counted in a capillary tube, with the volume analyzed being estimated with the aid of a ruled eyepiece. In 1877, Gowers described a counting chamber comprised of a glass slide with a grid ruled on its surface, and ridges at the edges of the observation area that supported the cover glass at a known height above the slide, simplifying the task of counting cells in a defined volume of diluted blood.[8] Variants of this hemocytometer design are still used for cell counting. In 1907, the statistician William Sealy Gossett, publishing under the pseudonym "Student" to prevent competitors of his employers at the Guinness Brewery from discovering the utility of statistics, showed that the theoretical minimum error of hemocytometer counts varies with the number of cells actually counted according to what are now generally known as Poisson statistics.[9] The same counting statistics also apply to other quantities encountered in cytometry, notably the photoelectrons generated by interaction of light emitted and scattered from cells with the detectors used in cytometric apparatus.

1.4 THE EARLY TWENTIETH CENTURY — ULTRAMICROSCOPY AND EINSTEIN

When Albert Einstein was awarded the 1921 Nobel Prize in Physics, he was cited "for his services to Theoretical Physics, and especially for his discovery of the law of the photoelectric effect." The emission of electrons from certain materials under the influence of light had been explained on the basis of light consisting of quanta in one of the four extraordinary papers Einstein published in *Annalen der Physik* in 1905; a second paper correctly attributed the Brownian motion of small particles in suspension to random atomic collisions (the third and fourth introduced relativity and the equivalence of matter and energy).

The motion of small particles was of great interest to chemists studying colloids, and one such chemist, Richard Zsigmondy, collaborating with Heinrich Siedentopf of the Zeiss works, had developed a highly sensitive darkfield microscope, termed an "ultramicroscope",[10] in 1903. This and Einstein's work allowed particles with dimensions of a few nanometers to be detected, and their masses estimated on the basis of the amplitude of excursions during their random motions, providing the most concrete evidence for the existence of atoms many scientists had encountered up to that time. Zsigmondy received the Nobel Prize in Chemistry in 1925 for his work, and ultramicroscopy became a useful tool for analysis of both colloidal solutions and aerosols. The apparatus itself was thereafter modified to permit an aerosol sample to be flowed through it intermittently, allowing particles to be observed and counted in a chamber of defined volume;[11] it was noted that, as "Student" had found,[9] the counts obeyed Poisson statistics.

Although photocells, i.e., devices that either produced electric current or changed their electrical properties in response to illumination, had been known since the late 1800s, Einstein's work probably contributed to their development and commercialization as devices for light measurement. He himself was the co-inventor of a camera in which exposure was automatically controlled by motion of a variable neutral density filter in front of the lens in response to an amplified signal from a photocell.[12]

A 1934 paper by Andrew Moldavan in *Science*,[13] while widely cited as the first publication on flow cytometry, appears to summarize an unsuccessful attempt to count cells flowing through a capillary tube using a photocell attached to the microscope eyepiece. Moldavan listed several obstacles to development of a practical apparatus, including the necessity to standardize capillary tubes and improve their optical characteristics, the limited sensitivity of available photodetectors, and the undesirable effects of cell clumping within the capillary.

1.5 WORLD WAR II TO VIETNAM — MAKING FLOW CYTOMETRY WORK

1.5.1 Gucker's Counter for Bacteria

The first working flow cytometer was described in 1947 by Gucker *et al.*[14] who reported success in flow cytometric detection of bacteria in aerosols; the work, done during World War II, was sponsored by the U.S. Army with the aim of rapid identification of biological warfare agents. The apparatus, clearly a descendant of the ultramicroscope, incorporated a sheath of filtered air to confine the air sample stream to the central portion of the flow chamber, in which it was subjected to dark-field illumination. The light source was a Ford headlight; a photomultiplier tube, then a newly developed device, was introduced as a detector. The instrument had about 60 percent probability of detecting a particle 0.6 μm in diameter. The Army has continued to fund work on flow cytometric detection of airborne microbial pathogens using more specific identification methods.

1.5.2 Optical and Electronic Blood Cell Counters

The sheath flow principle used in the Gucker aerosol counter was adopted by Crosland-Taylor[15] in the early 1950s for a blood cell counter in which cells in sheath flow were detected by light scattering with dark-field illumination. During this period, several industrial organizations in England, Germany, and the United States developed or attempted to develop similar photoelectric counters.

One American electrical engineer pursuing this goal (Wallace Coulter, personal communication) encountered some problems with optics and explored another means of cell detection, based upon the fact that the electrical conductivity of cells is lower than that of saline solutions. Coulter reasoned that blood cells, suspended in a saline solution and passing one at a time through a small (<100 μm) orifice, would be detectable by the transient increases in the electrical impedance of the orifice produced as the nonconducting cells passed through, displacing the conducting saline. The Coulter counter[16] proved accurate for counting[17] and sizing[18] blood cells and other particles and apparatus based on the principle is now used worldwide.

In 2009, when computers a thousand times as powerful as the multimillion-dollar mainframes of the 1960s are smaller than a phone book and cost only a few hundred dollars, with memory prices below US$10 per gigabyte and hard drives available for less than $100 per terabyte, it may not be easy for most readers of this book to contemplate the practical difficulties encountered in producing even a relatively simple flow cytometer, such as a blood cell counter, that could be used in laboratories by technicians without an engineering background. Both optical and electronic counters produced electrical pulses as cells passed through the measurement system; early experimenters would amplify the detector output and use it to drive a chart recorder, through which a known length of paper would pass per unit time. The volume of blood analyzed per unit time could be computed from the flow rate of diluted sample and the dilution factor, and the number of cells per unit volume could then be obtained by counting the number of pulses visible on the chart trace. Gucker *et al.*[14] had used an electromechanical counter to eliminate the labor-intensive process just described; the earliest Coulter Counters instead employed cascaded "Dekatron"-type vacuum tubes, each of which could both accumulate a count between zero and nine and indicate the count by the position of a glowing spot on the circular end face of the tube. "Nixie" tubes, which could accept a decimal digital output from an electronic counter and illuminate an internal electrode in the shape of the corresponding digit, became available in the mid-1950s; both forms of display were subsequently replaced by single- and multidigit LED- and LCD displays, which have themselves largely been supplanted by LCD screens.

The original blood cell counters performed red cell counts on blood diluted with isotonic saline. A microliter of blood typically contains approximately 5 million red cells, accompanied by 100,000 to 200,000 platelets and 5,000 to 10,000 leukocytes. Platelets are considerably smaller than red cells and leukocytes; the instrument threshold could therefore be set high enough so that they would not be counted. Leukocytes could be counted with red cells, since this would normally yield a result only a few tenths of one per cent above the true value. It had been established in the early days of hemocytometry that red cells could be lysed, and leukocytes preserved, by diluting a blood sample with a hypotonic medium or with chemicals such as acids or detergents, and the same dilution procedure was adapted to perform leukocyte counts in flow cytometric counters.

By the time these counters had become available, however, hematologists were aware that there was some variation in size even in normal red cells, and that cells from anemic patients might be smaller or larger than normal, depending on the cause of the anemia. A reasonably accurate value for average red cell size could be derived from the red cell count and the hematocrit, a measure of the fraction of the blood volume occupied by red cells, obtained by centrifuging a small volume of blood in a capillary tube and taking the ratio of the distances between the bottom of the tube and the top of the overlying column of packed red cells and between the bottom of the tube and the plasma meniscus at the top of the sample.

1.5.3 Approaches to Cell Heterogeneity: Pulse Height Analysis

The procedure described above, however, could not provide any information about the degree of heterogeneity of cell size within the red cell population, which had to be estimated by visual examination of blood smears under the microscope. Since the amplitudes of pulses produced by the passage of cells through a Coulter Counter were known to be proportional to cell volume, it was possible to obtain a cell volume distribution by connecting the output of the detector electronics to a pulse height analyzer, a special purpose computer that would convert pulse amplitude to a digital value and increment a corresponding memory location, with the result that, after a sample had been run, the distribution stored in memory could be displayed on an oscilloscope screen or written out on a chart recorder. The pulse height analyzer used for the original demonstration of electronic blood cell sizing[18] had been designed for atomic nuclear spectroscopy; Coulter Electronics soon brought out its "Channelyzer", designed to work with Coulter Counters, while other flow cytometer manufacturers would incorporate analyzers made by providers of nuclear instrumentation.

1.5.4 Pap Smears and Diff Counts: Scanning Approaches

By the 1940s, it had become established that examination of exfoliated cells from the cervix stained with a multicomponent dye mixture developed by George Papanicolaou in the late 1920s could reveal the presence of cancer.[19] A biochemical basis for identification of premalignant and malignant cells was also established, beginning in the 1930s, by Torbjörn Caspersson,[20] who used ultraviolet (UV) microspectrophotometry to demonstrate that such cells typically contained more DNA and RNA, both of which have strong absorption peaks at 260 nm, than did normal cells. In 1951, Mellors and Silver[21] proposed construction of an automatic scanning instrument for screening "Pap" smears, and such an apparatus, the Cytoanalyzer, was described by Tolles in 1955.[22] The Cytoanalyzer measured absorption of Papanicolaou-stained cells, rather than fluorescence, as had been suggested by Mellors and Silver; although results were encouraging enough for the American Cancer Society and the National Cancer Institute to continue funding research on cytology automation; false positive and false negative rates were too high for the instrument to be suitable for clinical use.

Because both Pap smear analysis and differential leukocyte counting (the "Diff Count" reports the percentages of neutrophils, eosinophils, basophils, lymphocytes, and monocytes in peripheral blood) utilized stains containing mixtures of relatively nonspecific dyes, identification of different cell types by either procedure required an observer to examine specimens under high magnification in order to determine sizes and shapes of cells and nuclei, the presence or absence of different types of granules in cytoplasm, etc. The obvious approach to automation involved computer analysis of high-resolution images and a number of groups were active in this area. The TICAS system, assembled in the late 1960s, interfaced Zeiss's commercial version of the Caspersson microspectrophotometer to a minicomputer, with the aim of automating Pap smears,[23] and Prewitt and Mendelsohn[24] and Preston and Ingram[25], among others, published work on differential counting.

In 1969, my colleagues and I at NIH built "Spectre II",[26] which incorporated a galvanometer mirror scanner and was the first microscope imaging system to incorporate interactive computer-controlled stepping motor-driven stage motion, focus, and illumination wavelength and intensity selection. The controlling computer was a Digital Equipment Corporation LINC-8 minicomputer, with only 12 K words of 12-bit memory, which would not have held data from a single 256 × 256 pixel, 8-bit scan; our high-resolution (0.2 μm pixels) cell image data were recorded on 9-track tape and transported (by "sneakernet") to a mainframe elsewhere on the NIH campus for analysis.[27] Although we could locate and mark the positions of cells of interest on a slide fairly rapidly, allowing the apparatus to scan them without operator intervention at a later time, it was obvious to us that collecting data at three wavelengths from several thousand cells, which we felt would be the minimum need for algorithm development, would occupy a great deal of time at two minutes per scan per wavelength. We were therefore careful to keep informed about work in flow cytometric analysis of cervical and blood cells that had begun at IBM Corporation a few years before our project started.

1.5.5 *Kamentsky's Rapid Cell Spectrophotometer; Cell Sorting*

In the early 1960s, promising results obtained with the Cytoanalyzer in attempts to automate reading of Pap smears[22] encouraged executives at IBM to look into producing an improved instrument. Assuming this would be some kind of image analyzer, they gave technical responsibility for the program to Louis Kamentsky, who had recently developed a successful optical character reader. He did some calculations of what would be required in the way of light sources, scanning rates, and computer storage and processing speeds to solve the problem using image analysis, and concluded that a different approach would be required.

Having learned from pathologists in New York that cell size and nucleic acid content could provide a good indicator of whether cervical cells were normal or abnormal, Kamentsky traveled to Caspersson's laboratory in Stockholm and learned microspectrophotometry. He then built a microscope-based flow cytometer that used a transmission measurement at visible wavelengths to estimate cell size and a 260 nm UV absorption measurement to estimate nucleic acid content.[28] Subsequent versions of this instrument, which incorporated a dedicated IBM 1130 16-bit minicomputer system, could measure light scattering and fluorescence as well as absorption and store values of as many as four cellular parameters for each cell.[29] A brief trial on cervical cytology specimens indicated the system had some ability to discriminate normal from abnormal cells;[30] it could also produce distinguishable signals from different types of cells in blood samples stained with a combination of acidic and basic dyes, suggesting that flow cytometry might be usable for differential leukocyte counting.

In the 1960s, even minicomputers, which cost tens of thousands of dollars at the time, were too expensive to be incorporated into scientific or medical apparatus by anyone not working for either the government or a computer manufacturer. My colleagues and I and Kamentsky were lucky enough to fall into the privileged group. Even at that time, it was obvious that the solution of nontrivial problems in cell discrimination would require measurement of multiple parameters, whether in imaging or flow systems, and that it would be necessary to examine more than one parameter at a time to reach a robust solution. The pulse height analyzer, which represented the most sophisticated data analysis system then in use in flow cytometry, was restricted to single-parameter analysis, although two-parameter versions, typically priced higher than many minicomputers, did become available by the end of the decade. The best that most people were able to do in terms of simultaneous display of two parameter values was to generate a dot plot, or "cytogram," on the screen of a storage oscilloscope as the sample was run, and preserve the data in a Polaroid photograph.

The minicomputer, which could store multiparameter data in what came to be called "list mode", provided a great deal more flexibility for data analysis. It did not, however, solve a central problem in developing cell identification methodology using flow cytometry, which was that cells normally went down the drain a fraction of a second after they were analyzed, which precluded their being examined to determine whether they had been correctly identified.

The first, and still most widely used, solution to that problem was described by Mack Fulwyler, then at Los Alamos National Laboratory, in 1965,[31] and was based on ink jet printer technology then recently developed by Richard Sweet[32] at Stanford. Following passage through the cytometer's measurement system, the saline sample stream was broken into droplets, and those droplets that contained cells with selected measurement values were electrically charged at the droplet breakoff point. The selected charged droplets were then deflected into a collection vessel by an electric field. Fulwyler's first sorter was built onto a Coulter volume measurement system; analyses of red cells had revealed a bimodal distribution, and the finding that cells sorted from the upper and lower peaks of the distribution and run a second time through the apparatus again yielded a bimodal distribution confirmed that the bimodal pattern was an artifact. Kamentsky's original approach to sorting used a syringe pump to withdraw fluid from the sample stream[33]; he later experimented with both electrostatic deflection and acoustic switching.

Leonard Ornstein and Regina O'Brien, who had worked on microspectrophotometry at various New York institutions, came into contact with Kamentsky in the early stages of the IBM project, and this interaction eventually led to the development of the first commercial flow cytometric differential counter, the Hemalog D, by Technicon Instruments in the early 1970s.[34–36] This instrument used relatively specific reagents and light scattering and absorption measurements made at different wavelengths in three different flow cytometers to classify leukocytes. Chromogenic enzyme substrates were used to identify neutrophils and eosinophils by the presence of moderate to high and very high concentrations of peroxidase, while another channel identified monocytes by their esterase content. Basophil identification was based on detection of glycosaminoglycans in basophil granules using Alcian blue. A single tungsten-halogen lamp served as light source for all three flow systems.

Although the Hemalog D employed cytochemical staining procedures that were well regarded by hematologists for such purposes as determination of lineage of leukemic cells, the apparatus, which demonstrably produced accurate counts, was initially regarded with a great deal of suspicion, at least in part due to the novelty of flow cytometry. The developers and manufacturers of image analyzing differential counters, which also reached the market in the early 1970s and, at best, performed no better than did the Hemalog D, did what they could to keep potential users suspicious of flow cytometry for as long as possible; the technology would eventually be legitimized by its dramatic impact on immunology, which was facilitated by the introduction of cell sorting and immunofluorescence measurements.

1.5.6 Flow Cytometry Meets Fluorescence and Goes Commercial

By the late 1960's, several groups had explored fluorescence measurement as a means of improving both quantitative and qualitative analysis in cytometry. In 1964, Hallerman *et al.*,[37] working with Leitz in Germany, conceived the idea of adding a fluorescence measurement channel to Leitz's optical blood cell counter. The original plan was to add acridine orange to samples, and use the green fluorescence of that dye to discriminate leukocytes and other nucleated cells from red cells; however, they found that when they measured red fluorescence instead, the signals from granulocytes were significantly larger than those from mononuclear cells. Although this would have made it possible for the instrument to do a two-part differential count, Leitz never commercialized the system. By 1969, Van Dilla *et al.*[38] at Los Alamos and Dittrich and Göhde[39] in Germany had described rapid fluorescence flow cytometric measurement of cellular DNA content, facilitating analysis of abnormalities in tumor cells and of cell cycle kinetics in both neoplastic and normal cells. The Los Alamos instrument used an argon ion laser source for excitation of a fluorescent Feulgen stoichiometric stain for DNA, and incorporated the orthogonal "body plan" now standard in laser-source instruments, with the optical axes of illumination and light collection at right angles to each other and to the direction of sample flow. Instruments that could make the same measurements were on the market by 1970; Kamentsky, who had left IBM to found Bio/Physics Systems, produced the Cytofluorograf, an orthogonal fluorescence flow cytometer that was the first commercial product to incorporate an argon ion laser; Göhde's Partec (Münster, Germany) Impulscytophotometer (ICP) instrument, built around a fluorescence microscope with arc lamp illumination, was distributed commercially by Phywe. Both instruments could use the dye ethidium bromide, which Dittrich and Göhde had introduced, for nucleic acid content quantification, eliminating the need to perform the Feulgen reaction.

Leonard Herzenberg and his colleagues[40] at Stanford, realizing that fluorescence flow cytometry and subsequent cell sorting could provide a useful and novel method for purifying living cells for further study, developed a series of instruments after exposure to a Kamentsky prototype lent by IBM.[41] Their original apparatus,[42] with arc lamp illumination, was not sufficiently sensitive to permit them to achieve their objective of sorting cells from the immune system based on the presence and intensity of staining by fluorescently labeled antibodies. The second version,[43] which used a water-cooled argon laser, was more than adequate, and was commercialized as the Fluorescence-Activated Cell Sorter (FACS) in 1974 by a group at Becton-Dickinson (B-D, now BD Biosciences [San Jose, CA]), led by Bernard Shoor.

Coulter Electronics (now Beckman Coulter, Fullerton, CA and Miami, FL), which by 1970 had become a very large and successful manufacturer of laboratory hematology counters, pursued the development of fluorescence flow cytometers through a subsidiary, Particle Technology, under Mack Fulwyler's direction in Los Alamos. The TPS-1 (Two Parameter Sorter), Coulter's first product in this area, reached the market in 1975. It used an air-cooled 35 mW argon ion laser source and could measure forward scatter and fluorescence.

Multiple wavelength fluorescence excitation was introduced to flow cytometry in apparatus built at Block Engineering during an abortive attempt to develop a hematology analyzer. The first instrument[44] derived five illuminating beams from a single arc lamp; the second[45] used three laser beams; both could measure eight parameters, analyze over 30,000 cells/second and, using hardwired preprocessors and integral Data General Nova minicomputers, identify cells comprising less than 1/100,000 of the total sample. The laser source system incorporated forward and side scatter measurements, influenced by work done at Los Alamos,[46] and, as did its predecessor, permitted implementation of elaborate multiparameter gating strategies.[47] Block also built a slow flow system intended for detection of hepatitis B virus and antigen in serum; it could discriminate scatter singles from large viruses[48] and could theoretically detect a few dozen fluorescein molecules above background. The Block cytometers were never sold commercially, but influenced the optical, electronic, and systems design of later instruments.

By the time the Society for Analytical Cytology (now the International Society for the Advancement of Cytometry, or ISAC) came into being in 1978, B-D, Coulter, and Ortho (a division of Johnson & Johnson that bought Bio/Physics Systems) were producing flow cytometers that could measure small- (forward scatter) and large- (side scatter) angle light scattering and fluorescence in at least two wavelength regions, analyzing several thousand cells per second, and with droplet deflection cell sorting capability. Ortho was also distributing the Partec ICP, which, by virtue of its optical design, could make higher precision measurements of DNA content than could laser-based flow cytometers. DNA content analysis was receiving considerable attention as a means of characterizing the aggressiveness of breast cancer and other malignancies, and, at least in part due to the results of a Herzenberg sabbatical in Cesar Milstein's lab in Cambridge, monoclonal antibodies had begun to emerge as practical reagents for dissecting the stages of development of cells of the blood and immune system. Loken, Parks, and Herzenberg had successfully performed a two-color immunofluorescence experiment, introducing fluorescence compensation in the process,[49] although it was clear that a great deal needed to be done in the area of fluorescent label development to realize the potential of monoclonal antibodies.

Since the 1980s, an increasingly wide range of fluorescent labels of different types have become available, including low molecular weight dyes, phycobiliproteins, phycobiliprotein-dye tandem conjugates, and, most recently, semiconductor nanocrystals, more popularly known as quantum dots; many of these labels can be used with ligands other than antibodies, e.g., nucleic acid sequence probes. A series of international workshops has defined several hundred "CD" (Cluster of Differentiation) antigens that are expressed on various cell types, and directly conjugated fluorescent antibodies to many of these are widely available. Perhaps the best known example of a CD antigen is CD4, expressed (with CD3) on the helper subset of T lymphocytes, which are the target cells for HIV, the human immunodeficiency virus; counts of patients' CD4-positive T cells are routine done to monitor therapy of HIV/AIDS wherever the technology (most commonly fluorescence flow cytometry) is supportable and affordable, and much current work on miniaturized flow cytometers has been motivated by the goal of producing CD4 counters suitable for use in resource-poor countries.[50]

1.6 BEHEMOTH TO BENCHTOP AND BEYOND: THINKING INSIDE THE BOX

Through the early 1980s, flow cytometer manufacturers represented the largest segment of the market for water-cooled multiwatt ion lasers, which typically cost tens of thousands of dollars and required both high-voltage, high current electrical power and cooling water flowing at rates of several gallons/minute. Today's benchtop flow cytometers achieve equivalent performance using solid-state and diode lasers, which are orders of magnitude smaller and lighter, considerably less expensive, run on line current and may not even need cooling fans. Paradoxically enough, the 488 nm solid-state lasers now in common use are no more powerful than was the air-cooled argon ion laser used in the Bio/Physics Systems Cytofluorograf of 1970, an instrument that sold for $20,000.

A flow cytometer, even in its simplest guise, is a relatively complex system. First and foremost, it must incorporate means of moving cells or other particles in a sample through the measurement system, preferably one at a time, and either regulating flow (and other factors) well enough so that the dwell time and measurement process are the same for all particles analyzed or monitoring variations in flow, etc. and compensating for them during measurement and/or analysis.

In an impedance measurement system, the sensing electrodes are incorporated into the flow system itself; in an optical flow cytometer, one or more illuminating beams must be collected from a light source and directed to a small region of the stream, providing sufficiently intense and uniform illumination to achieve set specifications for measurement sensitivity and precision. Light scattered by and/or emitted from particles passing through the illumination beam or beams must be collected and transmitted to one or more detectors, which may require separation of the collected light into two or more discrete wavelength regions.

Signals from each detector must be amplified and conditioned so that the peak amplitude, area or integral, and/or duration of the width of each pulse resulting from a particle's passage through a measurement station can be captured and recorded. Finally, in cell sorters, there must be a means of diverting particles meeting preset criteria for one or more measurement into one or more collection vessels.

The subsystems of a flow cytometer interact. Water-cooled lasers came into use when it was noted, circa 1970, that neither the arc lamp-illuminated, microscope-based instruments nor the low power laser-illuminated instruments then available could readily resolve cells stained with fluorescent antibodies from unstained cells. As it happens, the major problem with the arc lamp systems was that particles were observed as they flowed toward the collecting lens (a high-N.A. microscope lens) along the lens's optical axis; this resulted in much higher background fluorescence than would have been generated had the flow been orthogonal to the optical axis. The major problem with laser source systems was inefficiency of the optics, variously due to low N.A. of the collection lenses, the use of obscuration bars to block scattered laser light from reaching fluorescence detectors, the inadequacy of optical filters used to define fluorescence bandwidths, and the inefficiency of photomultiplier tubes at wavelengths in the upper half of the visible spectrum. By the early 1980s, at least one arc source system (the B-D FACS Analyzer) capable of immunofluorescence measurements had appeared; it had an orthogonal flow geometry as well as high-efficiency optics. Also, several manufacturers had introduced instruments with low power laser sources in which fluorescence was measured in cuvettes using higher-N.A. objectives, providing more than adequate sensitivity for immunofluorescence work. Although it is tempting to speculate that more rigorous systems analysis could have completely eliminated the water-cooled laser period of cytometric history, there were other reasons why benchtop instruments did not come into their own until after 1985.

Until monoclonal antibodies became commercially available in 1981, it was extremely difficult to measure immunofluorescence from samples treated with more than one antibody. Although monoclonal antibodies made it relatively simple to stain with multiple antibodies, there were, initially, no combinations of antibody labels with well- separated fluorescence emission compatible with 488 nm excitation. The short-term solution to this problem required the addition of a second laser, either a water-cooled krypton ion laser or a dye laser, to the instrument, making it substantially more complex, power-hungry, water-thirsty, and expensive. The appearance of phycobiliprotein and tandem conjugate labels eliminated the need for the second laser, but it was not practical to measure four (two scatter and two fluorescence) or more parameters until computers for data analysis became readily available and affordable.

The emergence of AIDS in the early 1980s created widespread demand for flow cytometers in both research and clinical laboratories, leading flow cytometer manufacturers, who, up to that time, had been losing money, to intensify their efforts to produce smaller, more user-friendly apparatus. Although cell sorters remained status symbols at many institutions, the prospect of generating infectious aerosols from patient samples created a niche for analyzers with closed fluidic systems. B-D's FACScan, introduced in 1985, was one of the most successful such systems; it used a low-power, air-cooled 488 nm argon laser, and could measured forward and side scatter and green (fluorescein), yellow-orange (phycoerythrin [PE]), and red (PE-Cy5 tandem) fluorescence. Cells were analyzed in a cuvette with an integral gel-coupled lens of higher N.A. (1.2) than had appeared in any previous laser source commercial system; the design essentially eliminated any need for regular optical adjustments by the operator. Data analysis was done with a dedicated Hewlett-Packard microcomputer. B-D's competitors also produced benchtop systems, and they and third parties began to offer data analysis systems based on microcomputers, typically IBM PCs and clones. Although manufacturers had previously made minicomputer data analysis and (in the case of Cytomation (now part of Beckman Coulter)) sort control systems available, their cost had impeded widespread adoption; from 1985 on, however, the microcomputer became an integral and indispensable component of a flow cytometer.

Since 1990, much of flow cytometer development has moved in the direction of systems of increased complexity, capable of measuring multiangle scatter and fluorescence in increasingly larger numbers of wavelength regions. B-D, which acquired Ortho's flow cytometry business in 1987, introduced the benchtop FACSCalibur, adding a second laser source, a red diode laser nominally emitting at 635 nm, to the 488 nm argon laser. This permitted measurement of red-excited far red fluorescence as well as fluorescence in the three regions measured by the earlier FACScan. Coulter countered with the EPICS XL, another benchtop, which used a single 488 nm laser and measured four fluorescence channels (adding the red-orange PE-Texas red tandem to fluorescein, PE, and PE-Cy5). The XL also introduced high-resolution (20-bit) analog-to-digital converters (ADCs), inaugurating a basic shift in signal processing philosophy.

A large percentage of flow cytometer users are interested in measuring as many immunofluorescence parameters as is practical. Immunofluorescence signal intensities tend to vary over a fairly wide dynamic range, traditionally taken as the four decades between a few hundred and a few million molecules of labeled antibody bound per cell. The ADCs in the first generation of microcomputer-based flow cytometry data analysis system had substantially lower resolution (8-10 bits) than would be needed to cover this dynamic range; as a result, the instruments provided the option of routing signals through analog logarithmic amplifiers (log amps), which compressed the intensity scale. Multicolor immunofluorescence measurements, however, required manipulation of signals to compensate for spectral overlap of the emissions of various probes among measurement channels, and such compensation had to be applied to linear, rather than log-transformed data.

Although the general n-color compensation problem can be solved analytically using relatively simple matrix algebra, early implementations for two and three colors were done using analog circuits that could add and subtract linear signals, with the instrument operator twiddling potentiometers until the two-parameter displays on the screen separated data into quadrants, approximating the analytical solution by eye. The number of active electronic devices needed in compensation circuitry increases rapidly, and the ability of the operator to compensate by eye decreases rapidly, with the number of colors involved; at the four-color level, the noise level of the electronics begins to narrow the effective dynamic range of the analysis system. By the time cytometry reached this level of complexity, the communications and consumer electronics industries had produced affordable high-resolution ADCs, and microcomputers fast enough to do analytically correct compensation on the fly. The Coulter XL's engineers approached the overall problem by developing high dynamic range analog peak detectors and integrators, allowing signals to be kept in a 20-bit linear form and eliminating thereby eliminating the need for log amps; data were converted to a logarithmic scale for display purposes. These electronics were adequate for an analyzer, but not fast enough to be used in a high-speed (>10,000 particles/second cell sorter).

High-speed sorter development was first driven by the need to separate relatively large numbers of individual human chromosomes to provide libraries for the Human Genome Project;[51] the apparatus is now used for a much wider range of applications[52]. In 1994, Cytomation introduced the MoFlo sorter, made under license from Lawrence Livermore National Laboratory, where it had been developed by Ger van den Engh and his associates. The approach to high-resolution digitization taken in the MoFlo, and in the newer Influx high-speed sorter, produced from 2000 on by van den Engh's company, Cytopeia (now part of BD), does not dispense with log amps but, instead, digitizes their output signals with 16-bit ADCs, allowing data to be converted from linear to log scale and back without significant loss of information. BD itself took a third route toward high-resolution digitization; signal processing in the DiVa analysis system used with their newer cytometers is almost completely digital, with conditioned detector preamplifier output streams being sampled every 100 nsec by 14-bit ADCs.

Until about 2000, the vast majority of fluorescence flow cytometers used at most two illuminating beams, one at 488 nm from a low-power argon laser and one from either a helium-neon (He-Ne) laser (633 nm) or a red diode laser (nominally 635 nm). Sorters, which have relatively inefficient optical systems, tended to use higher-power, water-cooled argon and krypton lasers, which could provide output wavelengths ranging from ultraviolet (UV) to green and, in the case of krypton, also yellow, red, and infrared; the high-power lasers were also used to pump dye lasers producing yellow-orange wavelengths. Some benchtop systems, built using more efficient optics, used mercury or xenon arc lamps instead of lasers as light sources, and air-cooled helium-cadmium (He-Cd) lasers were occasionally used as sources of UV and deep blue light. Since 2000, a wider range of diode and solid-state lasers have found their way into flow cytometers; although not inexpensive, these sources are at least an order of magnitude smaller, lighter, and more energy-efficient than other lasers, and are now available at many wavelengths from UV to near-infrared.[53] Although light-emitting diodes (LEDs) have also been used as light sources in some flow cytometers, their inability to deliver substantial power to small regions of space limits their utility.

1.7 MICROFLOW CYTOMETRY — A PERSONAL NOTE

By the mid-1980s, the success of the first generation of personal computers had demonstrated that making advanced technology smaller, simpler, cheaper, and more energy-efficient would greatly expand a user community, with resulting benefits to developers and manufacturers. Although microcomputer-based data analysis systems were rapidly becoming standard adjuncts to flow cytometers, there were few options then available that could have substantially decreased the size, cost, and/or complexity of other subsystems of the apparatus.

In 1982, I heard a meeting presentation by Jonathan Briggs that described what still may be the simplest fluorescence cytometer ever conceived.[54] Blue laser light was coupled into an optical fiber through a long pass dichroic filter, which diverted fluorescent light returning from the other end of the fiber into a photomultiplier fitted with a bandpass filter. The tip of the fiber was moved slowly through a dilute suspension of fluorescent particles. The relatively low (0.19) N.A. of the fiber and the divergence of illuminating light combined to define a relatively small volume within which the illumination intensity and the fraction of fluorescence emission collected from a particle would be sufficient to produce a signal above background from the photomultiplier. The apparatus was used to detect fluorescein-labeled antibody binding to red blood cells.

Briggs's presentation led me to imagine that a somewhat more sensitive and precise simple flow cytometer could be made by positioning the ends of four fiber optics in such a way as to define a small orifice through which particles could be passed in single file, with illuminating light coming through one fiber, an extinction signal being collected through the coaxial fiber opposite, and fluorescence and/or side scatter signals being collected through the two fibers with faces perpendicular to those of the illumination and extinction collection fibers. The notion seemed too simplistic to be practical, but, when I checked with a colleague, Mike Hercher, who was much better versed in optics than I was, he thought it might work, and we brainstormed our way to the "flow cytometer-on-a-chip" shown in Figure 1. The Figure shows part of a poster presentation I made in 1984; a more formal publication appeared in early 1986,[55] by which time we had been unsuccessful in attempting to obtain a broad patent and decided to give away the idea. The next publication on microflow cytometry of which I am aware appeared in 1993; Dan Sobek, working at MIT with Martha Gray and Steve Senturia, among others, successfully generated sheath flow in a micromachined flow cell[56] and cited my earlier work.

By the late 1990s, it was clear that, although putting the flow cytometer optics in a solid block decreased subsystem size, cost and complexity and eliminated the need for alignment and adjustment, an instrument that was to use sheath flow and the standard repertoire of fluorescent dyes would still have to include a sheath tank, waste reservoir, and a laser of substantial size. It was also becoming apparent that waveguides typically collect light from larger volumes than do the lenses conventionally used in flow cytometers, resulting in higher fluorescence background and decreased instrument sensitivity, although precision comparable to that of conventional instruments could be obtained when measuring relatively high-intensity signals. Finally, since I realized that I would be unlikely to make further progress in developing microflow systems with the relatively crude microfabrication capabilities I had available, I elected to keep up with what others were doing and rethink cytometry entirely.

During the past decade, flow cytometry, and cytometry in general, have diversified considerably. Whereas ten years ago, most cell sorting was done for the purpose of isolating cell populations bearing specific antigens, the best estimates now available suggest that selecting cells bearing specific fluorescent proteins inserted by genetic manipulation now accounts for more sorting time. The most complex flow cytometers can measure light scattered at two or more angles and fluorescence in eighteen or more spectral regions, using as many as five illuminating beams, and the highest speed sorters can process tens of thousands of cells per second. One might argue that it is wasteful to tie up more than half the work schedule

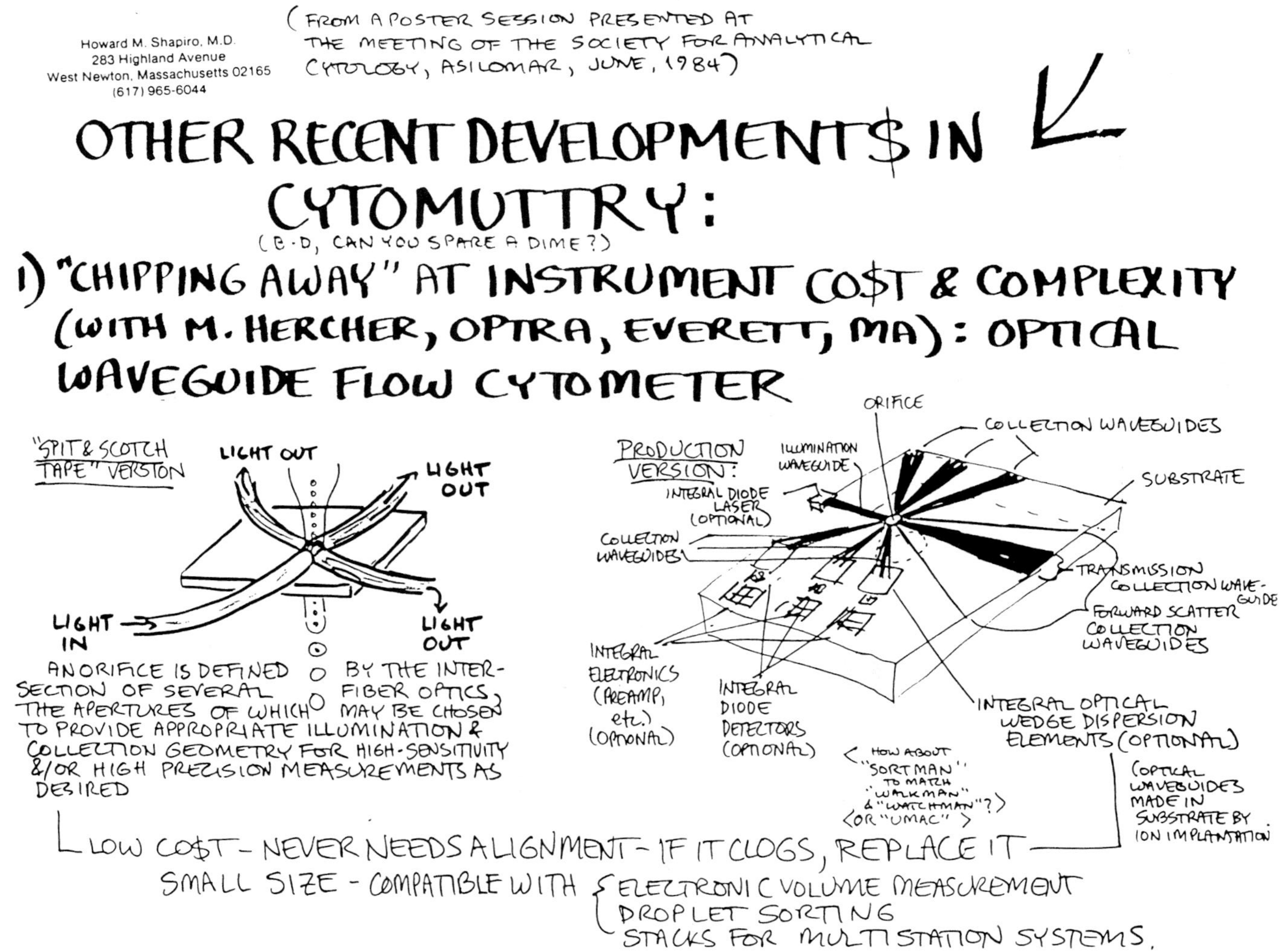

Figure 1.1. The "flow cytometer-on-a-chip" concept, 1984.

of such sophisticated, expensive ($500,000) instruments doing a task that typically requires only a single fluorescence measurement channel, especially when a simpler sorter adequate for most work with fluorescent proteins could probably be sold for a relatively small fraction of the cost of a high-end system. No one has yet built the simple system, and, if one is built, it is unlikely to be a microflow system.

In general, microflow cytometers process smaller volumes of sample per unit time than do conventional instruments. The actual volume flow rate of a sample in a high-speed sorter is typically no more than 2 μL/sec; although the instrument can theoretically process 100,000 cells/sec, analysis at this rate requires a cell concentration of 5 x 10^7/mL, which makes for a relatively viscous suspension. The microfabricated sorter described by Fu *et al.*,[57] built using soft lithography, was able to analyze and sort a few hundred *E. coli* per second, but the volume flow rate was so low that achieving this analysis rate required a bacterial concentration of 10^9/mL. This concentration would be impossible to achieve with mammalian cells; moreover, attempting to use a flow system of this type to detect enterohemorrhagic *E. coli* in a single unconcentrated food sample, in which they may be present at concentrations of less than a few hundred/mL, would take days. One could increase the analysis rate by building a massively parallel system, but that would increase cost and complexity. Failure to do the math is, unfortunately, prevalent in both the high-speed sorting and the microflow literature.

One of the simplest flow cytometers now in widespread use, the BD FACSCount,[50] is a relatively small instrument designed solely for the purpose of counting CD4-positive T cells and a few other relevant cell types in the peripheral blood of HIV/AIDS patients in resource-poor areas. The light source in the FACSCount is a 543 nm He-Ne laser emitting less than 1 mW, and the apparatus measures only two parameters, yellow-orange (PE) fluorescence and red (PE-Cy5) fluorescence, using relatively inexpensive photomultiplier detectors. For CD4 counting, the criterion used is the presence of both PE-Cy5-labeled antiCD3 and PE-labeled anti-CD4 on a cell; fluorescent beads added to the sample at known concentrations are used to derive the cell count per unit volume. The FACSCount has several competitors; among them are flow cytometers from Guava (Hayward, CA), Partec, and Pointcare Technologies (Marlborough, MA). All of these instruments cost at least a few tens of thousands of dollars, and a reasonably steady flow of manuscripts, grant applications, etc. across my desk suggests that there are a lot of other people trying to get into the CD4 counter business, many of them proposing to do so with microflow systems.

An increasing number of others, myself included, think that flow cytometry, micro- or otherwise, is the not best way to do a CD4 count in a really resource-poor area. As I noted above, a flow cytometer is a relatively complex instrument, even in its simplest form. The FACSCount still incorporates a flow system, laser, photomultipliers, associated electronics, and a computer; it is, however, now possible[50,58,59] to build a much smaller, simpler, cheaper cytometer by using one or more high-intensity LEDs to illuminate a large area of a slide, large-volume hemocytometer, or other static substrate and a digital camera chip, costing a few tens of dollars, to make low-resolution fluorescence images of all of the cells in a specimen at once. Such an apparatus can readily detect low-level fluorescence signals from cells stained with fluorescent antibodies, as well as the substantially stronger signals associated with cells stained with nucleic acid stains, fluorescent physiologic probes and enzyme substrates, fluorescent proteins, etc., and at least two companies (Inverness Medical, Waltham, MA and LabNow, Austin, TX) have shown prototype CD4 counters using this technology, which might also be useful for diagnosis and drug sensitivity testing in tuberculosis and malaria, both infectious diseases that interact with HIV/AIDS in resource-poor areas, taking millions of lives, and both still diagnosed by microscopy using techniques many decades old and long overdue for now-affordable automation.[60]

1.8 CONCLUSIONS

Nobody is quite sure who first opined that "when all you have is a hammer, everything looks like a nail," but many of us in the flow cytometry field seem to have acquired the mindset described. Cytometry exists to get information from cells; you put cells into a cytometer and you get numbers out, and, if they're the right numbers, you shouldn't care what's in the box. Flow cytometry got us from the point at which we had to characterize cells based on high resolution imaging and analysis of morphology to where we can do a much better job by using multiple reagents and making whole-cell measurements of emitted and/or scattered light. Flow cytometry will remain the method of choice for the most complex analytical tasks and for sorting; but the jobs now done by the simpler, relatively inexpensive flow cytometers may soon be done by even simpler and less expensive boxes in which flow systems are conspicuous by their absence.

There is, nonetheless, good news for the microflow diehards. A visit to http://www.particlecounters.org shows that the tradition of Zsigmondy and Gucker is now maintained by at least six manufacturers who offer battery-operated handheld aerosol particle counters that report counts of three to eight size classes of particles as small as 0.2 μm in sampled air. Some of these devices are sold for less than $2,000. If history repeats itself, the microflow cell analyzers should appear in a few years.

References

[1] H. M. Shapiro, *Practical Flow Cytometry, 4th ed.* John Wiley & Sons: Hoboken, NJ, pp. 1–681 (2003).

[2] H. M. Shapiro, The evolution of cytometers. *Cytometry* **58A**, 13–20 (2004).

[3] H. M. Shapiro, Cytometry and cytometers: Development and growth. In: Doležel, J., Greilhuber, J., Suda, J., Eds. *Flow Cytometry with Plant Cells.* Weinheim, Wiley-VCH: Weinheim, Germany, pp. 1–17 (2007).

[4] H. Harris, *The Birth of the Cell.* Yale University Press: New Haven, CT, pp. 1–212 (1999).

[5] R. Hooke, *Micrographia*. John Martyn & James Allestry, Printers to the Royal Society: London, England, 1665, pp. 1-246. (Reprinted with the Index from the 1745 and 1780 editions; Dover Publications: New York, NY, 2003.)

[6] T. Cowles, Dr. Henry Power's poem on the microscope. *Isis* **21**, 71–80 (1934).

[7] P. De Kruif, *Microbe Hunters* (originally published 1926; reprinted with a new introduction). Harcourt, Brace: San Diego, CA, pp. 1–357 (1996).

[8] W. R. Gowers, On the numeration of blood-corpuscles. *Lancet* **110**, 797–798 (1877).

[9] "Student" [Gossett, W. S.]. On the error of counting with a haemacytometer. *Biometrika* **5**, 351–360 (1907).

[10] R. Zsigmondy, *Colloids and the Ultramicroscope* (translated by J. Alexander). John Wiley & Sons: New York, pp. 1–245 (1909).

[11] E. H. M. Badger, Particle counts in the ultramicroscope. *Nature* **157**, 480 (1946) .

[12] G. Bucky and A. Einstein, Light intensity self-adjusting camera. *US Patent 2,058,562,* October **27**, (1936).

[13] A. Moldavan, Photo-electric technique for the counting of microscopical cells. *Science* **80**, 188–189 (1934). M. R. Flannery, *Adv. At. Mol. Opt. Phys.* 117 (1994).

[14] F. T. Gucker, Jr., C. T. O'Konski, H. B. Pickard and J. N. Pitts, Jr. A photoelectronic counter for colloidal particles. *J. Am. Chem. Soc.* **69**, 2422–2431 (1947).

[15] P. J. Crosland-Taylor, A device for counting small particles suspended in fluid through a tube. *Nature* **171**, 37–38 (1953).

[16] W. H. Coulter, High speed automatic blood cell counter and cell size analyzer. *Proc. Natl. Electronics Conf.* **12**, 1034–1042 (1956).

[17] G. Brecher, M. Schneiderman and G. Z.Williams, Evaluation of electronic red blood cell counter. *Amer. J. Clin. Path.* **26**, 1439–1449 (1956).

[18] C. F. T. Mattern, F. S. Brackett and B. J. Olson, Determination of number and size of particles by electrical gating: Blood cells. *J. Appl. Physiol.* **10**, 56–70 (1957).

[19] G. N. Papanicolaou and H. F. Traut, The diagnostic value of vaginal smears in carcinoma of the uterus. *Amer. J. Obstet. Gynecol.* **42**, 193–206 (1941).

[20] T. O. Caspersson, *Cell Growth and Cell Function*. Norton: New York, NY (1950).

[21] R. C. Mellors and R. Silver, A microfluorometric scanner for the differential detection of cells: application to exfoliative cytology. *Science* **114**, 356–360 (1951).

[22] W. E. Tolles, The Cytoanalyzer: An example of physics in medical research. *Trans. N.Y. Acad. Sci.* **17**, 250–256 (1955).

[23] G. L. Wied and G. F. Bahr, Eds. *Automated Cell Identification and Cell Sorting*. Academic Press: New York, NY (1970).

[24] J. M. S. Prewitt and M. L. Mendelsohn, The analysis of cell images. *Ann. N.Y. Acad. Sci.* **128**, 1035–1053 (1966).

[25] M. Ingram and K. Preston Jr. Automatic analysis of blood cells. *Sci. Amer.* **223**(5), 72–78 (Nov. 1970).

[26] P. G. Stein, L. E. Lipkin and H. M. Shapiro, Spectre II: general-purpose microscope input for a computer. *Science* **166**, 328–333 (1969).

[27] H. M. Shapiro, S. D. Bryan, L. E. Lipkin, P. G. Stein and P. F. Lemkin, Computer-aided microspectrophotometry of biolgical specimens. *Exptl. Cell Res.* **67**, 81–85 (1971) .

[28] L. A. Kamentsky, M. R. Melamed and H. Derman, Spectrophotometer: New instrument for ultra rapid cell analysis. *Science* **150**, 630–631 (1965).

[29] L. A. Kamentsky and M. R. Melamed, Rapid multiple mass constituent analysis of biological cells. *Ann. N.Y. Acad. Sci.* **157**, 310–323 (1969).

[30] S. H. Koenig, R. D. Brown, L. A. Kamentsky, A. Sedlis and M. R. Melamed, Efficacy of a rapid cell spectrophotometer in screening for cervical cancer. *Cancer* **21**, 1019–1026 (1968).

[31] M. J. Fulwyler, Electronic separation of biological cells by volume. *Science* **150**, 910–911 (1965).

[32] R. G. Sweet, High frequency recording with electrostatically deflected ink jets. *Rev. Sci. Instrum.* **36**, 131–136 (1965).

[33] L. A. Kamentsky and M. R. Melamed, Spectrophotometric cell sorter. *Science* **156**, 1364–1365 (1967) .

[34] L. Ornstein, Tenuous but contingent connections. *Electrophoresis* **8**, 3–13 (1987).

[35] L. Ornstein, H. R. Ansley, Spectral matching of classical cytochemistry to automated cytology. *J. Histochem. Cytochem.* **22**, 453–469 (1974).

[36] H. P. J. Mansberg, A. M. Saunders and W. Groner, The Hemalog D white cell differential system. *J. Histochem. Cytochem.* **22**, 711–724 (1974).

[37] L. Hallermann, R. Thom and H. Gerhartz, Elektronische Differentialzählung von Granulocyten und Lymphocyten nach intravitaler Fluochromierung mit Acridinorange. *Verh. Deutsch Ges. Inn. Med.* **70**, 217–219 (1964).

[38] M. A. Van Dilla, T. T. Trujillo, P. F. Mullaney and J. R. Coulter Cell microfluorometry: A method for rapid fluorescence measurement. *Science* **163**, 1213–1214 (1969).

[39] W, G Dittrich and W. öhde, Impulsfluorometrie bei Einzelzellen in Suspensionen. *Z. Naturforsch.* **24b**, 360–361 (1969).

[40] L. A. Herzenberg, R. G. Sweet and L. A. Herzenberg, Fluorescence activated cell sorting. *Sci. Amer.* **234**(3), 108–117 (Mar 1976).

[41] A. M. Saunders and H. R. Hulett, Microfluorometry: Comparison of single measurements to a rapid flow system. *J. Histochem. Cytochem.* **17**, 188 (1969).

[42] H. R. Hulett, W. A. Bonner, J. Barrett and L. A. Herzenberg, Cell sorting: Automated separation of mammalian cells as a function of intracellular fluorescence. *Science* **166**, 747–749 (1969) .

[43] W. A. Bonner, H. R. Hulett, R. G. Sweet and L. A. Herzenberg, Fluorescence activated cell sorting. *Rev. Sci. Instrum.* **43**, 404–409 (1972).

[44] R. Curbelo, E. R. Schildkraut, T. Hirschfeld, R. H. Webb, M. J. Block and H. M. Shapiro, A generalized machine for automated flow cytology system design. *J. Histochem. Cytochem.* **24**, 388–395 (1976).

[45] H. M. Shapiro, E. R. Schildkraut, R. Curbelo, R. B. Turner, R. H. Webb, D. C. Brown and M. J. Block, Cytomat-R: A computer-controlled multiple laser source multiparameter flow cytophotometer system. *J. Histochem. Cytochem.* **25**, 836–844 (1977) .

[46] G. C. Salzman, J. M. Crowell, J. C. Martin, T. Trujillo, A. Romero, P. F. Mullaney and P. M. LaBauve, Cell identification by laser light scattering: Identification and separation of unstained leukocytes. *Acta Cytol.* **19**, 374–377 (1975).

[47] H. M. Shapiro, Fluorescent dyes for differential counts by flow cytometry: Does histochemistry tell us much more than cell geometry? *J. Histochem. Cytochem.* **25**, 976–989 (1977).

[48] M. Hercher, W. Mueller and H. M. Shapiro, Detection and discrimination of individual viruses by flow cytometry. *J. Histochem. Cytochem.* **27**, 350–352 (1979) .

[49] M. R. Loken, D. R. Parks and L. A. Herzenberg, Two-color immunofluorescence using a fluorescence-activated cell sorter. *J. Histochem. Cytochem.* **25**, 899–90 (1977).

[50] G. Janossy and H. Shapiro, Simplified cytometry for routine monitoring of infectious diseases. *Cytometry B Clin. Cytom. 74 Suppl 1*, S6–S10 (2008) .

[51] J. W. Gray, P. N. Dean, J. C. Fuscoe, D. C. Peters, B. J. Trask, G. J. van den Engh and M. A. Van Dilla, High-speed chromosome sorting. *Science* **238**, 323–329 (1987).

[52] S. F. Ibrahim, G. van den Engh, High-speed cell sorting: fundamentals and recent advances. *Curr. Opin. Biotechnol.* **14**, 5–12 (2003).

[53] H. M. Shapiro and W. G. Telford, Lasers for flow cytometry. *Curr. Protoc. Cytom.* (2009), in press.

[54] J. Briggs, M. L. Fisher, V. E. Ghazarossian and M. J. Becker, Fiber optic probe cytometer. *J. Immunol. Methods* **81**, 73–81 (1985).

[55] H. M. Shapiro and M. Hercher, Flow cytometers using optical waveguides in place of lenses for specimen illumination and light collection. *Cytometry* **7**, 221–223 (1986).

[56] D. Sobek, A. M. Young, M. L. Gray and S. D. Senturia, A microfabricated flow chamber for optical measurements in fluids. *Proc. IEEE* **2**, 219–224 (1993).

[57] A. Y. Fu, C. Spence, A. Scherer, F. H. Arnold and S. R. A Quake, microfabricated fluorescence-activated cell sorter. *Nat. Biotechnol.* **17**, 1109–1111 (1999).

[58] H. M. Shapiro, "Cellular astronomy" — A foreseeable future in cytometry. *Cytometry Part A* **60A**, 115–124 (2004).

[59] H. M. Shapiro, Perlmutter, N. G. Personal cytometers — Slow flow or no flow? *Cytometry Part A* **69A**, 620–630 (2006).

[60] H. M. Shapiro and N. G. Perlmutter, Killer applications: toward affordable rapid cell-based diagnostics for malaria and tuberculosis. *Cytometry B Clin. Cytom. 74 Suppl 1*, S152–S164 (2008).

Chapter Two

Analysis of Single Cells Using Lab-on-a-Chip Systems

Tobias Preckel

Agilent Technologies R&D and Marketing GmbH,
Hewlett-Packard-Strasse 8, 76337 Waldbronn, Germany
tobias_preckel@agilent.com

Lab-on-a-chip (LOC) technology is capable of performing liquid-phase analysis with a dramatic reduction of sample and reagent volume. It also automates complex laboratory processes. Initially the use of the technology was focused on molecular analysis based on electrokinetic- or pressure-driven flow. Recently, there has been a growing interest to utilize LOC technology for analysis of cellular parameters. One of the major benefits is the low consumption of cells which makes the technology particularly useful for tests with primary cells. At the same time small subpopulations can be reproducibly measured.

2.1 INTRODUCTION

Lab-on-a-chip (LOC) technology relies on movement of fluids and particles in microfluidic channels fabricated into a glass, polymer or metal structure. It has benefited from achieving a dramatic reduction of sample and reagent volumes compared with traditional macroscopic analysis techniques. The proximity of components and short distances provide significant increases in speed of analysis. In addition, complex laboratory processes can often be automated. The first of these complex applications, based on traditional capillary electrophoresis, have been for separations of biomolecules such as nucleic acids and proteins,[1] where only low amounts of sample are available. In contrast to capillary electrophoresis, LOC technology lends itself to routine applications where open platforms and complex method development capability are seen as hindering. Hence, LOC technology has excelled at quality control applications, particularly for analysis of biomolecules, e.g., RNA, which is easily degraded by endonucleases within laboratory settings.[2]

The Microflow Cytometer by J S Kim & F S Ligler

www.panstanford.com
978-981-4267-41-0

In the past few years, the interest in flow cytometry performed in microfluidic devices has grown.[3–5] On the one hand, standard flow cytometry equipment is expensive and complicated, and therefore accessible primarily in central laboratories with specially trained operators. On the other hand, routine applications, such as checking the expression of a transfected protein in a cell line prior to starting a fermentation process, do not utilize the full functionality and the sophisticated functions of a complex flow cytometer. Hence, there is an obvious need for easy-to-use flow cytometers with a limited set of features at a reasonable cost.

To manipulate cells in microfluidic channels, either pressure driven- or electrokinetically driven systems are typically used. Traditional flow cytometry is based on pressure driven flow. Usually, cells are moved with a speed of several meters per second and are focused hydrodynamically before they pass the detection area. Several detectors either in line with the illuminating light beam or perpendicular to it read fluorescence and scatter-related signals. While the fluorescence is correlated with the concentration of a fluorescent chemical within or on the cell the scatter parameters yield information relating to size or volume (forward scatter) and "roughness" or inner structure (sideward scatter).

In many applications, only the fluorescent signals are required. A commercial microfluidic chip-based platform for the flow cytometric measurement of fluorescent parameters has been successfully used to evaluate antibody staining, green fluorescent protein transfection efficiency and apoptosis in cultured cells.[6] The instrument is equipped with two light sources, and emitted fluorescence is measured in single cells as they pass through the illuminating beams in individual channels of the chip. Data acquisition and analysis can be automated, allowing unattended measurement of the samples once the microfluidic chip is loaded on the instrument. As a small number of cells are consumed per sample, it is particularly suitable for working with cells of limited availability, e.g., primary cells.

The applications are based on the controlled movement of cells by pressure-driven flow inside networks of microfluidic channels. Cells are hydrodynamically focused and pass the fluorescence detector in single file. Initial applications are the determination of protein expression and apoptosis parameters. Results obtained with the microfluidic chips show good correlation with data obtained using a standard flow cytometer.

2.2 INSTRUMENT AND CELL-ASSAY CHIP

The 2100 Bioanalyzer was the first available product using microfabricated chip-based technology. On this platform, approaches for a variety of separation techniques for nucleic acids and proteins have been introduced. The system is capable of laser fluorescence detection and uses disposable microfluidic glass chips. For the detection of cellular fluorescence parameters, the system was modified to use vacuum to move cells through the microfluidic channels of the chip having channel dimensions of $25 \times 75\ \mu m$. Fluid flow is controlled by a peristaltic pump driven by

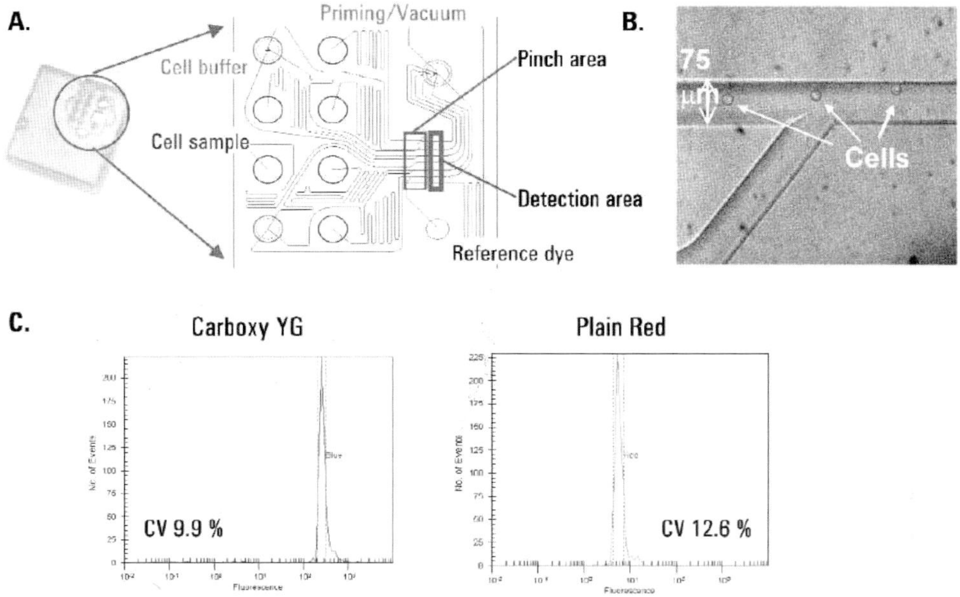

Figure 2.1. Chip features and layout. (A) Microfabricated glass chips for handling nanoliter amounts of liquid were glued into a plastic caddy which accommodates six sample wells, two buffer wells, one well for a reference dye, and a well for a vacuum interface and collection of fluid waste. (B) Each sample channel is joined by a buffer channel in close proximity to the detection area. Here each sample is focused to a portion of the microchannel in order to generate a single file cell stream. (C) Performance of the chip design and detection system was tested with commercially available calibration beads red Fluoresbrite™ Plain Red microspheres and blue Fluoresbrite™ Carboxy YG microspheres (Polysciences Inc., Warrington, PA). Beads were measured on the microfluidic system and frequency histograms are shown. CVs measured at half height of histograms were compared to the supplier's specifications (percentage for blue, percentage for red beads). Color reference – pg. 337.

a stepper motor and the vacuum is tightly controlled by a pressure sensor (range 0-140 mbar), thus guaranteeing a constant flow speed within the microfluidic channels. A special cartridge designed for the 2100 Bioanalyzer is used to interface the pressure control with the chip. The cartridge also contains a filter assembly to prevent fluid from entering the system in case liquid should overflow the waste well of the chip.

For optical detection, cells are prestained with fluorescent markers. Following staining, cells are resuspended in an isobuoyant buffer to guarantee that they remain in suspension during the course of a chip run (approximately 25 minutes). Only a few μl of this suspension are loaded per sample well (corresponding to $\sim$20,000 cells). Before the cells are loaded, the chip is primed with an aqueous solution, which fills the channels by capillary forces within one second. The chip can then be loaded with up to six cell samples. A fluorescent reference dye is also added to a separate well on the chip. This dye fills one channel after vacuum is applied to the chip, which acts as a reference point for the optical detection system of the instrument before cell measurements actually start. To achieve single cell analysis, cells pass the detector in single file. The chip was designed to hydrodynamically focus cells to one side of the microfluidic channel before passing the detector after vacuum is applied (Fig. 2.1A). This is achieved by a junction of the channel carrying the cells and a buffer channel shortly before the detection area (pinch area, Fig. 2.1B).

The performance of the chip design and detection system can be validated with calibration beads. Red and blue fluorescently labeled beads were analyzed on the microfluidic system (Fig. 2.1C). Histogram coefficients of variation (CVs) measured at half height of frequency histograms were 9.9% for blue beads and 12.6% for red beads while the supplier's specifications for the bead preparations were 7% and 10%, respectively. The data confirmed that the error due to the number of cases when more than one bead or only a part of a bead pass the detector at a given time is not significant under the experimental conditions used. If particle aggregation were to occur, one would expect significant broadening of the fluorescence intensity distribution.

During an experiment, fluid flow in all channels needs to be tightly regulated. A change of vacuum directly leads to a change in cell speed, which will have an effect on the cell's signal intensity as the detector samples at a specific frequency. Pressure regulation is performed by monitoring the applied vacuum or pressure.

In the case of the described instrument, only one channel is read by the optical system at a given time. However, in an interconnected network of channels, fluid will flow in all channels once pressure is applied. A chip can be run for a maximum time until its sample and buffer reservoirs have been depleted. Therefore, the total run time of a chip needs to be split between the number of samples. For the experiments described below, a constant flow rate was used (the average migration speed of the cells through the channel was 3-4 mm/s) and approximately 500-1000 cell events of 20,000 were analyzed per sample in 240 seconds. It is conceivable that, with higher flow rate, analysis time on such instrumentation could be reduced.

For most simple applications, such as measurement of expression of a given protein, a minimum of two signals are needed. First, the instrument requires a signal for confirming the presence of the cell in the detection area. Second, a specific signal is needed to indicate the protein expression (with the signal intensity correlating to its expression level). In the case of the Bioanalyzer, the instrument uses two independent excitation light sources: a blue LED and a red diode laser which excite at 458–482 nm and 625-645 nm, respectively. The light is focused on the microfluidic channel and fluorescence emission from the cells is detected at 510-540 nm (blue) and 674–696 nm (red). The light must be focused so that the channel is homogenously illuminated across the detection area perpendicular to the flow direction. At the same time, the illumination spot should be slit-like across the channel in order to illuminate only one cell at a time. These requirements can present challenges in the spacial limitations of a microscale fluidic chip.

2.3 DATA ANALYSIS

Cells are stained such that each cell carries a fluorescence intensity value in two colors. As a cell passes the instrument optics in single file, the fluorescence intensities in the two detection wavelengths are detected and recorded. Each cell that emits fluorescence above a threshold value is counted as an event, and the intensities of the two different fluorescent signals (red and blue) are recorded. The intensity of the fluorescent signal depends on the amount of fluorescent probe bound to or incorporated into the cell. The data can be analyzed in two ways: Events are either plotted against their fluorescence intensity in frequency histograms (number of events vs. fluorescence intensity in one color, Fig. 2.3(A)) or they are plotted in a two-dimensional dot plot where the fluorescence intensity of each color is depicted along one axis (Fig. 2.3(B)). Cell populations of interest can be gated by selecting regions (in dot plots) or markers (in histograms). Using these regions of interest, the software calculates percentages of subsets within a given population, e.g., percentage of antibody-stained or Annexin-positive cells within the live cell population and gives statistical information. In principal, if the data can be exported in common flow cytometry data formats and analyzed by third party software.

2.4 APPLICATIONS

Typical applications that benefit from the LOC cytometry approach are the monitoring of protein expression or apoptotic processes in eukaryotic cells. Monitoring cellular protein expression is a critical step for characterization of cell populations or assay optimization and can be achieved by staining the protein of interest with specific antibodies. Fluorescently labeled antibodies are used to detect cells bearing specific antigens (lipids, proteins or carbohydrates). The antibody may be directly conjugated to a fluorescent probe or a fluorescent secondary antibody may be used. The detection of proteins generally depends on the availability of a suitable, specific antibody. When using fluorescently labeled antibodies, cells can often be stained directly on the chip, eliminating time consuming washing steps.[7]

Tumor cell lines or immortalized cells have been standard models for the elucidation of biochemical pathways as well as drug testing. However, indeterminate changes in the phenotype may occur during immortalization of a cell.[8] Primary cells on the other hand, prepared directly from fresh tissues or fluids of an organism, often display most of the differentiated properties of the original source (e.g., isolated fibroblasts continue to secrete collagen). However, extended work with those cells is hindered by the limited number of cell divisions experienced by primary cells in culture as well as the inherent limitation in cell numbers. Flow cytometry offers extensive analytical opportunities but often requires high cell numbers for an experiment. For primary cells, the LOC approach is particularly useful due to the low numbers of cells required per analysis.[9]

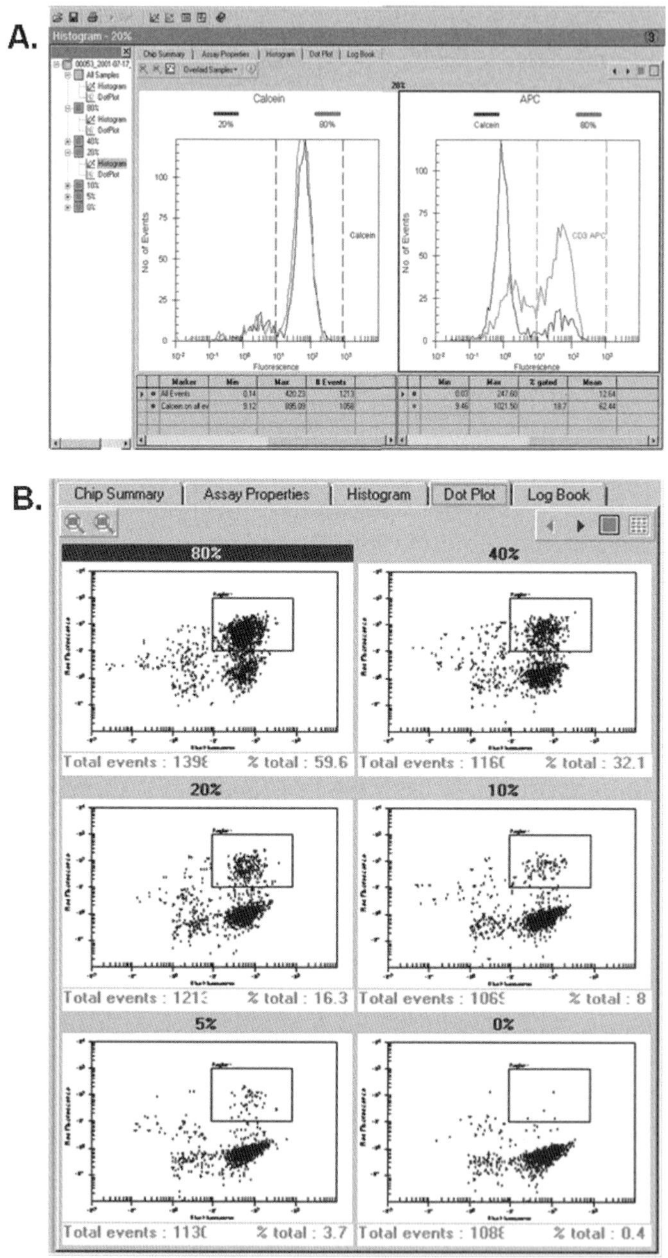

Figure 2.2. 2100 Bioanalyzer's user interface and software. The software used to analyze the raw data allowed to display the data either in a histogram view (A) or a dot plot view (B). Analysis was done by setting the appropriate gate and marker in the histogram view according to standard procedures. Color reference – pg. 338.

Figure 2.3 shows the results of a typical experiment using normal human dermal fibroblast (NHDF) cells. In order to analyze how many cells are required at a minimum to reproducibly detect a protein's expression level decreasing numbers of primary cells were stained with a live cell marker, calcein, and fluorescently labeled antibodies against HLA-A, B, C. As shown in Fig. 2.3(A) NHDFs can be stained and detected on-chip down to cell numbers of 625 cells per sample. the live cells were 100% HLA-positive and therefore were expected to be double positive, i.e. staining with both dye and antibody.

Figure 2.3(B) shows the impact of decreasing the the cell number per sample on the percentage value of a subpopulation. Here, a non-purified, impure preparation of lymphocytes was stained on-chip. As a control the same cells were stained with calcein-AM and an antibody against the non-expressed CD86 protein. The instrument detects a population of 82–84% double-positive, 16–18% single-positive cells with an acceptable reproducibility as expressed in standard deviation down

A

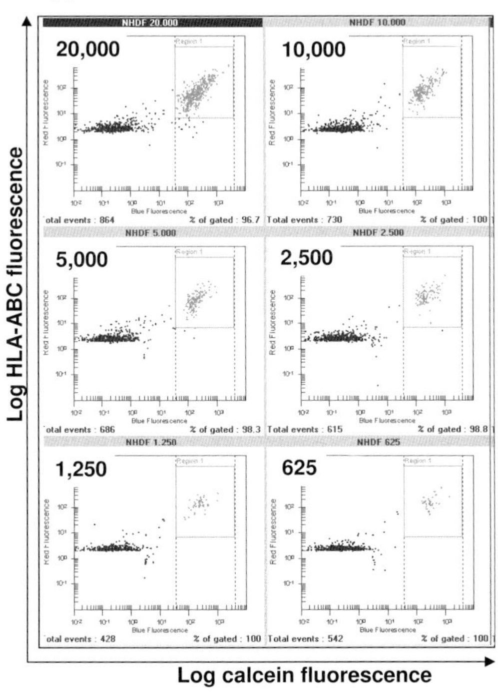

B

# cells in well	CD3 %gated	STD (n=4)
20000	83.7%	3.5%
10000	85.6%	4.1%
5000	87.7%	4.2%
2500	84.0%	3.0%
1250	89.8%	6.5%
625	90.0%	9.3%

# cells in well	CD86 %gated	STD (n=4)
20000	1.8%	2.3%
10000	1.2%	0.8%
5000	1.0%	0.2%
2500	0.1%	0.2%
1250	0.0%	0.0%
625	0.0%	0.0%

Figure 2.3. Staining of few cells and measured using the 2100 Bioanalzyer. (A) Different numbers of NHDFs were loaded into the sample wells of cell chip (10 μl) and stained on-chip with Calcein-AM (x-axis) and anti-human HLA-A, B, C antibodies (y-axis). Numbers of cells used per well are indicated. (B) Evaluation of protein expression with low cell numbers. Different numbers of lymphocytes were stained on chip with Calcein-AM and anti human CD3-APC or with anti-CD86-APC antibodies. Standard deviations (STD) of the percentage of CD3 gated live cells (CD3/calcein double-positive cells within the calcein-positive cell population) are listed.

to 2500 per sample. With lower numbers of cells per sample, the reproducibility deteriorates.

In this case a concentration of 20,000 cells/μL in the chip sample well per sample represents the optimum between reproducibly detecting subpopulations of as little as 5% and using as few cells as possible. The application of antibody staining is not restricted to the use of antigens that are expressed on the cell surface. Following fixation and permeabilization of cells, intracellular antigens can be easily detected.[6]

One of the most important applications for flow cytometry in the area of cancer research is induction of apoptosis in tumor cells. Upon induction of apoptosis, phosphatidylserine (PS), a membrane constituent that is actively confined to the inner leaflet of the cell membrane becomes displayed on the outer leaflet of the membrane. Annexin-V is a member of the family of calcium- and phospholipid-binding proteins with high affinity for PS and can be used as a sensitive probe for PS. The measurement of annexin-V binding to the cell surface can easily be performed in conjunction with the live cell marker calcein, which is a specific indicator for cells with an intact membrane.[10]

In addition to changes in the cell membrane, apoptosis involves the active participation of endogenous cellular enzymes. A family of cysteine proteases (caspases) seems to represent the effector arm of the apoptotic program. Caspase-3 is

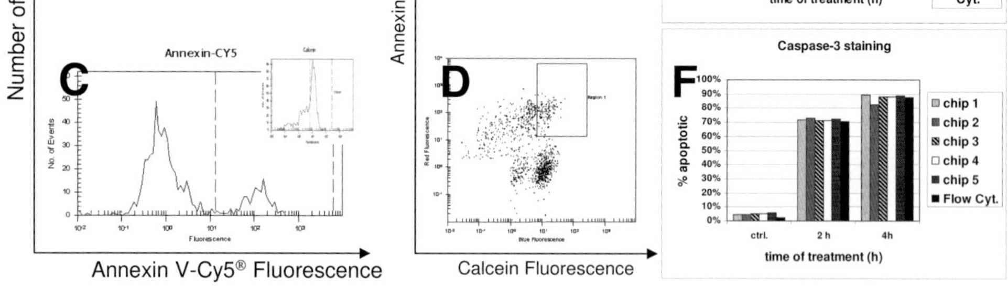

Figure 2.4. Detection of apoptosis in Jurkat cells by annexin-V and active caspase-3 staining. Cells were treated with camptothecin, subsequently stained with either the live cell stain calcein and annexin-V/biotin/Cy5-streptavidin or SYTO16 and active caspase-3/Cy5-labeled antibodies. Cells were washed and analyzed on the microfluidic chip-based system. Annexin-Cy5 histogram of untreated (A) and 24 h treated (C) sample. Dot plot of untreated sample (B) and 24h treated (D) annexin-Cy5 stained sample. Reproducibility of annexin-V (E) and active caspase-3 (F) staining. Cells were treated with camptothecin for different timepoints and stained. Data of 5 chips was compared with measurements of the same samples on a standard flow cytometer (Flow Cyt).

a key protease that is activated during the early stages of apoptosis and is synthesized as an inactive proenzyme. It is cleaved and thereby activated in cells undergoing apoptosis. The expression of active caspase-3 can be studied in apoptotic cells by staining the cells intracellularly with antibodies.

Figure 2.4 shows that data obtained using the microfluidic system in apoptosis experiments compares well with data obtained using a standard flow cytometer. Jurkat cells were treated with camptothecin at different time points to induce apoptosis. The percentage of apoptotic cells detected was almost identical at a given timepoint of induction independently of staining for caspase-3 or PS (compare ~70% apoptotic cells in Fig. 2.4 E, F at 2h). This indicates a nearly simultaneous induction of active caspase-3 and display of phosphatidylserine on the outer leaflet of the cell membrane. A concern of LOC-based flow cytometry is the small channel dimensions which could lead to clogging and possible cell-channel wall interactions. For microfluidic cell analysis, a variety of cell lines have been tested ranging from yeast and fungal spores[11] to mammalian cells[9] without any adverse effects. To avoid cell-channel interactions the microfluidic channels can either be covalently or dynamically coated with charge-neutral compounds (as in the case of glass chips) or manufactured from an inert material (polymer chips).

2.5 CONCLUSIONS

Microfluidics holds the promise to overcome many limitations of today's research. The reduction of scale saves reagents and sample, and often allows an enhanced degree of automation. Flow cytometry has become a method of choice for rapidly analyzing large numbers of cells individually using light-scattering, fluorescence, and absorbance measurements. The power of this method lies in the wide range of cellular parameters that can be determined and in the ability to obtain information on how these parameters are distributed in the cell population. Recently, several groups have demonstrated that chip-based systems can be used for flow cytometry.[3–5] Researchers have either used pressure-driven or electrokinetic flow to pump fluids and cells in microfluidic devices. Pressure driven flow seems more suitable for the analysis of eukaryotic cells since it does less damage to the cells than an electric field and is less dependent on ion content of the buffer.[12]

Flow cytometric analysis of primary cells can present a challenge for researchers due to limited availability and life span of these cells. However, our results show that the microfluidic chip-based technology allows the analysis of very few cells per sample, while maintaining consistent results compared to conventional flow cytometry. This is especially true if the technology is used in conjunction with on-chip staining protocols, i.e., performing the cell staining in the sample wells of the chip. For experimental designs with low statistical significance requirements (e.g., detection of a protein in a clonal cell population by antibody staining), a reduction of cell consumption of up to 80% may give acceptable results with the use of a simple assay protocol. Samples where only a low percentage of cells express the characteristic of interest (e.g., a protein) require not more than 20,000 cells per sample for the analysis. We have shown that subpopulations as small as 5% are reproducibly detected in a sample population of 500–1,000 measured cells.[6] Thus, microfluidic technology achieves a dramatic reduction in the amount of sample needed for analysis. Alternative technologies requiring low cell counts for analysis are laser scanning and imaging cytometry.[13] These technologies yield different types of information as the localization of subcellular staining is possible and additional cell features can be extracted from the data. The typical number of cells that is analyzed per sample is very similar to that needed for microfluidic analysis, as we described here. Using laser scanning and imaging cytometry, the cell samples can be archived and reanalyzed, as cells on the slides are usually fixed. However, the complexity of data analysis, assay development and equipment setup is substantially higher.

The microfluidic system described here was developed for purely analytical purposes. The particular advantages are ease-of-use coupled with easy sample preparation and low cell consumption. Multi-color staining, analysis of large cell populations, and high acquisition rates — hallmarks of standard flow cytometers — are not possible with current commercial LOC cytometry equipment. However, microfluidic technology can not only be used for analysis of cells but also for preparative sorting of bacteria using voltage-driven electrokinetic flow.[14] Future work will focus on the development of integrated reagents and staining procedures on chip to facilitate fewer sample preparation steps and to minimize time-to-result. In addition, improvements of the integrated detection systems will increase speed of analysis and data precision and reproducibility.

ACKNOWLEDGMENTS

The instrument and the microfluidic chips were jointly developed by Caliper Life Sciences and Agilent Technologies. The author thanks his colleagues at Agilent Technologies and Caliper Life Sciences for their work on microfabrication, assay development and engineering.

References

[1] M. Kuschel, C. Buhlmann and T. Preckel, High throughput protein and DNA analysis based on microfluidic on-chip electrophoresis, *Journal of the Association for Laboratory Automation* **10**, 319–326 (2005).

[2] O. Mueller, S. Lightfoot and A. Schroeder, RNA Integrity Number (RIN) – Standardization of RNA Quality Control, *Agilent Technologies Application Note* [online] (2004).

[3] D. P. Schrum, C. T. Culbertson, S. C. Jacobson and J. M. Ramsey, Microchip flow cytometry using electrokinetic focusing. *Anal. Chem.* **71**, 4173–4177 (1999).

[4] M. A. Unger, H. P. Chou, T. Thorsen, A. Scherer and S. R. Quake, Monolithic microfabricated valves and pumps by multilayer soft lithography, *Science* **288**, 113–116 (2000).

[5] S. R. Quake and A. Scherer, From micro- to nanofabrication with soft materials. *Science* **290**, 1536–1540 (2000).

[6] T. Preckel, G. Lüdke, S. D. H. Chan, B. N. Wang, R. Dubrow and C. Buhlmann, Detection of cellular parameters using a microfluidic chip-based system, *Journal of the Association for Laboratory Automation* **7**, 85-89 (2002).

[7] S. Chan, G. Luedke and T. Preckel, Flow cytometric analysis of human primary cells using the Agilent 2100 bioanalyzer and on-chip staining, *Agilent Technologies Application Note* [online] (2002).

[8] T. R. Yeager and R. R. Reddel, Constructing immortalized human cell lines, *Curr. Opin. Biotechnol.* **5**, 465–469 (1999).

[9] S. D. H. Chan, G. Lüdke, M. Valer, C. Buhlmann and T. Preckel, Cytometric analysis of protein expression and apoptosis in human primary cells with a novel microfluidic chip-based system, *Cytometry* **55A**, 119–125 (2003).

[10] X. M. Wang, P. I. Terasaki G. W. Rankin, D. Chia, H. P. Zhong and S. A. Hardy, New microcellular cytotoxicity test based on calcein AM release, *Hum. Immunol.* **37**, 264–70 (1993).

[11] Z. Palková, L. Váchová, M. Valer, T. Preckel, Single cell analysis of yeast, mammalian cells and fungal spores with a microfluidic pressure driven chip-based system, *Cytometry* **59A**, 246–253 (2004).

[12] P. C. H. Li and D. S. Harrison, Transport, manipulation, and reaction of biological cells on-chip using electrokinetic effects, *Annal Chem.* **69**, 1564–1568 (1997).

[13] G. E. Benito, M. L. Sanchez, J. del Pino-Montes, J. J. Calvo, P. Menendez, M. A. Garcia-Marcos, P. Osdoby and A. Orfao, A new cytometric method for the immunophenotypic characterization of bone-derived human osteoclasts, *Cytometry* **50**, 261–266 (2002).

[14] A. Y. Fu, C. Spence, A. Scherer, F. H. Arnold and S. R. Quake, A microfabricated fluorescence-activated cell sorter, *Nature Biotechnol.* **11**, 1109–1111 (1999).

Chapter Three

Personal Flow Cytometers — Luminex

Wayne D. Roth
Chief Engineer & Luminex Fellow Carrage Return
Luminex Corporation, 12212 Technology Blvd, Austin, Texas, USA
wroth@luminexcorp.com

Luminex® Corporation employs flow cytometry to measure color-encoded polystyrene microspheres for detection of multiple analytes in a single sample. This chapter describes the original Luminex instrumentation and later enhancements. Alternate measurement techniques employing two dimensional imaging are also discussed.

3.1 LUMINEX, CYTOMETRY, AND MULTI-ANALYTE MEASUREMENTS

Flow cytometers were originally designed to measure multi-parameter cellular assays. In the 1970s, scientists also started using flow cytometers to analyze micron-sized spherical bead based assays. The magnitude of light scattered by a bead as it passes through a cytometer's excitation laser is proportional to bead diameter, enabling discrimination of multiple bead size populations. Thus, the scatter profile made it possible to identify different bead sets; each set having a unique surface-coupled analyte.

There are considerable obstacles to perfecting the size-multiplexing method. The corresponding surface area change with particle diameter affects the kinetics of the surface reaction, making it desirable to keep size variation to a minimum across the population superset. Yet, each bead diameter must be different enough to discriminate. Further, light scatter is also used to gate out beads that simultaneously pass through the laser spot, and these aggregate signatures must be separable from the individual bead set responses. These complex aspects of using scatter as a classifier limited the achievable analyte count to fewer than 10.

In 1997, Luminex Corporation abandoned the crowded scatter-space discriminator in favor of fluorescently dyed, same-sized microspheres in a system they called FlowMetrix® The kinetic differences among populations were resolved, and the multiplex count was raised from 10, to 64.

The Microflow Cytometer by J S Kim & F S Ligler

www.panstanford.com
978-981-4267-41-0

3.1.1 Internal Dyes and Instrumentation

The FlowMetrix system used polystyrene beads 5.6 micron in diameter and impregnated with two fluorescent dyes at multiple concentrations. The resultant fluorescent emissions from the internal dyes were measured by independent photodetectors, the resultant electrical signals were digitized and identified each particle's class, in real-time, as it passed through the instrument's flow cell.

Having no flow cytometer of its own, Luminex modified a Becton Dickinson's FACScan to read its beads. Luminex created a custom digital signal processing board, which was inserted in the FACScan general purpose I/O (GPIO) bus. The DSP board interpreted the real-time data flow from the FACScan fluorescent detectors. A Microsoft Visual Basic application on an external PC hosted the FlowMetrix user interface, interpreted the data, and also controlled the FACScan acquisition sequence.

3.1.2 Bead Classification Using Internal Dyes

The FACScan's red (633 nm) laser excited both of the bead's internal dyes, which emitted peaks intensities at orange (585 nm) and deep red (650 nm). These wavelengths matched the FACScan's FL2 and FL3 filter responses, and were used as "classification" channels to identify the individual analytes. Multiple beads passing through the interrogation zone too closely together to be individually identified were eliminated from processing by gating on the 90 degree side scatter channel. A two dimensional logarithmic dot plot of the FlowMetrix "bead map" is shown in Fig. 3.1 below.

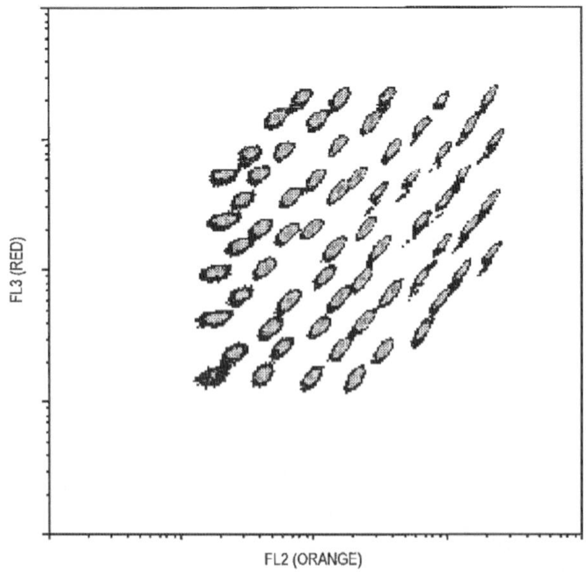

Figure 3.1. Fluorescence dotplot of Flow Metrix microspheres.

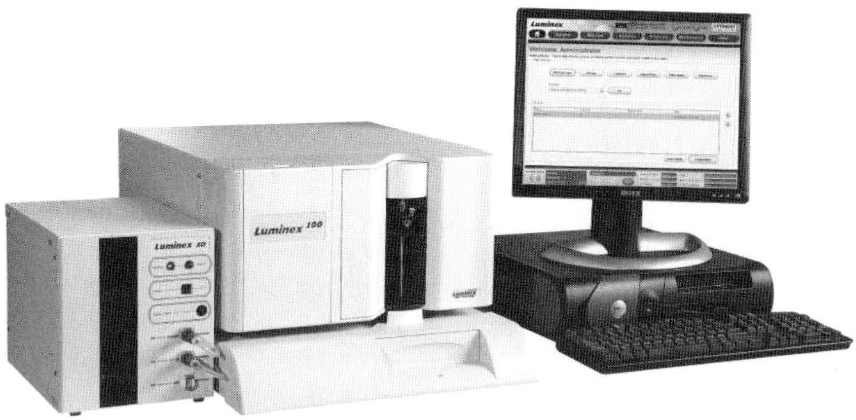

Figure 3.2. The Luminex 100 multiplex analyzer, 96 well plate transport, and sheath delivery system. Color reference – pg. 339.

The number of discrete populations in FlowMetrix was limited to 64 by the distribution of each population (CVs), the separation between populations required to prevent misclassification, dynamic range of the system, and the spectral nature of the internal dyes. The long wavelength tail of the orange dye spilled into the FL3 detector; an increase in the orange dye resulted in a corresponding (but smaller) FL3 increase. Hence, real estate in the extreme lower right and upper left corners of the map was unusable.

3.1.3 Reporter Response

The reaction between antibodies coupled to the bead surfaces and the target analytes were labeled with a fluorescent dye spectrally distinct from the internal dyes.[2] The FACScan's 488 nm laser excited the surface reporter tags, and the resultant emissions at 532 nm were measured by the FL1 PMT. Common reporter dyes included boron-dipyrromethene (BODIPY) and fluorescein iscothiocyanate (FITC), but any green emitting dye with high quantum yield was a candidate.

3.2 THE LUMINEX 100 FLOW CYTOMETER AND xMAP® TECHNOLOGY

In 1999, Luminex marketed the Luminex® 100™. The new instrument employed a number of novel techniques to reduce cost, decrease footprint, and extend the analyte limit from 64 to 100. Shortly after product introduction, an automated plate transport mechanism and sheath delivery pump were offered, enabling Luminex customers to analyze samples in large batches using the 96-well plate format popular on ELISA platforms.

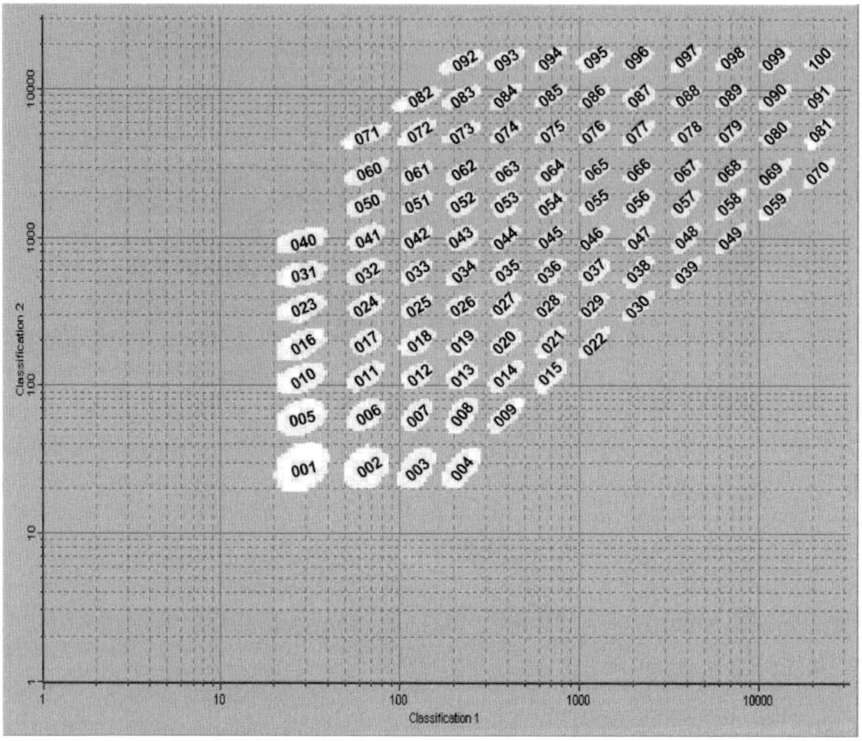

Figure 3.3. The xMap Technology bead map employed by the Luminex 100 system to classify each bead in real time.

3.2.1 Optical Design

The two internal classification dyes previously used on FlowMetrix were modified slightly to enhance photostability without materially affecting the emission signature. The spectral responses of the Luminex 100's photodetectors were matched to the dye using multi-layer interference filters. The exact filter edges were chosen to maximize channel to channel orthogonality, yet minimize signal attenuation, which resulted in the greatest multiplex count. Refer to Fig. 3.3 for a graphical log-log representation of the 100 region bead map used for classification.

3.2.1.1 Classification Excitation and Detection

In the Luminex 100 optics, an inexpensive 10 milliwatt 635 nm diode laser excited the internal classification dyes. Low cost avalanche photo diodes (APD) replaced the photomultiplier tubes (PMT) previously used in the FlowMetrix. While the photon-to-electron conversion gain of an APD is orders of magnitude less than a PMT, their high quantum efficiency (QE) and red spectral sensitivity made them a viable choice to measure the relatively bright emissions from the internal dyes.

A complication in the use of APDs in any application is extreme gain sensitivity to temperature, succinctly described by Su.[3] APD modules that electrically

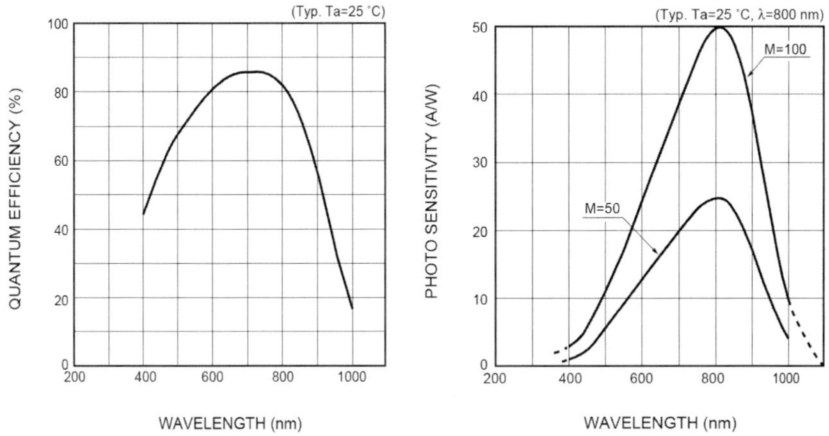

Figure 3.4. Quantum efficiency and spetral sensitivity of an avalance photodiode versus wavelength, peak sensitivity in the red to near-IR regions matches the emission from the days used in the xMap beads.

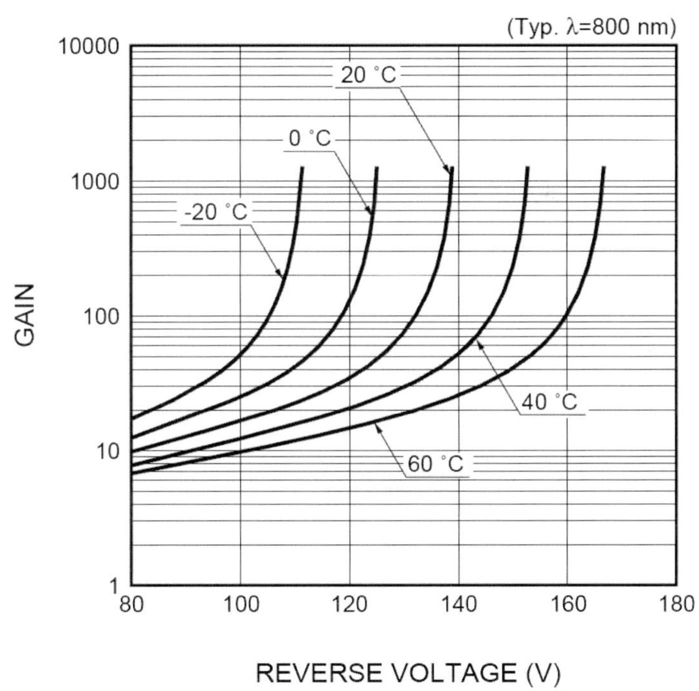

Figure 3.5. APD gain vs temperature and bias voltage.

correct for temperature are available today, but were not during the Luminex 100 design cycle. Although the module effectively compensates gain over temperature, the high cost and large physical form factor of these modules have prevented their use in follow-on Luminex designs.

The reverse bias necessary to achieve a given gain is different from one device to the next and complicates compensation. Each APD is specified by the manufacturer by the bias voltage at a reference temperature (typically 25°C) where the current gain is 60 times that of a silicon photodiode with equivalent surface area. In addition to needing a family of curves defining gain versus temperature, the response of any one curve is not easily described by a polynomial, making a closed form solution difficult.

Luminex chose to develop a temperature compensation technique based on empirical measurements of a family of APDs with different gain of 60 bias voltages. A multidimensional table of gain, temperature, bias voltage and "gain 60 voltage" was constructed, and is consulted by an embedded microprocessor at the beginning of each data acquisition to determine the optimum APD bias at any temperature. The method is described by Roth in US patent 7,318,336[4] and controls gain variance to less than 1/2 percent over the instrument's operating temperature range.

3.2.1.2 Reporter Excitation

Initially, an attempt was made to move the Luminex 100 reporter excitation from the argon excited 488 nm line used in FlowMetrix, to 780 nm, where the ubiquitous laser diodes used in CD players emit, which would save thousands of dollars over the cost of an Argon laser. Unfortunately, the QE of infrared dyes excitable at 780 nm was not sufficient to achieve the desired assay sensitivity.

At the time, 532 nm Nd:YAG single-mode lasers were just beginning to be affordable. While more expensive and shorter lived than a monolithic laser, the green light they emit is readily absorbed by the high QE of commercially available R-Phycoerythrin (PE) dye. The Nd:YAG—PE combination resulted in a relatively bright signal from only a few molecules, and the choice was solidified. In the decade since the introduction of the Luminex 100, competition amongst laser vendors and manufacturing process improvements have again lowered the cost of Nd:YAG lasers by more than a factor of two and simultaneously extended the lifetime by an order of magnitude. Although advances in solid state 488 nm lasers have evolved as well, primarily due to the argon laser replacement market, their cost tends to be higher for the same power levels as Nd:YAG lasers, and Raman scattering by water molecules can be problematic at that wavelength.

3.2.1.3 Reporter Detection

While semiconductor-based photon counting technology, such as Hamamatsu's multi-pixel photon counter (MPPC), shows promise today for application in flow cytometery, in the 1990s, the 8 decade amplification capability and low dark current of a PMT made it the logical choice to measure the potentially low light levels emitted from fluorescent tags attached to the bead surface. Luminex chose a Hamamatsu HC120 series PMT module for the reporter detector. The module includes a Cockcroft-Walton voltage multiplier, and with a 0 to 1.0 volt input range;

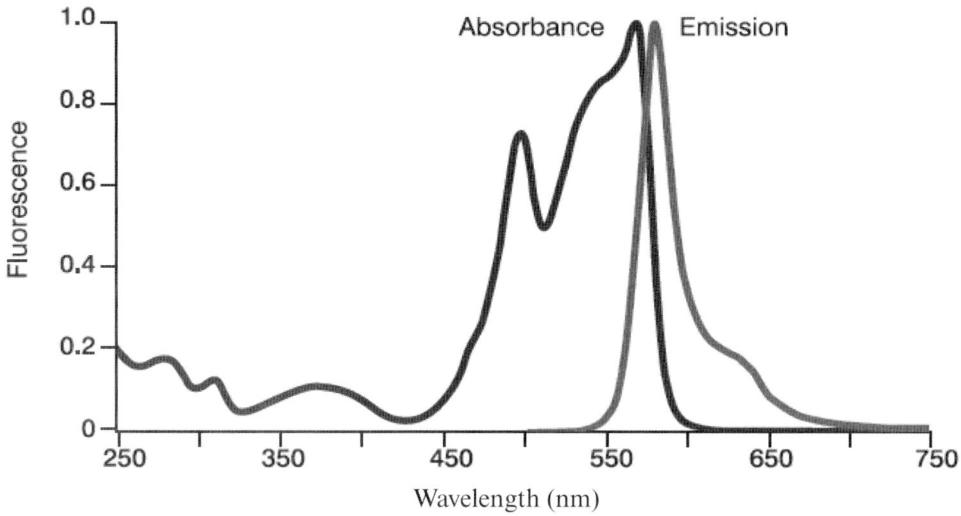

Figure 3.6. R-phycoerythrin absorbance and emission versus wavelength.

the HC120's bias is easily controlled by low voltage electronics. Other advantages to the module include magnetic shielding and compact size. A variety of tube options are available; Luminex chose a low dark current tube with a peak QE that best fit the spectrum of R-Phyrcoerythrin.

3.2.1.4 Background Light Reduction

The chief inhibitor of a low limit of detection in a flow cytometer is often the amount of stray light entering the detectors. The resultant shot noise overwhelms small signals in proportion to the stray light passed by the optical filters. Thus, one of the most critical components considered during the Luminex design was the width and blocking dynamic range of the filters that modify the light impinging on the detectors. The most deleterious wavelength is that of the excitation laser, since a bead's scattered light is typically brighter than the largest fluorescent emission by several orders of magnitude. Custom multi-layer filters in the Luminex 100 attenuate the laser lines by more than 6 decades.

Another place where it is important to reduce stray excitation light is at the cuvette, where changes in index of refraction also give rise to light scatter. In a flow cytometer that uses a faster-moving liquid sheath fluid to hydrodynamically focus the sample of interest, there are three major interfaces to be concerned with; air to glass, glass to sheath, and sheath to sample core. Luminex matched the index of the sheath fluid to the sample, utilized a flat surfaced cuvette to intentionally direct the air to glass reflection in a benign direction, and placed a physical mask at the PMT aperture to block the images of the laser scatter at the sheath to cuvette interface.

3.2.1.5 Laser Geometry

The red and green lasers illuminate the square cuvette from opposite sides, saving on the cost of a combining beamsplitter and allow the orange and red fluorescent emissions to be collected from the remaining sides. The Gaussian beam profile of the red and green lasers was stretched to an elliptical shape, with major axis perpendicular to the direction of particle flow. The width of the major axis was designed to minimize variation in bead-to-bead illumination as it varies horizontal position in the sample core, while not so wide as to significantly dilute the applied energy density. Another consideration in horizontal spot size was keeping the $1/e^2$ width less than that of the cuvette channel to minimize the magnitude of the scatter. The spot's minor axis lies along the axis of particle flow path and determines the temporal width of the electrical pulses out of the photodetectors. As the minor axis is reduced, the pulse width decreases, resulting in the need for more electrical bandwidth to preserve signal intensity, which increases noise by the square root of the bandwidth. Conversely, an increase in the minor axis increases the illumination of unbound fluorochrome molecules in the sample core, resulting in a higher non-specific background light level. A small minor axis is most important to keep the limit of detection low in unwashed assays. Given the 15 micron wide core in the Luminex 100, Luminex went with a 75×25 micron spot geometry. The red and green laser spots were spatially separated, primarily to minimize undesired emissions stimulated by excitation wavelengths far from the respective dye absorbance peaks.

3.2.2 Electrical and Algorithm Design

Log amps were employed in early flow cytometers to enable downstream analog circuitry to operate over a mutli-decade dynamic range. Unfortunately, log amps are noted for their drift over temperature and time, as well as frequent calibration requirements. During the 1990s, advances in analog-to-digital (A/D) converter technology gave rise to inexpensive 14 bit A/D converters with effective dynamic ranges in the order of 4 decades, and sampling rates faster than 1 megasample per second. This made it possible for the Luminex 100 to abandon log amps, and measure the reporter and classification channels using linear amplifiers and A/D converters. The Luminex 100 was the first production flow cytometer to implement direct digital sampling as described in several US Patents by Chandler.[5]

Rather than using the traditional analog peak-hold circuit to capture the maximum instantaneous pulse magnitude and then digitize the result, the analog signal from each photodetector was band limited, amplitude scaled, and presented directly to an A/D converter. A 12-bit device provided sufficient dynamic range for the classification and side-scatter channels, while the reporter utilized a 14-bit converter to improve the signal-to-noise ratio and gain more dynamic range. All converter outputs were fed to a digital signal processor (DSP) that computed the area of each pulse; effectively in real time. Direct digital sampling also allowed the

DSP to provide DC restoration directly from the signal itself, eliminating the cost of extra signal conditioning circuitry.

Another benefit to avoiding the analog peak hold is what Luminex calls a "zero dead time" architecture. A peak-hold circuit, typically implemented by storing charge on a capacitor must be measured, and then discharged, before the circuit is available to capture another peak. Often, a second particle arrives in the flow stream before the discharge process is complete and the subsequent event is lost. In the Luminex 100 implementation, the DSP quickly saves the waveform of any event of interest to a small circular memory buffer, and queues the buffer segment for the more computationally intensive pulse area measurement, which can be done between pulses. In the Luminex 100, the downstream processing by the PC, not the DSP hardware or algorithms, was found to be the limiting factor in system throughput. More information on the technique can be found in the Chandler patents.

The output of the DSP is a continuous stream of calibrated fluorescent intensities (SS, CL1, CL2, and RP1), which is transmitted over a RS232 or USB cable to a personal computer. A continuous process runs on the PC that uses a predetermined discriminant function table (the bead map) to classify each bead event in real time and route the reporter measurement to the appropriate analyte database. One significant advantage of Luminex's real time classification, as opposed to follow-on bead-based systems, is the ability to determine when enough events have been collected to generate acceptable statistics for each analyte. This permits early termination of the acquisition when conditions are met, allowing the system to move on to the next sample andgreatly improve system throughput. Details of the classification algorithms can be found in additional patents by Chandler.[6]

3.2.3 Luminex 100 Fluidic Design

The Luminex 100 particle flow is a driven by a constant pressure system. The sheath supply is driven by a diaphragm-based air pump and is governed by spring-loaded mechanical regulator. Variations in cuvette geometry require a slightly different pressure setting for each instrument to achieve identical flow rates and keep the characteristics of the hydrodynamic focusing equivalent among instruments.

An industry standard syringe pump aspirates sample fluid into a 250 microliter sample loop. The sample is subsequently injected into the cuvette at 1 microliter per second. To minimize sample carryover, the maximum aspiration volume is restricted to less than the volume of the loop, which keeps sample out of the syringe, where it can contaminate the seals.

To prevent sample debris from clogging the cuvette, a laser drilled mechanical filter is appended to the sample probe inlet. The pores of the filter are 125 microns in diameter, which is less than the diameter of the cuvette. Thus, if an obstruction does occur, it is located at the easily accessed sample probe, rather than deep within the bowels of the instrument.

Since the Luminex 100 is typically used to process sample batches from a 96-well plate, the time between sample preparation and acquisition can be up to an hour at the last well measured. Microspheres are heavier than the sample buffer, and tend to settle out of solution over time. In order to mitigate the resultant drop in bead concentration in an aspirated sample, the syringe pump performs a series of rapid aspirate-dispense cycles to create eddies in the sample volume, which resuspends the beads from the well bottom before each draw into the sample loop,. Cycling of the sample also tends to expel objects that get stuck in the sample probe filter pores.

3.3 TECHNOLOGY ENHANCEMENTS POST LUMINEX 100

In addition to the Luminex 100, Luminex has offered three flow-based products using the same electronics, and a similar optical configuration. First out was the Luminex® 100e™ embedded detector that is the heart of Bio-Rad's BioPlex 2200 system. The Luminex® HTS™ system, designed for high throughput users (such as those in the pharmaceutical industry), was coupled to a multi-probe sampling fluid handler that could move samples in parallel and had faster sheath and sample flow rates. In 2005, Luminex launched the Luminex®200™, which maintained the 100 analyte limit, but improved the user experience with a more robust sample probe mechanism and other enhancements to increase reliability and ease maintenance.

Several improvements to the Luminex 100 family of products were rolled into the next generation flow cytometer FLEXMAP™ 3D first sold in November 2008. Many of the enhancements described below are employed in the FLEXMAP 3D design and are generally applicable to traditional flow cytometry.

3.3.1 *Increasing Multiplex Capability*

Luminex's core competency is multiplexing via internal dyes. The most logical method to increase the discrete analyte count was addition of another internal dye and an APD spectrally matched to that dye's emission profile. The new dye fluoresces at a higher wavelength than the others, but is excitable at the same 635 nm wavelength, negating the need for another laser. FLEXMAP 3D is capable of discriminating at least 500 individual bead populations, a five-fold improvement over earlier instruments.

Similar to the Luminex 100 two-dye system, the additional channel is not 100% orthogonal to the others, resulting in a classification space that is far from cube shaped. A three-dimensional map was constructed based on the original map to preserve compatibility with legacy two-dye bead assays. The addition of another dimension required the new classification "regions" to be an ellipsoidal shape that was more complex to implement, and the sheer number of them also made it difficult to optimally tailor the region positions and shapes using legacy empirical methods. This time, an algorithm was designed and programmed in the

C# language, which computed optimal ellipsoid axis lengths and orientations and also defined the spacing between ellipsoids, in order to maximize region count and minimize misclassification.

3.3.2 Increasing Throughput

FLEXMAP 3D acquires data from a sample plate faster than legacy systems via two main mechanisms; addition of a second syringe pump, and doubling the sample injection rate. The added pump enables a sample to be aspirated while the other is dispensing the previous sample through the cuvette, thus the cuvette is not dormant while sample is moved in and out of the tubing. Matching the doubled sample rate with the sheath flow was not necessary for acceptable CVs, and keeps generated waste fluid to a minimum. The combination more than halves the time to read a plate in most assays.

3.3.3 Improving the Signal

To increase the signal-to-noise ratio of the pulse measurements from dim beads, the resolution of the A/D converter was increased from 14 to 16 bits, the sample rate increased by a factor of 8 to 10 MHz, and a moving sum search algorithm was employed in a FPGA. Details of the algorithm can be found in US patent 7,274.316 by Moore.[7]

3.3.4 Viscosity Compensation

A factor complicating the performance of flow cytometers over temperature is variation of sheath viscosity. In Luminex's air over sheath system, viscosity changes by an impressive 42% over the modest operating range of 15 to 30°C. In a constant pressure system, flow rate is inversely proportional to viscosity, and thus the particle dwell time in the laser spot and transit time between lasers, changes proportionally. Transit time is a dominant effect to counter, as the Luminex system triggers on bead scatter as it transitions first through the red laser spot, and if the velocity changes, it is difficult to know exactly when to sample the spatially separated reporter signal. For beads with bright reporter values, time-searching for a pulse maximum eliminates measurement inaccuracy. However, this does not work for the dimmer beads that are partially buried in noise and artificially increases the reported signal.

One way to keep the flow rate steady is to monitor the sheath temperature and adjust the pressure via a viscosity look up table. The table can be built empirically or computed from Poiseuille's equation for a fluid flow through a uniform straight pipe[11] shown in Eq. 3.1.

$$\Delta P = \frac{8Q\mu L}{\pi R^4} \tag{3.1}$$

where,

ΔP = pressure drop
Q = volumetric flow rate
u = viscosity
L = tube length
R = tube radius

Sheath fluid is composed primarily of water, the viscosity of which has been determined empirically and can found in table form, or computed from a second order polynomial that was later created and fit to the measured data.[8]

3.3.5 Extending Dynamic Range

Many users have found that the Luminex 100's limit of detection (LOD) can be improved in many assays by increasing the original default gain of the photomultiplier tube that collects the light from the fluorescent molecules on the bead surface. While the LOD is improved, the ability to measure high concentration analytes accurately is impeded due to the instrument's fixed dynamic range. The signal from the brighter beads begins to clip, and the measurement becomes non-linear.

There are a host of ways to increase dynamic range. The most straightforward approach modifies the analog electronics to accept larger signals before clipping. This approach would be expensive to implement and at most, could increase the dynamic range in the neighborhood of a single decade. Additionally, the implementation necessary is likely to increase the electronic noise of the system, which would then hamper the LOD.

A brute force method that can extend dynamic range by as much as 4 or more decades is to split off a portion of the fluorescent light emitted by the beads with a beamsplitter and feed it to a second PMT, or APD. The downside is the added expense of the beamsplitter, detector, and electronics necessary to process the additional signal.

Another approach is accomplished on the excitation side via a very simple and inexpensive modification to the laser focusing optics. As shown in Fig. 3.7 below, a flat of optically clear material the thickness of a glass microscope slide is inserted at an angle between the green excitation laser and the cuvette through which the beads flow. Since the index of refraction η of the flat plate is greater than 1 in air, internal reflection splits the laser beam into a series of decreasing intensity beams. The vertical distance between the beams is proportional to the plate angle, and the amount by which each successive beam decreases in intensity is inversely proportional to the index of refraction. Sapphire is a good choice for the plate because of its relatively high η of approximately 1.77 at 532 nm, which results in a reduction ratio of 167 for each successive beam as it exits the plate. The beam series illuminates different zones on the cuvette, and an image of all the zones fills the active area of a single PMT. A digital signal processor measures each resultant electrical pulse from the succession of spots and picks the brightest one that has not driven the signal into the non-linear region of the electronics. Realistically, only the main beam and the first reflection are of any value. An example implementation that enables detection of more than two decades of additional dynamic range is described in detail below.

Referring to Figs. 3.7 and 3.8, as each bead traverses through the cuvette, it is first illuminated by the attenuated green laser spot in zone 1. The fluorescence emitted during this time period is captured by the reporter channel PMT, and is digitized by the A/D converter. Next, the bead exits zone 1 and enters zone 2, which is illuminated by the much brighter primary laser beam. Similarly, the resultant zone 2 PMT signal is also digitized and stored by the same A/D converter. The area under the zone 1 and zone 2 signals is computed by the digital signal processing electronics and is available to a downstream software algorithm which decides which of the two signals best represents the fluorescent intensity of the bead. For

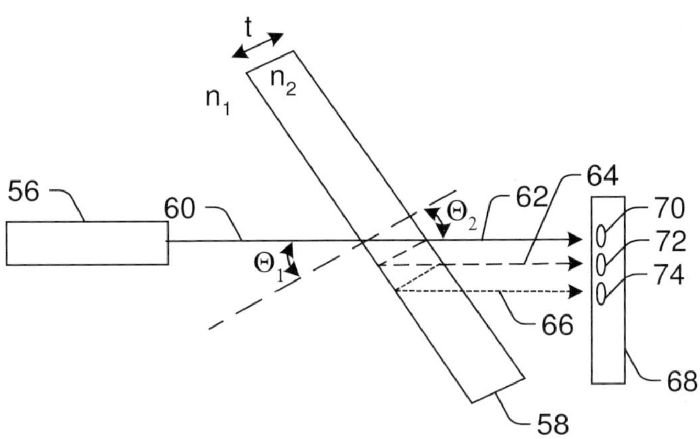

Figure 3.7. Diagram showing laser sport (60) reflecting at glass plate (58) surfaces resulting in main beam (62) and attenutated beam (64). As the bead traverses the cuvette (68) it is illuminated with two different intensity spots at zone 1(72) and zone 2(70).

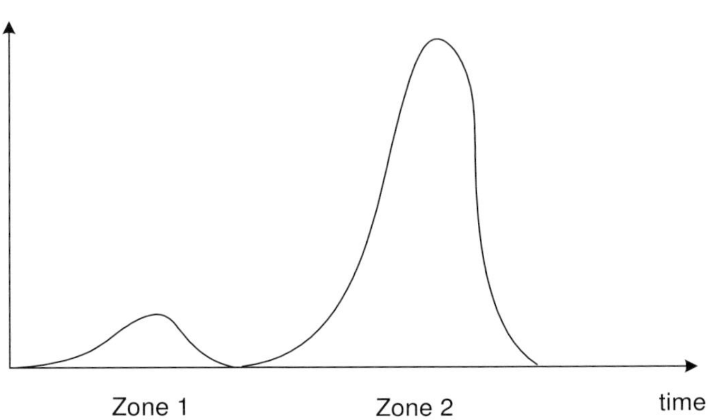

Figure 3.8. Time domain pulsetrain from the photodetector showing the main pulse at zone 2 and attunuated at zone 1.

very bright beads that the electronics would clip signal peaks, the smaller zone 1 signal is chosen. Conversely, for very dim signals where the zone 1 measurement is small, and likely embedded in electronic noise, the larger zone 2 signal is utilized. In this manner, the dynamic range is effectively extended by more than 2 logs, enabling the PMT gain to be increased for better LOD without the linearity problems previously encountered at high analyte concentrations. Note that the order of the illumination is important; to minimize photobleaching, it is best to illuminate the bead first with the dim beam, and then the bright beam. For more detail, refer to US patent 7,362,432 by Roth.[9]

3.4 FUTURE TECHNOLOGIES FOR MULTIPLEXED ANALYTES

Luminex is currently investigating alternate techniques to benchtop flow cytometery. Most of the readers of this book will be interested, and potentially well versed, in the evolution of miniaturized flow systems. Luminex has performed research on semiconductor flow systems — the details can be found in a US patent application by Schilffarth.[10] However, in the near term, the most promising approach for Luminex to produce an instrument with a lower cost and complexity is to divert from the continuous flow concept in favor of static imaging.

3.4.1 Static CCD Imaging of Beads

Cost can be a significant obstacle to adoption of any new technology. Since Phycoerythrin has very high QE, it has proven to be the fluorescent label of choice for Luminex bead-based assays. Moving the excitation wavelength far from PE's optimal absorbance of 488-532 nm to enable use of an inexpensive semiconductor laser is not feasible. Since coherence is not a requirement, the broadband emission of a light emitting diode (LED) and equally wide absorbance spectra of the dye, makes LEDs a viable excitation source.

Unfortunately, given a flow cytometer's typical sub-ten-microsecond dwell time, the intensity of LEDs available today is insufficient to excite the fluorescent molecules enough to result in emission of sufficient number of photons to overcome the dark noise of photodetectors. Also, focusing an array of LED emitters to a spot just tens of microns in width will not result in a higher energy density than is emitted at any one LED surface, effectively killing simple parallelism as a solution. But, if you can keep the beads in the light long enough, the increased dwell time does provide enough photons for an accurate measurement. Essentially, time is traded for intensity.

The resultant dwell time difference in flow systems that would implement today's LEDs instead of lasers would be measured in decades, and that just is not yet practical. What is practical is distribution of the beads on a flat surface, epi-illuminating them with multiple LEDs with combined emission area less than that of the bead surface area, and then capturing images at the wavelengths of interest with a 2-dimensional CCD or CMOS sensor. In essence, this is an epi-illuminated fluorescence microscope that uses inexpensive LEDs for the light source.

The concept is simple enough, if you can keep the beads far enough apart and within the depth of field of the lens, you can easily measure individual beads by finding them in the image and integrating the pertinent intensities at each wavelength. The key to keeping the beads separated and on a plane was made possible when Luminex offered MagPlex™magnetic beads in 2007, often used to make wash steps easy during assay preparation. A simple permanent magnet can be used to draw the beads out of solution and onto a flat plane. As an added benefit, the final wash necessary to remove free fluorochrome from the assay can be done right in the system. A requirement necessary to achieve high bead count (for better statistics) and a wide dynamic range, is to use a lens with sufficient resolution, often measured by its modulation transfer function (MTF), such that the lens doesn't smear the bead image and pollute dimmer beads with light from a nearby bright bead. For more information on Luminex's work in this area, refer to US patent applications by Roth.[11,12]

3.5 CONCLUSIONS AND OUTLOOK

This article has discussed the evolution of the bead-based multiplexing techniques employed by Luminex Corporation. The Luminex platform offers a cost effective means to simultaneously measure multiple analytes in a single sample. In just a little over a decade, Luminex has increased the analyte count possible by early flow systems by more than a factor of 50. The company has also begun to investigate alternate bead measurement techniques, including two-dimensional imaging, but flow cytometry still offers significant advantages and will not be retired any time soon.

ACKNOWLEDGMENTS

The design of the Luminex 100, and many of the follow-on products was successful largely to the following individuals: Dr. Howard Shapiro as professor and consultant, whose unmatched broad knowledge base made it possible to educate us in the art of flow cytometry; Van Chandler, a prolific inventor who's experience in digital signal processing has solidified our measurement accuracy and delivered an extensive patent portfolio; Ted Calvin, who created and perfected our first optical platform and also helped us work through many important conceptual details; and Paul Pempsell, who's extreme attention to detail resulted in an almost noise-free analog design that has required little modification over the years. I would also like to thank Dr. John Carrano, a brilliant (and tough) leader who has helped me to install some measure of discipline, and also Dr. Sherry Dunbar, Dr. Don Chandler, and Dr. Kurt Hoffacker, who have helped a simple engineer understand the pertinent concepts of biology and chemistry as they relate to flow cytometry. And most importantly, I'd like to thank my wife Theresa, who has graciously allowed me to divert much of my attention (in excess, I will admit) to Luminex over the past decade.

References

[1] J. R. Kettman, T. Davies, D. Chandler, K. G. Oliver and R. J. Fulton, Classification and properties of 64 multiplexed microsphere sets, *Cytometry* **33**, 234–243 (1998).

[2] R. J. Fulton, R. L. McDade, P. L. Smith, L. J. Kienker and J. R. Kettman, Advanced multiplexed analysis with the FlowMetrix system, *Clin. Chem.* **43**(9), 1749–1756 (1997).

[3] Y. K. Su, C. Y. Chang and T. S. Wu, Temperature dependant characteristics of a PIN avalanche photodiode/APD/ in Ge,Si and GaAs, *Optical and Quantum Electronics* **11**, 109–117 (1979).

[4] W. D. Roth and D. E. Moore, Methods for controlling one or more parameters of a flow cytometer type measurement system. U. S. Patent 7, 318, 336, January 15 (2008).

[5] V. S. Chandler, Zero Dead time architecture for flow cytometer. US Patents 6,411,904 6,658,357 and 7,047,138, May 13 (1999).

[6] V. S. Chandler, J. R. Fulton and M. B. Chandler, Multiplexed analysis of clinical specimens apparatus and methods. US Patents 5,981,180 6,524,793 and 6,939,720, November 9 (1999).

[7] D. E. Moore, System and method for managing data from a flow analyzer. US Patent 7,274,316, November 16 (2005).

[8] The Viscosity of Water 0°C To 100°C; Weast, R. C., Ed. CRC *Handbook of Chemistry and Physics* 61st ed; pp F-48–F49 (1980).

[9] W. D. Roth, Method and systems for dynamic range expansion. US Patent **7**, 362–432, April 22 (2008).

[10] A. R. Schilffarth, W. R. Deicher, J. C. Carrano and J. C. Phillips, Chip-based flow cytometer type systems for analyzing fluorescently tagged particles. *US Patent Application* 20070281311A1, November 22 (2007).

[11] W. D. Roth, Methods and systems for image data processing. *US Patent Application* 20070064990A1, March 22 (2007).

[12] W. D. Roth, Systems and methods for performing measurements of one or more materials. *US Patent Application* 20070281311A1, December (2007).

Chapter Four

The Accuri C6 Flow Cytometer® — A Small Revolution

Collin Rich* and Grant Howes†

Accuri Cytometers, Inc. 173 Parkland Plaza, Ann Arbor, Michigan 48103, USA
**crich@AccuriCytometers.com*
†*ghowes@AccuriCytometers.com*

Flow cytometers are typically expensive, high-maintenance instruments requiring extensive training and specialized operators. As a result, most flow cytometer users have to share instruments or submit samples to a core facility for analysis. The recent introduction of affordable, compact systems — such as the Accuri C6 Flow Cytometer — has increased the number and type of laboratories that can adopt flow cytometry and its applications, to the extent that flow cytometers can now be taken out of the laboratory and into the field. The C6 represents a major step forward in realizing the concept of miniaturizing and increasing the accessibility of a commercially available flow cytometer. Instrument specifications have been chosen to match the performance of larger instruments and to afford backwards-compatibility with experiments already being performed in the life sciences. Focusing on the interfaces between sub-systems common to all flow cytometers, namely fluidics, optics and electronics, the components of the C6 were optimized to maximize system performance, including creating intuitive, user-friendly software. The resulting innovations miniaturized the entire instrument, a major step in fulfilling the definition of a Microflow Cytometer.

4.1 INTRODUCTION

The term "microflow cytometry" can mean many things to many people; however, it typically implies some form of small, disposable module manufactured using microfabrication techniques (hence "microflow"). The simplest might be nothing more than a convenient sample carrier, whereas a very complex module could include active sample preparation (mixing, incubation) using pre-measured reagent contents, and incorporate valves and flow channels.

The Microflow Cytometer by J S Kim & F S Ligler

www.panstanford.com
978-981-4267-41-0

Until recently, entrenched patterns in the microflow cytometry industry have tended to restrict, rather than foster, growth and mainstream acceptance of the technology. Historically, the industry has incrementally developed new hardware and software features based on existing designs, rather than innovating new technology to reduce complexity, size, and/or cost. Conventional flow cytometers are typically expensive, large, high-maintenance instruments that require extensive training and specialized operators. As a result, most flow cytometer users have to share instruments or submit samples to a core facility for analysis. Centrifugation and PCR systems, by contrast, have experienced much more widespread adoption, enabled by many user-friendly design innovations.

Although microflow cytometry has been gaining technical momentum, this approach requires widespread adoption in order to be commercially sustainable. Specifically, for a cartridge-based system to be competitive, the per-cartridge cost to the user must be less than the cost and time of using conventional sample management methods (e.g. a pipette and standard sample tube and/or well-plate). It is important to recognize that microfabrication is typically cost-effective only when executed in high volume (e.g. tens to hundreds of thousands of units), and this becomes more significant the more complex the cartridge. Accordingly, the market for a cartridge-based system must be able to support a sufficiently large consumables revenue stream to maintain acceptable production costs. Furthermore, disposable cartridges are often promoted as beneficial because of their application-specific nature and convenience; this, in turn, means that a particular cartridge design may be dependent on a single market having sufficient volume to be cost-effective.

In order to expand the market to the point where mass adoption of microflow cytometry is commercially viable, several barriers must be overcome industry-wide:

i. Users must recognize opportunities to improve their research effectiveness using cytometry;
ii. Cytometry instruments must be sufficiently user-friendly and self-maintaining to mitigate the need for extensive operator training;
iii. Cytometry instruments must be sufficiently affordable that interested users can obtain them; and,
iv. Cytometry instruments must have sufficiently flexible configuration (e.g. particle type, filter bands, software features) that users can explore new applications without requiring extensive, technician-intensive customization.

4.2 DESIGN GOALS

Even if the market is not yet large enough to support widespread adoption of cartridge-based microflow systems, there is certainly room for technological innovation to overcome the four aforementioned barriers and aggressively proliferate cytometry as a research technique using more conventional sample-handling methods. This unaddressed market potential was, in fact, the primary impetus for development of the Accuri C6 flow cytometer. Based on the results of extensive, field-based market research, the founders of Accuri Cytometers set about building an entirely new type of flow cytometer. They created their designs from scratch, taking advantage of leading-edge technology when possible, and inventing new technology when necessary. The resulting design of the C6 was based on several core principles:

i. Customers first: Researchers' needs and requirements are the main focus;
ii. Simplicity: Use the simplest design that can possibly work
iii. Performance: Product specifications should meet or exceed those of more expensive instruments;
iv. Innovation: Question all existing design assumptions; build new products from the ground up using advanced engineering theory; and
v. Advanced technology: Employ recent advances in technology, and if needed, invent new ones.

The outcome of this design philosophy is the C6 Flow Cytometer System: a low-cost, full-featured instrument that is easy-to-use, is back-compatible with standard protocols, produces results that are comparable with existing instruments and requires minimal maintenance.

When it began shipping in February of 2008, the Accuri C6 Flow Cytometer System represented a major step forward in realizing the concept of miniaturizing and increasing the accessibility of a commercially available flow cytometer. For the first time, researchers had access to a full-function analytical flow cytometer in a system covering only two square feet of bench space, at an affordable price, making cytometry easily accessible to a much wider range of life scientists than had previously been possible. Standing in the gap between large, complex legacy instruments of the past and ultra miniaturized "microflow cytometer" technology of the future, the small size and user-friendly features of instruments like the C6 have begun to transform the scientific community's perceptions of what cytometry means – similar to the transformation experienced by society at large as mainframe computing gave way to the PC.

In addition to expansion of the present cytometry market through instruments like the Accuri C6, continued miniaturization of the requisite technology subsystems is also critical to the future viability of microflow cytometry. The term "microflow cytometry" by definition focuses on miniaturization. And while much microflow cytometry research has addressed miniaturization of the particle path, doing so is of little benefit unless the balance of the system is also miniaturized. Visionary concepts of bench top, handheld, or even wearable cytometry systems require miniaturized electronics, fluidics, optics and power supply.

In the process of miniaturization, it is often tempting to seek integration of all components into a single module or substrate as the pièce-de-résistance of engineering accomplishment; in most cases however, this is not the best approach from a manufacturability or economic perspective. Just because something can be integrated does not necessarily mean it should be — as the Microelectromechanical Systems [MEMS] community has often demonstrated. By contrast, wise parsing of the system into a balance of integrated and discrete modules will typically yield the most cost-effective solution.

The miniaturization of optics, fluidics, and electronics systems, as well as simplification of user interface features, has been critical to the development of the Accuri C6 instrument. It would not be unreasonable to envision retooling the Accuri platform into a cartridge-based bench top system, given that it already incorporates substantial miniaturization in its optics, fluidics, and electronics relative to legacy systems. As such, the Accuri development effort already contributes to the advancement of microflow technology even if the Accuri system is not a microflow cytometer *per se*.

The balance of this chapter will explore the many innovations in the Accuri C6 to illustrate how they facilitate expansion of the existing cytometry market, as well as contribute to foundational technologies for future microflow cytometry systems.

4.3 DEVELOPMENT PROCESS

Development of the Accuri C6 was uniquely market-driven, as opposed to being built around a preexisting cytometry technology. Although this presented significant R&D challenges, it had the benefit of allowing the design to be largely dictated by market needs. Several of the more prominent qualitative functions/features (along with examples) for the ideal instrument are as follows:

i. User-friendly: Minimize or eliminate controls that contribute to user confusion (e.g. PMT voltage gain). Minimize the number of steps required to collect data (e.g. allow the user to open the software, turn on the instrument, and click "run" with little or no setup required). Always preserve the original data so that post-processing steps like gating and compensation can be freely changed and/or undone.
ii. Low-maintenance: Have the instrument automatically run cleaning cycles on startup and shutdown. Make any user-serviceable parts easy to reach and replace. Eliminate any need for user alignment of optics. Only provide design flexibility when useful to the user (e.g. user-changeable emission filters).
iii. Full backward-compatibility: Provide sensitivity, laser wavelengths, and default filter selection that match legacy instruments so that a user's existing protocols can be easily transferred to the C6.
iv. Affordable: Use engineering creativity and innovation to design a quality, high-performance instrument that is affordable to most labs likely to use cytometry.
v. Rugged: Design the instrument to ship in a simple cardboard box via commercial shippers, with no on-site alignment, so that instruments can be easily

transported and/or serviced from a central repair depot. Eliminate the need for on-site technician servicing.

In order to fulfill the aforementioned vision of the ideal instrument, a number of standards were established. In addition to traditional quantitative specifications for function and performance, more qualitative tools like *user personas* and *time trials* have played an important role in guiding development. A summary of key standards are thus:

4.3.1 User Personas

One of the seminal steps in Accuri's development effort was the establishment of several fictional but relevant user *personas* for its C6 instrument. A persona is a picture and description characterizing a targeted user of the instrument including experience, skill levels, work responsibilities, and even a little lifestyle background. Below are two of the user personas for the C6. Personas were used as test scenarios to validate specifications, user interface features, required operator skills, and assumptions made about users when designing the C6.

Persona 1: Brad Shaw, Second year graduate student in the Biological Sciences Program, University of Wisconsin.

Goals:

- Wants to publish 2 or 3 times in order to finish his dissertation and be a good candidate for a prime post-doctorate position.
- Wants to "do good science"
- Wants to understand flow cytometry thoroughly as a tool so that he can confidently design his ground-breaking immunology experiments
- Wants a short learning curve so he can stop practicing and start collecting data
- Likes to learn from colleagues, but wishes he could teach himself
- Wants to get today's work done quickly so he can leave lab and go home
- Not really sure how the flow cytometer works, but is curious

Level of Expertise:

- Brad is just beginning to learn flow cytometry from a senior graduate student in his lab.
- He knows that he must soon use flow cytometry independently as a central tool for his dissertation research. He was introduced to flow cytometry in his graduate courses that reviewed topics and methods in molecular and cellular biology.
- He has read multiple articles which report flow cytometry results, so he understands in a general sense the utility of flow cytometry, but he is not a power user (yet!).

Job Context:

- Academic

- Good school with typical funding for research.

Persona 2: Emily Menendez, PhD., Post Doctoral position, Immunology Department, University of Wisconsin.

Goals:

- PUBLISH! Ensure experiments work, gather data as fast as possible
- Do good science and obtain clean data from her experiments.
- No time to waste on anything that does not bring her closer to landing a good job
- Network with an eye toward a permanent position (still exploring academic track as well as industry)
- Have a life

Level of Expertise:

- 2nd or 3rd year post doctoral researcher
- A flow cytometry "power user". Flow cytometry was central to her dissertation work and her current job.

Job Context:

- Academia
- She has been awarded a NRSA fellowship to support her salary.
- Doing own research and managing (baby-sitting) a 1st year rotating graduate student working with her for the quarter — wants to be a good teacher but worries that the student is slowing her down
- Married two years with no children, but worried about work/life balance

As an example of how personas contribute to instrument design, one can consider the issue of PMT voltage setting on a cytometer. Conventional cytometers require the user to set the high-voltage for each PMT, which in turn determines the PMT's relative sensitivity. Experienced cytometrists fully understand how to set voltages; however, the novice user — perhaps a "Brad" — barely understands how to properly prepare his reagents, let alone the nuances of setting proper PMT sensitivity and why high-voltage should have anything to do with biology (outside of a primordial soup experiment!). Consequently, Accuri pushed the limits of available technology, increasing dynamic range and sensitivity to the point where all of the PMTs in the C6 have a single, fixed voltage that allows them to capture the full range of fluorochrome intensities. Now, Brad does not even need to know that PMT voltages exist, let alone understand how to set them. Ironically, while Brad has no need to adjust instrument settings, users of conventional flow cytometers have difficulty accepting this apparent loss of control.

4.3.2 *Instrument Specifications*

The instrument specifications have been chosen to match the performance of and be backward-compatible with the types of experiments typically run on larger

Table 4.1 Summary of key specifications of the Accuri C6 Flow Cytometer System.

Specification	Description
Excitation wavelengths	488 nm (rated at 20,000hr life)
	640 nm (rated at 20,000hr life)
Laser profile	15 × 75 microns
Scatter detection	Forward (0 degrees, +/−15)
	Side (90 degrees, +/−15)
Emission detection	4 colors, user swappable optical filters
Standard filter set	* FL1 530 / 30 nm (FITC/GFP)
	* FL2 585 / 40 nm (PE/PI)
	* FL3 >670 nm (PE-Cy5, PE-Cy5.5, PerCP-Cy5.5,PE-Cy5, PE-Cy7)
	* FL4 675 / 25 nm (APC)
Optical alignment	Fixed alignment, no maintenance required
Flow cell	200 micron internal diameter quartz capillary
Minimum particle size	1 μm
Minimum sample size	25 μL
Nominal flow rate - variable	10 to 100 μL/min
Recommended Sheath Fluid	0.2 μm filtered DI water
Florescence sensitivity	<750 MESF FITC
Scatter resolution	Resolves human peripheral blood erythrocytes, lymphocytes, monocytes & granulocytes
Fluorescence precision	<3% CV for CEN
Speed	10,000 events/second maximum
Absolute count capability	Yes
Warm-up time	Less than 5 minutes
Instrument size	11"H × 14.3"W × 16.5"D (27.9 × 36.3 × 41.9 cm)
Foot print with fluid tanks	17.3"W × 16.3"D (43.9 × 41.4 cm)
Weight	30 lbs (13.6 kg)
Signal processing	24-bit A/D conversion
Computer interface	USB 2.0

legacy cytometers. The abridged specifications (Table 4.1) below illustrate some of the key elements of the Accuri C6.

4.3.3 Standardized Intensity Bead Set

Accuri regularly uses 8-peak, 488 nm excitation and 6-peak, 633 nm excitation bead sets from Spherotech for assessing instrument performance during manufacturing

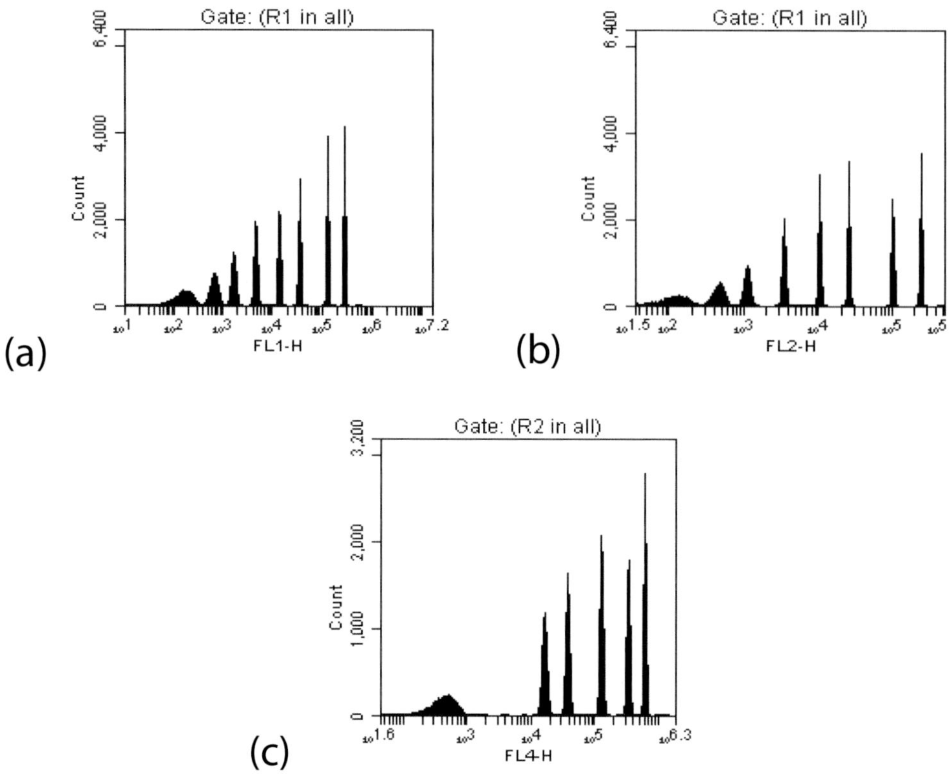

Figure 4.2. Data from Accuri C6 flow Cytometer System showing Spherotech SPHERO™ Rainbow Calibration Particles: (a) un-zoomed data for FL1 with 8 peak beads (b) FL2 data on same bead set showing the Zoom feature, allowing specific portions of the full dynamic range to be displayed (c) zoomed data for FL4 with 6-peak beads.

and to ensure the system is fully functional after delivery. These bead sets are recognized as an industry standard for evaluation of instrument sensitivity and stability. Figure 4.2 shows plots of typical runs using the Spherotech beads. The ability of the C6 to produce data similar to that seen on conventional systems was a key design requirement and the ability to separate the discrete peaks in such a quality control material was a significant achievement.

4.3.4 *User Time Trials*

As part of its user-friendly design evaluation, Accuri sets specific time targets for certain defined user processes. For example, a user with minimal training must be able to set up a new instrument out-of-the-box within a half-hour. Likewise, a 96-well microplate sampler accessory must be set up within fifteen minutes. Note that Accuri instrument set-up and use is simple enough to be successfully captured in a Quick-Start guide, as is now common practice for consumer electronic products.

4.4 MAJOR SYSTEM COMPONENTS

All flow cytometers are composed of three major, interdependent sub-systems: fluidics, optics and electronics. Focusing on the interfaces between the sub-systems, Accuri's team designed all components to optimize overall system performance, including creating intuitive, user-friendly software. This resulted in several interlocking innovations that set the industry on a path to miniaturizing not just the flow cell or capillary, but the whole instrument. While it may not fulfill the final definition of a microflow cytometer, the C6 is certainly a major step in the right direction.

4.4.1 *Fluidics*

Conventional flow cytometers typically use either syringe pumps or pressurized, air-over-water fluidics. Historically these systems have been selected because they can provide smooth, non-pulsatile flow of the sample, which is essential to achieve high performance. These two approaches to fluidics also have significant downfalls, however: they add cost, complexity and size to instruments, and require regular maintenance — a time consuming nuisance. Additionally, a choice can be made between coaxial sheath-sample vs. sheathless sample flow. A system without sheath flow has the advantages of reduced overall complexity and final instrument size, zero bulk reagent cost, and a reduction in potentially bio-hazardous waste production. Disadvantages, however, include an inability to provide the required flow precision for applications such as DNA analysis, or the small core size necessary to minimize coincident cells in the detection area when running concentrated samples.

With the aim of maintaining a sheath based design if possible, the C6 design effort sought an alternative to conventional fluidics designs. An ideal approach would use primarily non-pressurized fluidics, avoid expensive syringe pumps, and still reduce the size and cost of the instrument. To that end, peristaltic pumps were very appealing: They are compact, inexpensive, low maintenance and offer direct drive fluid flow. The challenge was to produce a smooth, pulse-free flow from an inherently pulsatile pump. The solution came as an Accuri innovation to regulate the fluid flow. Novel fluidic pulse dampeners, combined with a sophisticated, microprocessor-controlled dynamic feedback system, were developed to produce a precise, pulse-free flow that is user adjustable. Adopting this technology has enabled the C6 to produce consistent, robust control of the velocity of both the sheath and sample fluids to within extremely tight tolerances ($< 0.01\%$).

The production C6 Flow Cytometer fluidics system employs two peristaltic pumps in a "push/pull" configuration (Fig. 4.3). One pump pushes sheath fluid into the flow cell, while the other pulls the combined sample and sheath fluid from the flow cell to the waste tank. The differential between the two pump pressures creates suction. This suction draws the sample up through the sample introduction probe and into the flow cell, where the sample fluid is focused using the

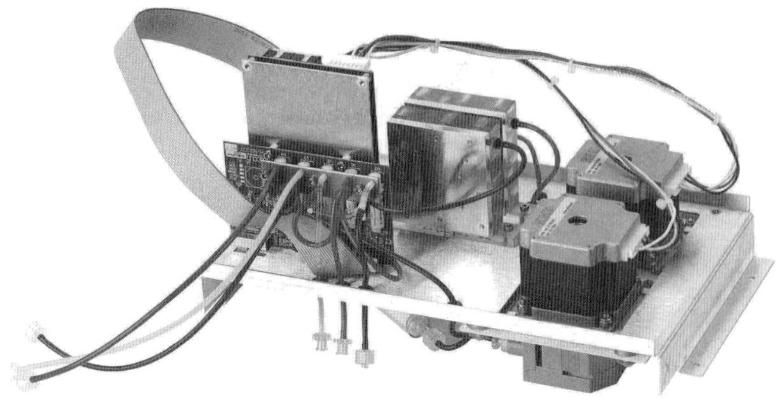

Figure 4.3. Accuri's patented pulse dampeners and pressure sensors allow the use of inexpensive, peristaltic pumps to provide a non-pressurized, zero pulsation, pull/push system with sophisticated microprocessor controlled dynamic feedback. Color reference – pg. 339.

hydrodynamic force applied by the sheath flow. The push/pull design allows for independent control of both the sheath and sample flow speeds, making it possible to precisely control the sample core diameter. This enables the user to quickly fine-tune the fluidics for each individual sample. The precision of the "direct drive" peristaltic pumps combined with the closed-loop controller offers the additional benefit of being able to meter the sample fluid uptake. As a result, the C6 Flow Cytometer can automatically calculate events per unit volume for each run, providing counts and particle concentration.

In summary, the fluidics design of the C6 has achieved the benefits of a hydrodynamically-focused, sheath-fluid-based system, but without the complexity of large pressurization mechanisms, and with minimal reagent cost and waste production.

4.4.2 Optics

The complexity of today's flow cytometers require vibration-isolated optical tables loaded with a complex array of photomultiplier tubes, photodiodes, lenses, filters, beam splitters or mirrors, each aligned in a precise manner to get optimum results. These component-dense setups are expensive to build and are extremely sensitive to jarring or bumping, which results in misalignment that requires service. By contrast, design goals for the Accuri instrument mandated decreasing complexity to lower costs, while simultaneously increasing stability to minimize service requirements.

To achieve these goals, the C6 optic table was designed with the shortest and simplest possible light path. Novel configurations of the optics subsystems were needed to make this possible. First, the system employs two solid-state lasers (488 nm and 640 nm) that are arranged to be co-linear, but modulated so that they are not co-temporal. In addition, a mirror-less, direct light detection scheme was

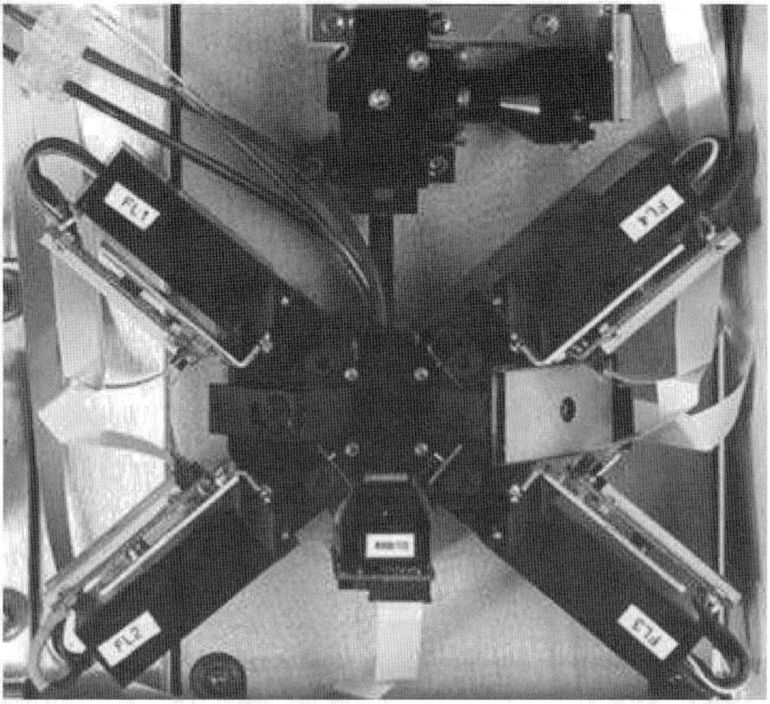

Figure 4.4. Clustered in a "pie" design around the flow cell, the C6 Flow Cytometers photomultiplier tubes maximize light collection and reduce alignment issues. Color reference – pg. 340.

employed. Photomultiplier modules are clustered in a pie-shaped design around the flow cell to maximize light collection and reduce alignment issues (Fig. 4.4). Focused on the same spot, these detectors sample data only when the desired laser combination is exciting the sample. This arrangement makes it possible to assign any of the four fluorescence detectors to read from either of the lasers. Most lenses and optics are self-aligned at the point of manufacture by using precisely-controlled machining tolerances (sometimes 0.0005 in or less). Likewise, the interference filters are self-aligning, yet designed to be easily accessed and swapped by the user. In fact, the entire optics train requires only three factory adjustments for permanent alignment.

4.4.3 Electronics

Analog front-end noise performance and the quality of the analog-to-digital converters used determine an instrument's dynamic range. Unfortunately, the four decades of dynamic range typical for conventional cytometers and sometimes, even five decades in advanced digital systems, is still less than the range of signals in many biological experiments. Until now, dynamic range limitations have been managed through the use of voltage gain controls on the photomultiplier tube (PMT) detectors. To observe faint signals, voltage gains are set higher; to detect bright signals, voltage gains are set lower. Although sufficient for most experiments, there are two major drawbacks to the use of variable voltage gain for overcoming dynamic range limitations:

i. Bright and faint signals cannot be resolved simultaneously; and
ii. The user must carefully set the gain in real-time or the desired data may not be detected. The complexity of managing dynamic range with gain control settings contributes greatly to the steep learning curve for novice flow cytometrists, as they must learn the "art" of setting voltage/amplification/gain levels.

In contrast to approaches that depend on giving the user control over PMT voltages, the C6 design focused on increasing the dynamic range to obviate the need for voltage control. Careful analog front-end design, the incorporation of ADC chips with 24-bit resolution (120 dB dynamic range), and sophisticated digital filtering have yielded an instrument with six decades of dynamic range, a first in flow cytometry history. This means the C6 has the ability to look at both faint and bright signals simultaneously and with great resolution. Data can be collected and analyzed later, without concern over data losses due to improper gain settings. The learning curve is greatly reduced because a common source of user error, the "art" of setting voltage gain levels, has been eliminated. While being sophisticated "under the hood", this approach is user friendly, encourages scientists unfamiliar with flow cytometry to try the technology, and reduces complexity, cost and size. Fixed voltage systems also result in more stable fluorescence compensation values as they are not dependant the effect of changing one detector voltage, independent of others.

4.4.4 Software

User interface complexity is a common complaint among users of legacy cytometry instruments and greatly limits the growth of the technology. Manufacturers typically offer multi-day training classes to compensate for the complexity of their software design, but classes do little to reduce the barriers to entry and can overwhelm users, especially novices.

Because the interface is critical to the user experience on any flow cytometer, large or small, the C6 companion CFlow software was designed from the ground up to be intuitive and easy-to-use. Rather than including every possible option or feature without thought, behavioral methods, including extensive observation of existing flow cytometry operators, ensured a user-focused design process and simple but efficient user interface. New users can be fluent in the software within 30 minutes.

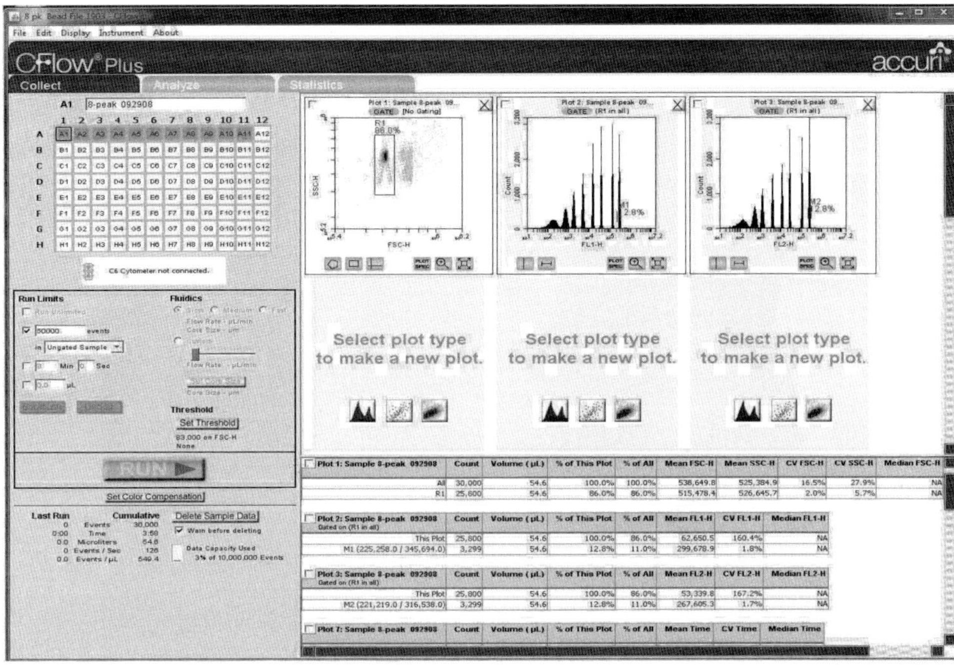

Figure 4.5. Intuitive CFlow software requires no training. Color reference – pg. 340.

4.4.5 *Enhancing the Manufacturing Process*

The Accuri C6 is presently shipping as a production instrument, supported by all of the key innovations discussed in earlier sections. One of the design principles has been to minimize the number of required adjustments at the time of instrument assembly, instead relying on self-aligning features and tight tolerances during the machining of critical components. This has required constant tradeoff between manufacturing complexity, part complexity, and tolerance capability. As production ramps up, quality control becomes more critical. Several major components in the C6 design have undergone over ten pre-production and several in-production iterations, each of which might take a month or more to evaluate, to further improve manufacturability and tolerancing. For example, the optics assembly initially had separate machined blocks holding the capillary and detector lenses. Since then, the detector lens mounts have been integrated directly with the capillary into the flow cell in order to improve alignment stability. This required significant retooling of the production line, and months of preproduction work, but has substantially increased production yield and assembly reliability.

4.5 CHALLENGES TO BE ADDRESSED

The single greatest shortcoming of the Accuri system, as originally launched in February 2008, was its lack of automation in the form of microplate handling.

Figure 4.6. The Accuri C6 on-board a research vessel, analyzing bacteria found in Lake Erie.

Although a substantial user base has been established around single-tube sample handling, which was the only option in cytometry for many years, a significant portion of scientists now prefer microplates for sample preparation and plate-based science is anticipated to become the norm. This includes high-throughput screening, sample preparation systems, and integration with robotic plate handler environments. To address this, Accuri developed a microplate sampler accessory for launch in 2009. In keeping with its user-friendly philosophy, Accuri's plate handler is designed for easy user installation (or removal, if so desired) from the standard Accuri C6 instrument in fifteen minutes.

4.6 THE FUTURE

The advent of affordable, compact systems such as the Accuri C6 has increased the number and type of laboratories that are able to adopt flow cytometry and its applications. Institutions undertaking research in areas as diverse as plant genetics or the analysis of organisms found in fresh or marine aquatic environments can now easily use the technique without having to receive special training. Miniaturization of the major subsystems found in flow cytometers means that in addition to immunophenotyping, a single small instrument can be used for cell counting in place of hemocytometers and multiplex bead analysis in place of ELISA methods.

Indeed, there is no reason why systems have to remain within the confines of the laboratory. Small flow cytometers can be taken into the field in small vehicles or even on-board ships (Fig.4.6). While this has been attempted in the past with mixed success, it is now becoming a practical proposition. It is possible that future systems will continue to be further simplified, with fewer detectors or simplified electronics, designed and built for specific applications rather than as general purpose tools.

Chapter Five

Progress in Capillary Flow Cytometry

David King,[1,*] Amedeo Cappione,[1] Fedor Ilkov,[1] Bruce Goldman,[1]
Ray Lefebvre,[1] Rick Pittaro[1] and G. J. Dixon[2,†]

[1]*Millipore Corporation, Inc. 25801 Industrial Boulevard Hayward, CA 94545-2991 USA*
[*]*dking@guavatechnologies.com*
[2]*Dixon Consulting PO Box 178 Socorro, NM 87801, USA*
[†]*lazrjoc@gmail.com*

5.1 INTRODUCTION

For much of its history the flow cytometry field has followed a 'basic research' model that placed sensitivity and measurement precision above all other considerations. The limitations of this approach became apparent as flow cytometry began to move out of the research laboratory and into applications where instrument cost and user-friendliness were important considerations. In many of these settings, users were less interested in a cytometer with *maximum* sensitivity and precision than they were in an instrument that combined *adequate* sensitivity and precision with a robust, low-cost package that could be reliably operated.

Until the early part of this decade, flow cytometry technology was dominated by hydrodynamically focused flow cells that confine the sample to a small region near the flow axis as it passes through the excitation beam. These cells were introduced during the infancy of the flow cytometry field[1] and continue to be used in a large majority of instruments.[2] Conventional designs inject the sample into a clear sheath fluid through a pressurized nozzle and subsequently reduce the flow diameter[3] as illustrated in Fig. 5.1.

In the nozzle approach to hydrodynamic focusing, axial sample confinement improves measurement sensitivity and precision. There are, however, a number of practical drawbacks. Foremost amongst these is the complexity and expense of the injection nozzle and fluidics system. Stable operation of a hydrodynamically focused flow requires the use of a carefully designed injector nozzle in combination with a complex and highly controlled fluidics system. These components are expensive and can malfunction in a variety of ways that make them poor choices

The Microflow Cytometer by J S Kim & F S Ligler

www.panstanford.com
978-981-4267-41-0

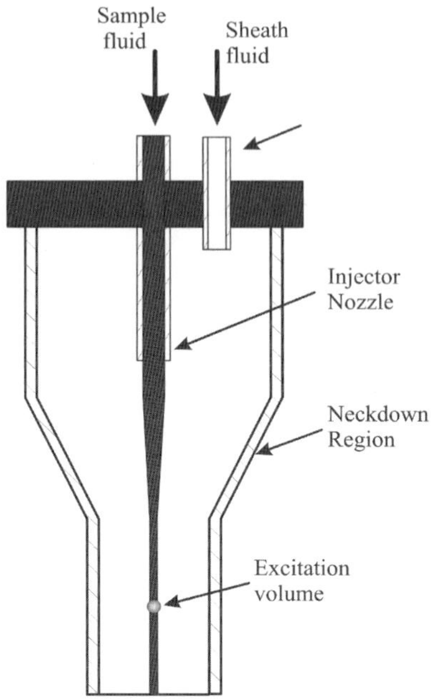

Figure 5.1. Schematic illustration of a traditional, hydrodynamically focused sheath flow cell. By injecting the sample fluid into the center of the sheath fluid and reducing the diameter of the flow tube, particles are confined to a narrow region near the flow axis as they pass through the excitation volume.

for a low- cost robust instrument. Pressurized sample injection also requires a significant volume of sample fluid and greatly complicates the direct measurement of particle concentration.

The drawbacks of the traditional, hydrodynamically focused sheath flow cells are largely overcome in capillary flow cells where the sample fluid completely fills the flow tube cross section. Configured as shown in the schematic of Fig. 5.2 the typical capillary cell acts, in some respects, like an ordinary soda straw and requires no sheath fluid or injector nozzle. The sample is drawn through the excitation volume by immersing one end of the capillary in the sample and applying a negative pressure to the other. This arrangement significantly reduces the complexity and cost of the flow system in addition to facilitating the direct measurement of particle concentration and minimizing the volume of sample required to make a measurement.

In the late 1990s, Philippe Goix and coworkers at Guava Technologies found that a cytometer with a capillary flow cell configured as in Fig. 5.2 could perform a number of common assays with an accuracy that was more than sufficient for routine laboratory application. Their work resulted in the introduction of the first commercial capillary flow cytometer in 2001 and the subsequent shipment of more than 1500 instruments to laboratories throughout the world. A continuing in-house R&D effort has led to significant improvements in capillary

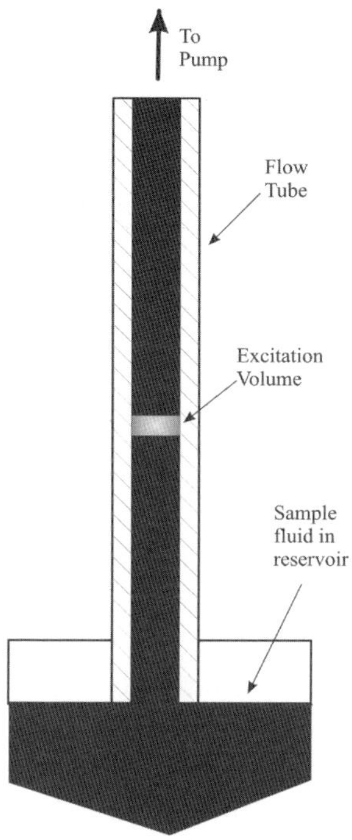

Figure 5.2. Schematic illustration of a capillary flow tube. In this scheme, the sample is drawn into the flow tube in 'soda straw' fashion and fills the entire tube cross section. Capillary flow has a number of practical advantages when compared to hydrodynamically-focused sheath flow, but the larger excitation volume leads to greater variation in a number of experimental parameters.

flow cytometer design and data analysis techniques. Currently, Guava's R&D effort is focused on improvements that will facilitate the movement of flow cytometry from central diagnostic and research labs to field locations where ease of use and robustness are of tantamount concern.

This chapter discusses the basic operation of Guava's capillary flow cytometers and some recent research results. Specific sections are devoted to the improved performance of capillaries with asymmetric cross sections, real-time particle velocity measurement and applications, and the dramatic simplification of analysis of complex flow cytometry data.

5.2 GUAVA CAPILLARY CYTOMETERS

Guava capillary flow cytometers share many features with conventional sheath flow instruments, differing primarily in the design of the flow tube and associated fluidics system. In our commercial product line, instruments are differentiated by the wavelength of the excitation sources and by the number of samples that can be interrogated in a single run. Excitation by 488 nm light is used in the EasyCyte product line with the EasyCyte Mini being configured for the manual insertion of single samples and the EasyCyte Plus for the computer-controlled interrogation of multi-sample plates and tubes. Guava's Personal Cell Analyzer or PCA products are also configured for single sample (PCA) or multi-sample plate (PCA-96) operation and typically incorporate a frequency-doubled, diode-pumped neodymium-doped yttrium vanadate laser at 532 nm. Both instrument lines can be configured for excitation at other wavelengths on a special order basis. Dual-source platforms that allow multi-color excitation of a single sample are in an advanced stage of development.

In all Guava instruments, assays are initiated by immersing the tip of the flow tube in the sample. The sample is then drawn through the excitation volume of the flow cell, where it is excited by the focused output of a frequency-doubled solid-state or semiconductor laser. Because the distance between the tube end and the excitation volume is small, the sample volume can be dramatically minimized relative to sheath fluid-based instruments and it is possible to utilize samples down to the sub-μL range.

Figure 5.3 is a schematic representation of the excitation and collection optics in a typical device. This optical design is qualitatively similar to that employed in conventional sheath flow instruments with the excitation and collection lens designs optimized to minimize variations in the excitation beam intensity and collection efficiency over the cross sectional area of the flow tube. The cross section of the 'standard' square capillary tube that is used in all current Guava instruments is shown in Fig. 5.4.

The operational simplicity of capillary flow cytometers is a direct consequence of a flow-tube design in which the sample fluid travels, without impediment, between the input and output ends. While it is possible to push or pull the sample through the flow tube, the latter arrangement is employed in all Guava instruments since it allows the sample to be drawn into an open end of the tube.

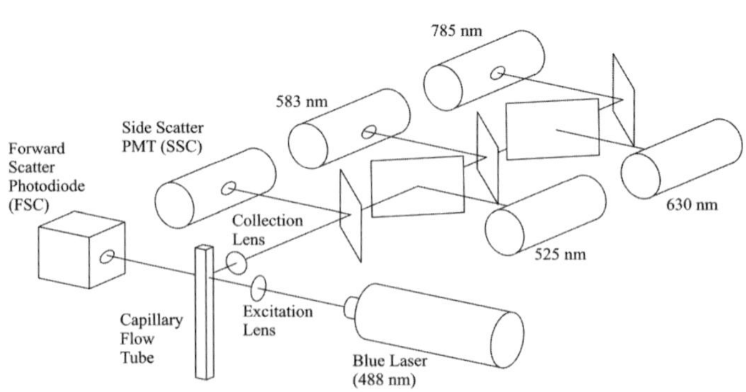

Figure 5.3. Optical excitation and collection systems of Guava EasyCyte product line. The 45-degree beam splitters transmit light with wavelengths longer than their cutoff and reflect shorter wavelengths. Bandpass filters, located immediately in front of the laser and all detectors, have been omitted from the drawing for clarity.

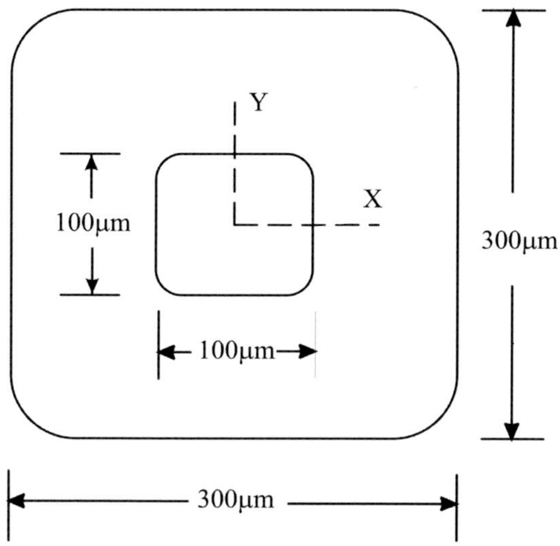

Figure 5.4. Cross sectional dimensions of 'standard' capillary in which a hollow, 100 m square bore is surrounded by a wall that is approximately 200 m thick. The bore is completely filled with sample fluid in a typical capillary flow system.

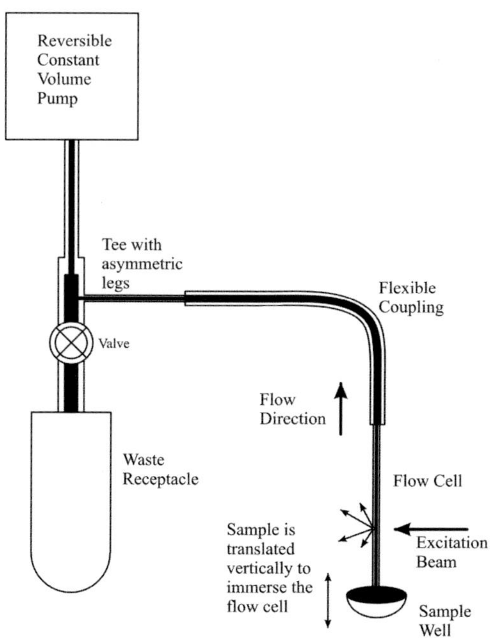

Figure 5.5. Fluidics system of typical capillary flow cytometer. A sample is drawn into the capillary flow tube by the pump when the open end of the tube is dipped into the sample well. The valve is opened and the pump direction reversed to deposit waste fluids in the waste receptacle.

Fig. 5.5 is a schematic illustration of a typical fluid handling system in which a syringe pump is used to draw the sample into the open end of the flow tube. Following interrogation by the excitation beam, the sample flows through the pumped end of the flow tube. The downstream valve allows fluids that have been drawn through the cell to be ejected into a dump.

The elimination of the sheath fluid and the associated sample injector nozzle are major advantages of the Fig. 5.5 fluidics system. In a hydrodynamically-focused system, particle concentration measurements typically require the addition of calibration beads to the sample or, in cases where this is not possible, a separate calibration run must be performed. In the capillary fluidics system of Fig. 5.5, the flow rate of the pump is typically known with great precision and the volume of sample fluid that is interrogated during a measurement cycle can be accurately determined from the measurement time. Concentrations are then ascertained by gating out the pulses with characteristics that correspond to the particle type, counting them, and dividing the result by the volume.

Capillary flow systems are also less prone to flow-related malfunctions than their sheath flow counterparts and, when malfunctions do occur, can be returned to operation with significantly less effort. In Guava cytometers, the capillary flow tube is attached to a carrier that is kinematically mounted on the optical system subchassis. A single fitting connects the flow tube to the downstream components of the fluidics system. In the event of tube breakage or clogging, the flow tube can be easily removed from the instrument and replaced. This is a much simpler process than replacing or unclogging the injector nozzle in a sheath flow instrument.

5.2.1 Asymmetric Capillary Designs

Spatial variations in the excitation and optical collection efficiencies of a flow cytometer have a deleterious effect on measurement precision as quantitatively described by the coefficient of variation or CV. This parameter decreases with increasing precision and is numerically equal to the standard deviation for a measurement divided by the arithmetic mean. In a hydrodynamically focused flow, CVs are reduced by confining the sample fluid to a region near the flow axis. In a capillary flow, the particles are also concentrated by fluid forces, but the magnitude of the effect is typically small and spatial variations must be reduced in other ways.

In the standard square capillary flow cell of Fig. 5.4, a large portion of the spatially dependent variation in large-angle collection efficiency is caused by refraction and reflection at the tube walls. As light rays pass from source points inside the tube to the collection lens, they encounter two steps in the index of refraction – a comparatively small step at the fluid/glass interface of the inner wall and larger step at the air/glass interface of the outer wall. Refractive effects occur principally at the outer wall and act to bend some rays into the collection aperture of the optical system as illustrated in Fig. 5.6. This has the largest effect on light rays that emanate from source points near the back and side walls of the tube. These effects can be effectively minimized by stretching the outer capillary wall nearest

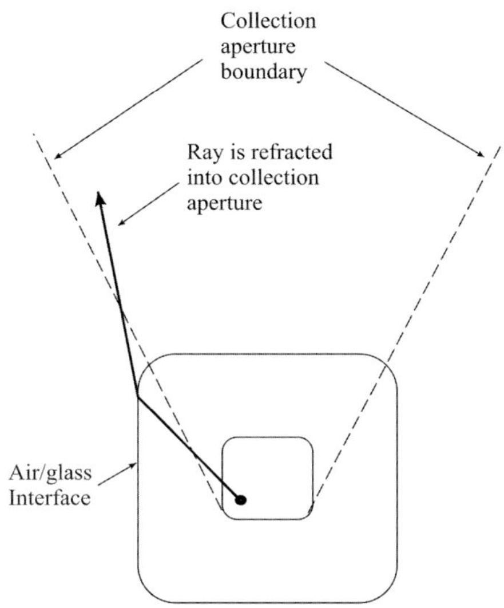

Figure 5.6. Some rays are refracted into the collection aperture by the outer capillary wall. This increases the collection efficiency by an amount that is dependent on the source point location.

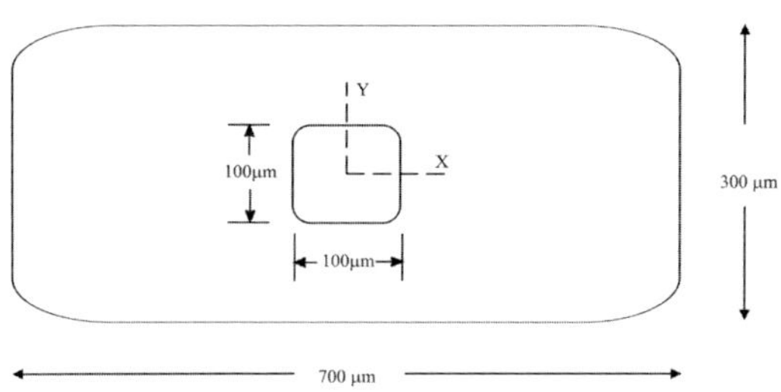

Figure 5.7. Cross-sectional detail of capillary with extended front wall and square bore. This design minimizes the contribution of outer wall refraction to the collection efficiency by moving the side walls and corners out of the collection aperture.

the collection lens as shown in Fig. 5.7. Although stretching the front wall does not eliminate refraction, it moves the corners and side walls far enough from the axis of the collection optical system that the refracted light is not collected.

For source points near the side walls of a square capillary, the refractive increase in collection efficiency is partially cancelled by the inner wall reflection of rays that strike the fluid/glass interface at a large angle of incidence. This effect changes the propagation direction of some rays so they lie outside of the collection aperture as illustrated schematically in Fig. 5.8. This effect can be reduced by increasing the angle of the inner bore side walls with respect to the axis of the collection optical system so that the reflected rays still lie within the collection aperture.

Because the refractive and reflective effects act in opposite directions, a capil-

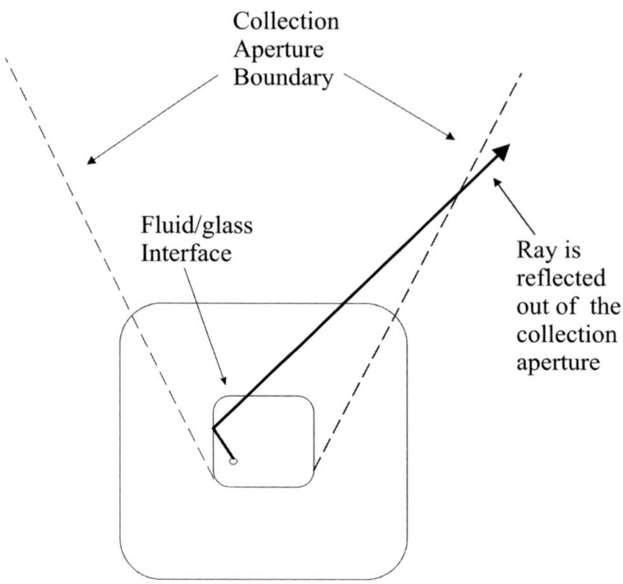

Figure 5.8. Light is reflected out of the collection aperture by the glass/fluid interface at the inner capillary wall. This effect reduces the light collection efficiency for some source points by an amount that depends on the source point location.

lary design must address both to effectively reduce spatial variations in collection efficiency. This is accomplished by capillary designs that combine an extended front wall and a bore with angled side walls. The capillary design of Fig. 5.9 is one such design and its effectiveness is illustrated by the calculated results in Tables 5.1 and 5.2.

Table 5.1 Light collection efficiency calculated by ASAP ray tracing model at representative source points in capillaries with square walls and square bore (Fig. 5.4), an extended front wall and a square bore (Fig. 5.7) and an extended front wall and a trapezoidal bore (Fig. 5.9). The parameters of the collection optical system are illustrated in Fig. 10. Source points are specified in coordinate systems with origins at the centers of the bores as illustrated in the figures. In all cases the y-axis and collection optical axis are coincident.

Source point location	Square wall Square bore	Extended wall Square bore	Extended wall Trapezoidal bore
(0 μm, 0 μm)	3.6%	3.6%	3.57%
(0 μm, −45 μm)	5.5%	3.5%	3.55%
(0 μm, 45 μm)	3.7%	3.6%	3.59%
(−45 μm, 0 μm)	4.3%	2.3%	3.58%
(45 μm, 0 μm)	4.3%	2.3%	3.58%
(35 μm, −35 μm)	3.6%	3.5%	3.59%
(−35 μm, 35 μm)	4.4%	2.7%	3.55%

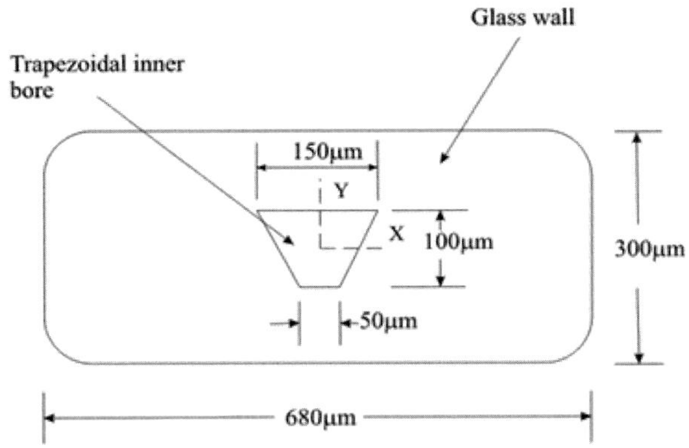

Figure 5.9. A capillary with an extended front wall and trapezoidal bore effectively minimizes the position-dependent variations in collection efficiency that are caused by refraction and reflection at the walls.

In Table 5.1, the calculated light collection efficiencies are tabulated for seven representative source points in the capillary designs of Figs. 5.4, 5.7 and 5.9. Calculations were performed for light collection by a lens with a numerical aperture (NA) of 0.5 positioned 5.85 millimeters from the outer front wall of the capillary using non-sequential ray tracing in the high-end optical design package, ASAP.[4] Source coordinates are specified in the coordinate systems shown in the Figures, all of which have an origin at the bore center.

Table 5.2 compares these various designs by tabulating the coefficient of variation (CV) in collected fluorescence intensity of a set of identical particles contained in a sample. In addition, CVs for linear translation in three directions are also shown. The Table 5.2 results were calculated under the assumptions that source points were uniformly distributed within the capillary cross section and

Table 5.2 Overall and directional coefficients of variation (CVs) from ASAP model of light collection in representative capillary cross sections using the optical system of Fig. 5.10. The cross sectional dimensions of the square wall/square bore capillary are given in Fig. 5.4, the extended front wall/square capillary in Fig. 5.7 and the extended front wall/trapezoidal bore in Fig. 5.9. Values are based on a model that only includes refraction and reflection of emitted light by the capillary walls and assumes unapertured light collection from an isotropic distribution of emitters.

Direction	Square wall Square bore	Extended wall Square bore	Extended wall Trapezoidal bore
Overall	15.8%	19.7%	0.48%
X-axis	9.5%	26.3%	0.16%
Y-axis	24.6%	2.0%	0.56%
Diagonal	11.9%	15.7%	0.56%

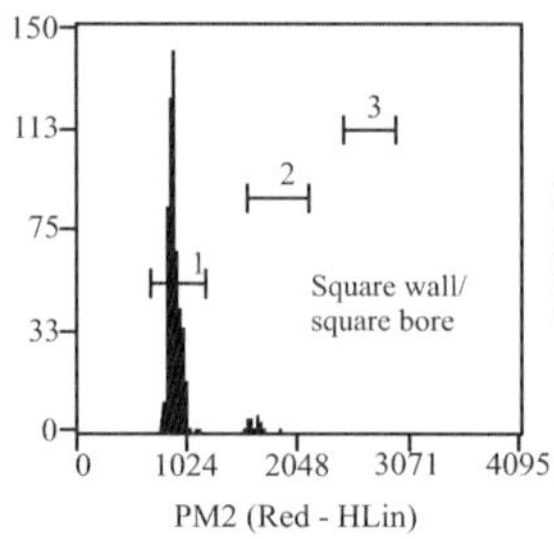

MARKER 1

Count	Cells/ml	% of Gated	x-Mean	%CV
2502	1.05e05	89.35%	904.65	5.54%

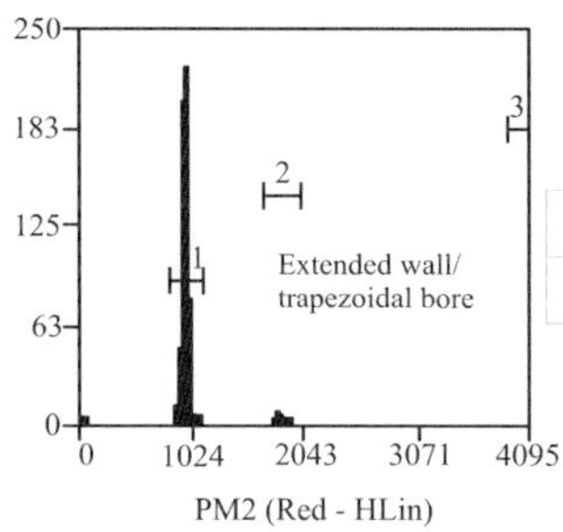

MARKER 1

Count	Cells/ml	% of Gated	x-Mean	%CV
1952	4.93e04	91.56%	963.11	2.82%

Figure 5.10. Histograms for chicken erythrocyte nuclei stained with propidium iodide. Samples were analyzed using square wall/square bore and extended wall/trapezoidal bore capillaries in Guava EasyCyte capillary cytometer. The CV reduction observed with the extended wall/ trapezoidal bore capillary is representative of its performance in other measurements.

that reflection and refraction at the tube walls were the only physical effects that contributed significantly to the CVs.

The regions of the capillary bore in which refraction contributed significantly to the variations in collection efficiency may be identified by comparing the first two columns of Tables 5.1 and 5.2. Extension of the front wall can be seen to have the largest effect for source points near the back and sides of the bore and, in the absence of reflective effects would significantly decrease the overall CV. In the presence of reflection, however, the reduction in refractive effects associated with stretching the front wall increases the CV since the reflective decreases in collection efficiency are no longer balanced by refraction.

Both reflection and refraction are decreased by extending the front wall and angling the sides, and this is verified by the values listed in the final columns of the tables. In Table 5.1, variations are small enough that an added decade of precision is required to see them and all of the CVs in the last column of Table 5.2 are significantly reduced from the values for the standard square wall, square bore design.

While the numerical results in Tables 5.1 and 5.2 qualitatively reflect the improvements afforded by improved capillary design, the numerical values differ from measured values in two respects. In an actual capillary cytometer, apertures are used to block stray light and axially directed hydrodynamic forces concentrate particles in certain regions of the flow. The resulting CVs are smaller than those calculated for unapertured light collection from an isotropic distribution of particles. Minimum values are also underestimated by the exclusion of other physical effects from the model.

The quantitative reductions that can actually be achieved with modified capillary designs have been experimentally determined using capillaries with different cross sections in a Guava EasyCyte Mini cytometer. In one experiment, fluorescence amplitude histograms were recorded for chicken erythrocyte nuclei (CEN) stained with propidium iodide (PI). The PI-stained nuclei were analyzed in capillaries with the cross-sectional geometries shown in Fig. 5.4 and 5.9. The pulse height histograms for the two designs are shown in Fig. 5.10 with the standard square capillary results in the upper half. The width of histogram peaks is indicative of CV and, ultimately, instrument resolution. The results shown in Fig. 5.10 clearly indicate that the resolution of the trapezoidal capillary is greater than that of the square capillary. CV values for the square and trapezoidal bores are 5.5% and 2.8%.

In another experiment, the CVs for a Guava Cell Cycle assay of fixed Jurkat cells were calculated using measured fluorescence intensities. The cell cycle assay is commonly used for determining the fraction of cells in the G0/G1, S and G2/M phases of the cell cycle. In the assay, cellular DNA is stained with PI and the amount of DNA in the nucleus of each cell is determined by measuring the intensity of PI fluorescence. Resting cells in the G0/G1 phase contain two copies of each chromosome and have the smallest fluorescent intensity. As cells progress toward mitosis they synthesize DNA (S phase) the fluorescent intensity increases until all chromosomes have replicated (G2/M phase) and the fluorescent intensity is twice that observed from a cell in the G0/G1 phase.

A histogram of pulse amplitudes recorded using a conventional square capillary (Fig. 5.4) is shown in the upper half of Fig. 5.11 with the corresponding extended front wall/trapezoidal bore data in the lower half. As in the chicken erythrocyte nuclei assay, the 2.4% CV for the G0/G1 peak of the extended front wall/trapezoidal data was significantly reduced from the square capillary value of 4.1%. The observed improvements support the accuracy of the ray tracing analysis and indicate that cross sectional geometries that minimize outer wall refraction and inner wall reflection can be expected to reduce the CV for a given measurement to a value that is 50% to 75% of that obtained with a conventional square capillary.

5.2.2 Particle Velocity Measurement

Like most modern cytometers, Guava's capillary flow instruments are 'all digital' in the sense that the optical detector outputs are connected to analog-to-digital converters (ADC's) that digitize the signals before any signal processing or analysis takes place.[5,6] Digital data from an ADC output is transferred to a buffer before it is stored in the memory of a personal computer that performs the signal

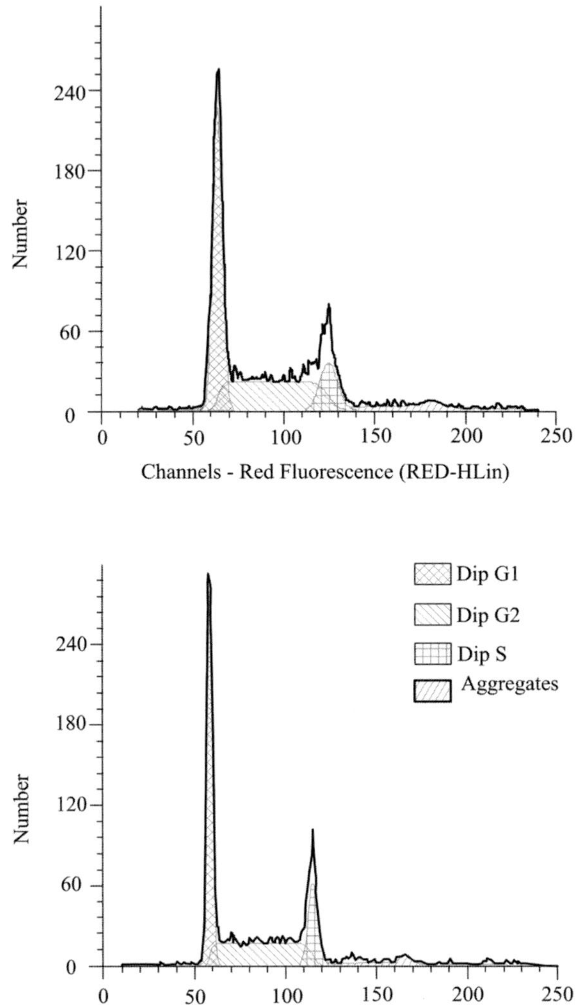

Figure 5.11. Histograms for cell cycle analysis assays of Jurkat cells in square wall/square bore (upper) and extended wall/trapezoidal bore capillaries using a Guava EasyCyte cytometer. The 4.1% CV for square wall/square bore capillary is reduced to 2.4% by the extended wall/trapezoidal bore design.

processing and analysis tasks. The fluidics system, sample translators (if present) and the detector amplifier gains are also controlled by the computer. With this architecture, the complexity of signal processing tasks and the speed with which they can be carried out has increased with personal computer power to a point where it is possible to remove instrumental artifacts from measured data in near real time.

Of particular interest for capillary instruments are the artifacts caused by the decrease of particle velocity with increasing radial distance in the flow tube.

Flow cytometers typically operate in a laminar flow regime where particle velocity varies in a parabolic fashion with distance from the flow axis. In a hydrodynamically focused flow where the sample is confined to a region near thepeak of the parabola, there is little variation in particle velocity. In a capillary flow where the sample fills the flow tube this is clearly not the case.

Significant variations in the shapes and widths of the fluorescent time pulses emanating from identical particles are one consequence of particle velocity variation in a capillary flow. Since they spend a longer time in the excitation beam, particles traveling near the walls of the tube give rise to pulses that are longer than those generated by those traveling near the axis. This effect is illustrated in the

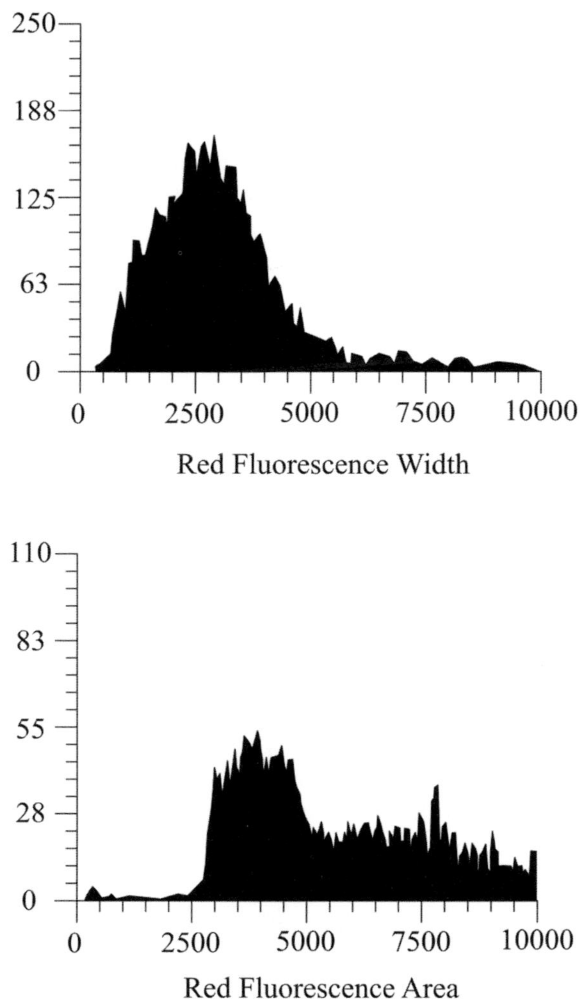

Figure 5.12. Histograms of uncorrected pulse width and areas recorded in an assay of uniformly-sized Guava Check beads with an EasyCyte cytometer. Both curves are significantly broadened by particle velocity variations within the excitation volume.

histograms of Fig. 5.12 which show the widths and areas of the fluorescence pulses recorded in an analysis of uniformly sized and stained Guava Check beads with a Guava EasyCyte cytometer. On a per-particle basis, the histograms would be much narrower in the absence of velocity variations. The observed velocity-smearing can limit resolution for a small number of applications and may complicate analyses that rely on fluorescence pulse area or width measurements.

In DNA cell cycle analyses, for example, doublet pulses generated by the simultaneous illumination of two particles are often identified using the width or area of the pulses. Doublet pulses are typically generated by cells in the G0/G1 stage but have fluorescent amplitudes that are characteristic of cells in the S or G2/M stages of the cycle. A failure to recognize and eliminate doublets leads to erroneous results in which the fraction of cells in the later stages is overestimated. Doublet pulses are longer than those generated by single cells and can be identified by measuring their width in addition to their area or amplitude because two cells are rarely illuminated at exactly the same time.[7] In the absence of velocity smearing, this is a straightforward task, and several different techniques have been developed for this purpose. When the data is velocity-smeared, however, it is necessary to use an alternative technique that is not based on pulsewidth or to correct the data for variations in particle velocity.

We have found that the estimation of derivative pulse shapes allows for the correction of velocity smeared data. Derivative pulse shapes can be calculated through numerical differentiation of pulse amplitude data Using appropriate numerical techniques, pulse shape differentiation can be carried out in near real time on a personal computer. We have also found that multiplets can be directly identified by counting the number of zeros in the derivative pulse.

Before numerically differentiating digitized amplitude data, it is useful to upsample using sinc interpolation[8] or a similar numerical technique. When carried out appropriately, upsampling increases the precision with which extreme values can be determined and improves the accuracy of all subsequent analyses. Numerical differentiation of the upsampled data is then carried out using one of the many numerical techniques that have been developed for this purpose. Because of its speed and stability, the Savitsky-Golay method[9] is well suited for velocity correction. This filter performs a local polynomial regression around each point in the digitized data and uses the polynomial fit to estimate the derivative of the original signal. It combines data smoothing (via the regression) with a derivative estimation filter and has the speed, accuracy and stability needed for the real-time analysis of cytometer pulses with a personal computer.

Estimated derivative pulse shapes may be used to calculate particle velocities for pulse shape correction or to identify doublet pulses in a more direct fashion. For similarly-sized particles, particle velocities are calculated by defining a measurement window (the $1/4$ intensity points of the pulse, for example) and determining the maximum and minimum values of the derivative pulse within the window. The particle velocity is then estimated using the following equation:

$$\mathrm{V} = (\mathrm{k}/2\mathrm{P})[abs(\max) + abs(\min)] \tag{5.1}$$

where V is equal to the particle velocity, *abs* is the absolute value function, (max) is the largest value of the derivative in the window, (min) is the smallest value of the derivative in the window, P is the pulse height, and k is a constant scaling factor used to express the velocity in physically-meaningful units. Velocity-smeared area and width data are corrected by multiplying the measured parameters by the particle velocity.

The histogram of Fig. 5.13 was obtained by applying a derivative velocity correction to the raw data of Fig. 5.12. The peaks in the corrected data are clearly much narrower than the velocity-smeared peaks and closely resemble histograms of data from a sheath flow instrument. This similarity is more than cosmetic, and we have found that multiplets may be identified in velocity-corrected cell cycle data using the techniques described in Wersto.[7]

It is also possible to discriminate doublet pulses directly by counting the number of derivative zero crossings in a single pulse. Fig. 5.14 shows the detector signal and first derivative of singlet and doublet pulses that were recorded by a forward scatter detector of a Guava EasyCyte capillary cytometer in an assay of a sample containing Guava Check beads. The pulsewidths are equal to the difference between the times T_1 and T_2 when the signals cross a preset threshold and the derivative signals cross zero at each maximum and minimum – once for the singlet pulse and three times for the doublet pulse. In the rare event that more than two particles are illuminated simultaneously, the derivative signal is expected to cross zero ($2n$-1) times where n is the number of simultaneously illuminated particles. With this technique, doublet and singlet pulses are easily separated by estimating the derivative pulse shape and counting the number of zeros in the derivative shape.

In general, derivative calculations may be used to determine individual particle velocities or as the basis for new analysis techniques. Velocity information can also be used to keep track of individual particles as they travel between multiple excitation beams or to determine the radial distribution of particles in a flow cell.

5.3 MULTISAMPLE DATA ANALYSIS

As flow cytometers are used with increasing frequency in drug screening, functional biology, and similar applications, the need to perform multiple analyses of plate-based data sets has increased. Conventional flow analysis software is designed to analyze data from single samples and is not well-suited for plate-based data analysis. Conventional analysis packages also fail to take full advantage of the computing and display capabilities of modern personal computers and are quite primitive when compared to the data analysis and visualization capabilities that are available in other fields.

We have addressed these limitations in a new software package called "In-Cyte" that allows a user to rapidly and very simply perform multiple analyses of plate-based data sets containing information from the interrogation of multiple samples and to visually compare the results. The package is named to reflect the primary objective of simplifying flow analysis to a point where it can be carried out by individuals with a wide range of educational and laboratory experiences. The focus of this effort is to enable researchers to quickly and easily arrive at biologically meaningful conclusions from complex data sets.

In contrast to the menu- and list-driven interface of conventional analysis packages, InCyte uses maps of the wells on a typical sample plate for data input and display operations. Drag and drop capability is enabled throughout the package and greatly speeds operation. All flow data is analyzed using data gates on one or more histograms or dot plots. In InCyte, this set of gates and graphs is called an analysis method, which is stored separately from data. Methods can be

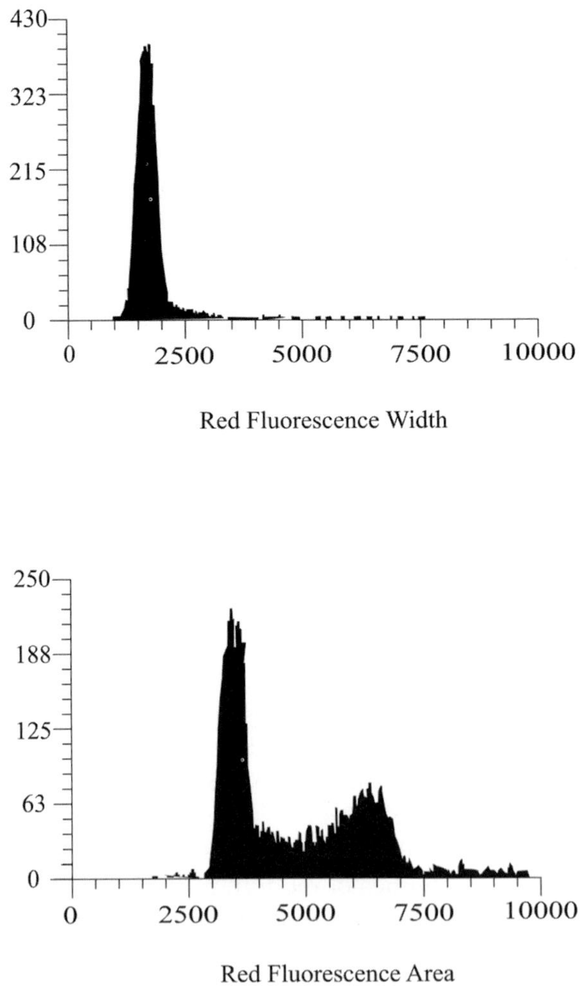

Figure 5.13. Histograms of the Guava Check Bead assay data from Fig. 5.12 that have been velocity-corrected using the derivative velocity technique.

recalled and applied to a data set enabling simultaneous analysis of all wells on a plate.

Analyzed results can be presented in a number of formats but the most useful is a multiplexed, location-specific heat map in which the wells on a sample plate are represented by subdivided icons. This approach enables the output data from multiple analyses from each well to be viewed simultaneously. Using this display, a user can quickly identify samples that satisfy multiple criteria. A linkage between the analysis definition window and the output heat map also allows a user

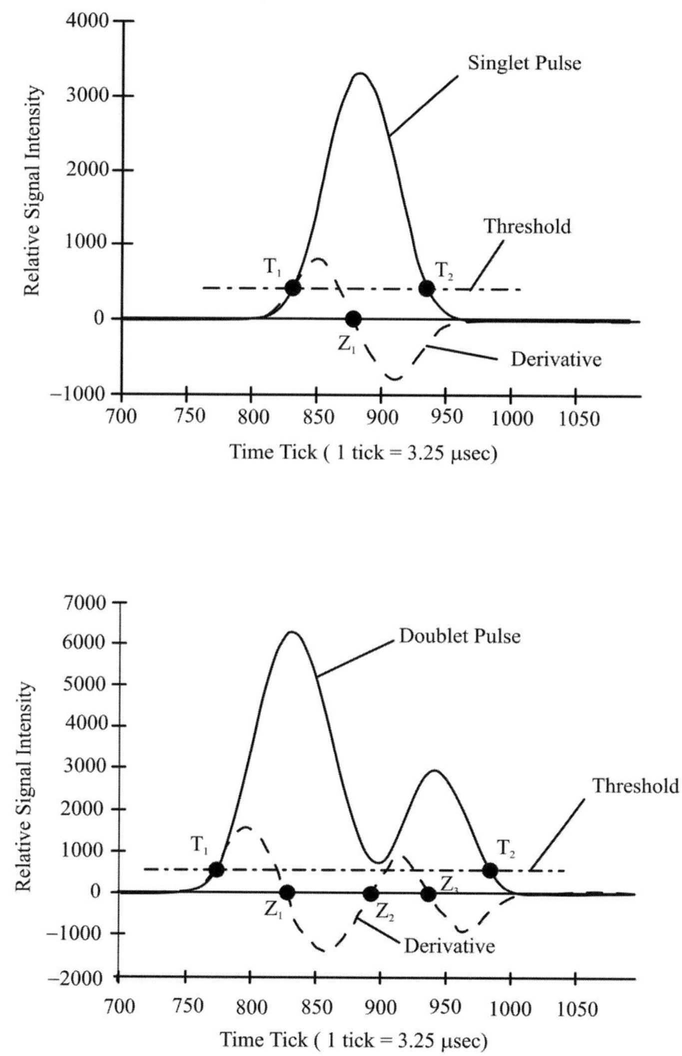

Figure 5.14. Representative amplitude and first-derivative pulse shapes for singlet and doublet pulses. The pulse crosses the detection threshold at points T1 and T2 while the zeros are labeled Z1 – Z3. The two shapes may be easily discriminated by counting the number of zeros in the derivative pulse shape.

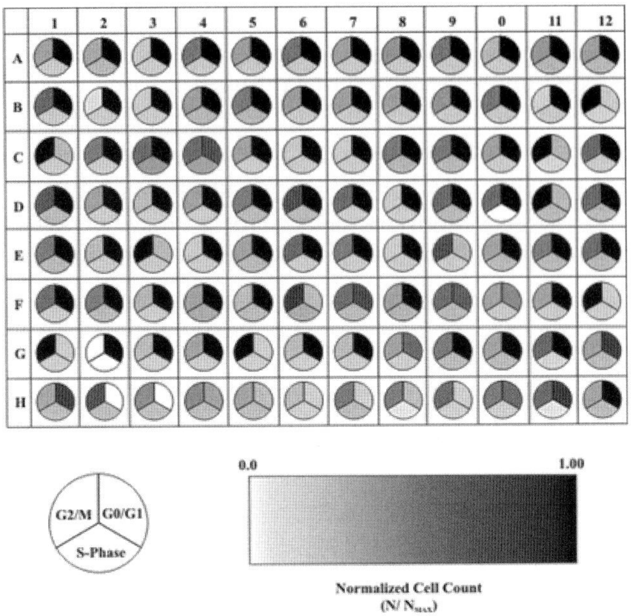

Figure 5.15. Multiplexed, location-specific heat map from InCyte output window. The sample locations on a standard 96-well plate are mapped with circular icons that have been subdivided into three regions to allow the display of cell cycle output data. Data sets are assigned to the regions in a clockwise fashion as shown in the circular inset and a tonal scale is used to represent the normalized cell counts in each region. The sample in location G2 has an absence of cells in the S-Phase and G2/M phases, indicating the cytotoxic compound added to this well was effective in blocking the later stages of the cell cycle.

to observe the effect of changing an analysis parameter (moving a gate boundary) on the entire output data set and its relationship to other output sets in real time. One example in which the InCyte output map has been used to advantage is the rapid identification of cytotoxic compounds that may block the growth of cancer cells. Under normal physiological conditions, the cell cycle is tightly controlled at key regulatory points to ensure a balance in cell number is maintained. In cancer, these regulatory checkpoints are lost or compromised, leading to the aberrant over-proliferation of the cancer cell and the ultimate formation of a macroscopic tumor. Consequently, compounds that are capable of blocking the cell cycle preferentially inhibit cancer cell growth and are highly effective chemotherapeutic agents.

In this example, Jurkat cells were exposed to a variety of cytoactive compounds ($n = 96$) with the ultimate goal being the identification of unique cell cycle inhibitors.[10,11,12] Following exposure, cells were harvested and then stained with propidium iodide, a cell-permeant DNA-intercalating dye used to measure DNA content.[13,14] Relative staining is visualized by flow as red fluorescence and the number of cells in each stage of the cell cycle counted using a histogram as illustrated in Fig. 5.11. An untreated culture has cells present in all 3 phases of the cell cycle while cultures that are blocked at entry to S-phase or M-phase show most cells stuck within the G0/G1 or G2-phases, respectively. Following staining, samples were collected on an EasyCyte then analyzed using the InCyte software package to determine which compounds caused phase-specific blockage. Data interrogation and comparative results are displayed at the high-level experiment level rather than on a single well basis.

The location-specific, multiplexed heat-map of Fig. 5.15 shows the results of this analysis. In the map, the icon representing each well (sample) is divided into 3 sectors, with each sector being assigned to one of the three cell-cycle phases in a clockwise fashion. Thus, for any well, the relative frequency of cells in a given cell phase can be easily determined from the heat-map display. For example, the cells of well G2 are predominantly in G0/G1 suggesting the compound acts to block progress at the entry to S-phase.

While the full capabilities of InCyte go far beyond those illustrated in this example, it clearly demonstrates the utility of the multiplexed heat map display. In its current release, the package allows icons to be subdivided into a maximum of six regions that may be used to represent sample locations on multiple plates or multiple analyses. Future releases are expected to significantly increase the ease with which input data groups and analyses are defined.

With continued progress in instrumental design and digital data processing, capillary cytometers and the InCyte software package are unrivaled in their potential for moving flow cytometry out of the core laboratory facility and into a wide range of clinical and diagnostic settings. At Guava we believe that flow cytometers will ultimately be as common in biological research and clinical diagnostics laboratories as FTIR and UV/VIS/NIR spectrometers are today.

References

[1] M. A. van Dilla, T. T. Trujillo, P. F. Mullaney and J. R. Coulter, *Science* **163**, 1213 (1969).

[2] H. M. Shapiro, *Practical Flow Cytometry, 4th Ed.*, Wiley-Liss, New York p.167 (2003).

[3] V. Kachel, H. Fellner-Feldegg and E. Menke, *Flow Cytometry and Sorting, 2nd Ed.*, Eds. M. R. Melamed, T. Lindmo and M. L. Mendelsohn, Wiley-Liss; New York, p. 27 (1990).

[4] ASAP Reference Guide, Breault Research Organization, Tucson, AZ (2008).

[5] S. Murthi, S. Sankaranarayanan, B. Xia, G. M. Lambert, J. J. Rodriguez and D. Galbraith, *Cytometry A* **66A**, 109 (2005).

[6] C. K. Snow, *Cytometry A* **57A**, 63 (2004).

[7] R. P. Wersto, F. J. Christ, J. F. Leary, C. Morris, M. Stettler-Stevensen and E. Gabrielson, *Cytometry* **46**, 296 (2001).

[8] T. Schanze, *IEEE Trans. Signal Process.* **43**, 1502 (1995).

[9] A. Savitsky and M. J. E. Golay, *Analytical Chemistry* **36** ,1627 (1964).

[10] B. E. Clurman and J. M. Roberts, *J. Natl. Cancer Inst.* **87**, 1499 (1995).

[11] G. DeCarcer, I. Perez de Castro and M. Malumbres, *Curr. Med. Chem.* **14**, 969 (2007).

[12] M. Malumbres and A. Carnero, *Prog. Cell Cycle Research* **5**, 5 (2003).

[13] A. Moore, C. J. Donahue, K. D. Bauer and J. P. Mather, *Methods Cell Biol.* **57**, 265 (1998).

[14] A. Krishan, *J. Cell Biol.* **66**, 188 (1975).

Chapter Six

Focusing Particles Without Sheath Flows in Microflow Cytometers

Sungyoung Choi, Eujin Um and Je-Kyun Park*

Department of Bio and Brain Engineering, KAIST, 335 Gwahangno, Yuseong-gu, Daejeon 305–701, Republic of Korea

**jekyun@kaist.ac.kr*

Miniaturization of flow cytometry is of fundamental importance in point-of-care and personalized diagnostics. Conventional flow cytometry is a standard analytical method in cell biology, but restrictive due to its high price, complexity, and large size. To address miniaturization needs, a number of investigators in microfluidics, lab-on-a-chip, and bio-MEMS (Micro-Electro-Mechanical Systems) have focused on the development of microflow cytometers using microfluidic technologies and miniaturized optical components. We herein present various sheathless, microfluidic particle focusing methods that replace conventional flow chambers with microfabricated devices capable, reducing both device size and sheath volumes.

6.1 INTRODUCTION: WHY FOCUS PARTICLES WITH OR WITHOUT SHEATH FLOWS?

Focusing particles in a fluid stream is critical to the performance of a flow cytometer for counting, analyzing, and sorting microscopic particles.[1,2] Before the development of hydrodynamic focusing, flow cytometers with wide channels suffered from the problem that multiple cells would pass through the detection laser beam simultaneously, distorting the cell analysis. Narrowing the channel to force particles to flow in single file often resulted in clogging. To solve these problems, hydrodynamic focusing utilized sheath flow to guide a sample stream.[3,4] Lamellar flow at the microscale enabled sheath streams to guide particles in a single file while utilizing a larger channel to prevent clogging. This confinement of particles in a tube also enabled a more accurate analysis because particles are located in a streamline of a parabolic velocity profile.

Microflow cytometers have been developed by replacing conventional flow chambers in flow cytometers with microfabricated devices.[5–8] The miniaturization

The Microflow Cytometer by J S Kim & F S Ligler

www.panstanford.com
978-981-4267-41-0

of the flow chambers can allow efficient particle sorting by incorporating innovative particle manipulation techniques such as dielectrophoresis (DEP), electro-osmosis, pneumatic microvalves, and optical switching. However, most microflow cytometers still rely on relatively complex fluidics for accurate control of the differential flow rates of sheath and sample flows.[9–11] The microchannel designs for two- or three- dimensional hydrodynamic focusing typically require at least four inlets and their individual control to fully sheathe a sample flow. This focusing approach makes microflow cytometers more complex, requiring well-controlled fluidics. In addition, large volumes of sheath fluid make it difficult to miniaturize the overall size of the system. Hydrodynamic focusing may also dilute the sample stream with sheath fluid. This dilution is inevitable, especially for blood analysis, to prevent interference from red blood cells that occupy almost the half volume of the blood. Therefore, microflow hemocytometers require both the on-chip implementation of sample concentration for further analysis and an additional storage reservoir for sheath buffer.

To overcome the limitations by conventional hydrodynamic focusing methods, microfluidic sheathless focusing methods have been developed for miniaturizing flow cytometers. Here, we provide a review of the methods for focusing particles without sheath flows in microfluidic devices. These methods include (1) field-based techniques that use external forces such as dielectrophoresis, acoustic, and optical forces, and (2) flow-assisted techniques that exploit hydrodynamic phenomena such as inertial lift forces, hydrodynamic filtration, deterministic hydrodynamics, and hydrophoresis. We describe details of the focusing principles and their application to flow cytometry, comparing their advantages and disadvantages. Finally, we present some challenges and future trends of microfluidic technologies for sheathless particle focusing.

Table 6.1 Performance of sheathless focusing devices.

Reference	Focusing method	Sample	Focusing variation[a]	Driving velocity
Morgan (12)	DEP	40–460 nm beads	–	2 mm/s
Holmes (13)	DEP	6 μm beads	–	1 mm/s
Yu (14)	DEP	10 μm beads	–	2.7 mm/s
		HL-60 cells	–	1.2 mm/s
Goddard (16)	Acoustic	10 μm beads	–	4 μL/min
Shi (17)	Acoustic	1.9 μm beads	–	6.7 cm/s
Aoki (20)	Flow	5 μm beads	4.8%	4 μL/min
Morton (24)	Flow	2.7 μm beads	–	–
Choi (27, 28)	Flow	10 and 15 μm beads	3.6% and 4.4%	2–9 μL/min
		Jurkat cells	1.7%	4 μL/min
Carlo (30)	Flow	10 μm beads	–	–

[a]The focusing variation denotes the coefficient of variation for each focused stream.
"Flow" refers to flow-assisted, sheathless focusing methods.

6.2 MICROFLUIDIC TECHNIQUES FOR SHEATHLESS PARTICLE FOCUSING

Sheathless particle focusing methods can be categorized into two groups — field-based and flow-assisted methods (Table 1). Field-based methods are those in which external physical fields such as electric, acoustic, and optical fields exert physical forces upon particles to induce focusing at a target position. Flow-assisted methods typically utilize steric hindrance mechanisms in which micro-channels or micro-structures form barriers to move particles out of their streamlines and into a desired focusing position. Here, we provide a detailed review of sheathless focusing methods realized in microfluidic channels.

6.2.1 *Dielectrophoresis*

Dielectrophoresis can be applied for the sheathless focusing of micrometer- and nanometer-scale biological particles in a flow stream. A non-uniform electric field drives motions of dielectric particles to a certain direction in a suspending

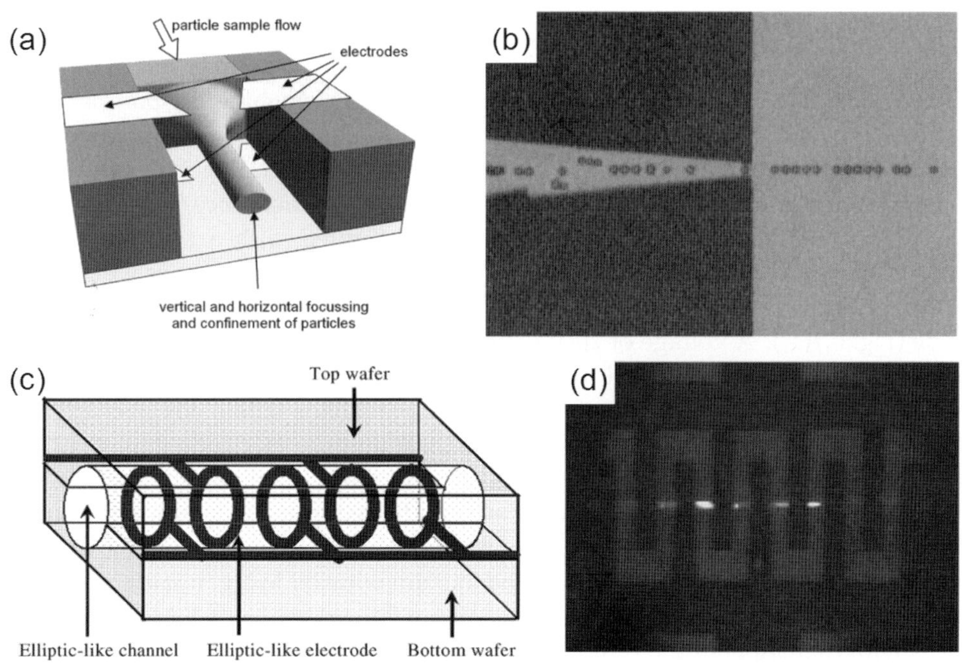

Figure 6.1. (a) Electrodes patterned on the top and bottom of the channel for dielectrophoretic focusing of particles in three dimensions and (b) the photograph of the result showing focusing of 6 μm diameter latex particles by Holmes *et al.* (2006). Reprinted from,[13] with permission from Elsevier. (c) Schematic illustration of microfluidic channel with a microelectrode array patterned on its circumference and (d) particles tightly focused at the center of the microfluidic channel described by Yu *et al.* (2005).[14] [Copyright 2005 IEEE.]

medium. The time-averaged dielectrophoretic force acting on a particle is expressed as the following equation,

$$\langle F_{DEP} \rangle = \pi a^3 \varepsilon_m Re\left[f_{CM}\right] \nabla \left|E\right|^2 \qquad (6.1)$$

where a is the particle radius, ε_m is the permittivity of the suspending medium, $Re[f_{CM}]$ is the real part of the Clausius-Mossotti factor, f_{CM}, and E is the electric field.[12] The direction of the particle movement is determined by the sign of the real part of the Clausius-Mossotti factor which depends on the permittivity and conductivity of the particle and the suspending medium as well as the frequency of the applied electric field. Particles move either towards the region of high-electric field strength (positive DEP) or to the minimum field gradient (negative DEP).

We can apply dielectrophoretic forces to the particles flowing in a channel with a few planar microelectrodes aligned on the top or bottom of the channel. The electric field gradients are generated between the electrodes, having the maximum field strength near the edges. Since the dielectrophoretic forces experienced by particles with various sizes and dielectric properties differ under the electric field of particular frequency, this scheme has been used for separation of distinct microparticles. For focusing of particles, however, one must determine the frequency region where the particles experience a large dielectrophoretic force, regardless of their other properties. Also, the force should be caused by negative DEP, because the focused particles must continuously flow in a stream and not be trapped near the electrodes. This repulsive force can also carry out two-dimensional focusing of particles with relative ease by pushing the particles into the center of the channel with appropriate electrode designs. To achieve negative DEP, the particles have to be less polarizable than the suspending medium.

Detection of 6 μm particles with three different fluorescence intensities was demonstrated with a DEP focusing channel integrated with a confocal optical detecting system by Holmes *et al.*[13] Two triangular-shaped electrodes were evaporated onto both the top and bottom surface of a channel 40 μm high and 250 μm wide (Fig. 6.1(a)). A voltage of 20 V peak-to-peak was applied to the top and bottom electrodes at 10 MHz and the maximum field gradients were created at the edges of the electrodes. All the particles could be focused within a region approximately 5 μm in diameter after they passed through the electrode gap of 10 μm (Fig. 6.1(b)). In order to achieve DEP focusing of particles of nanometer size, the dimensions of the gap and the channel should be reduced to increase the magnitude of the dielectrophoretic force. The funnel-shaped electrodes having the gap of 10 μm and the channel 30 μm high and 110 μm wide by Morgan *et al.* induced a dielectrophoretic force sufficient to focus 40 nm latex particles.[12] Focusing of particles with diameters in the range of 40-460 nm was demonstrated in the channel with 1-to-30 V peak to peak at a frequency of 18-20 MHz.

Yu *et al.* investigated the electrode arrays on the circumference of a round microchannel for the purpose of three-dimensional focusing of microparticles.[14] The elliptical channel was fabricated by bonding of two soda-lime glass wafers after chemical etching and electrode deposition (Fig. 6.1(c)). The electric field gradient was generated in the radial direction from the electrode pattern and was minimal at the center of the channel. Therefore, particles were directed towards the center from all directions by negative DEP. The etched channel of 50 µm in depth, 250 µm in width, and 100 µm center-to-center distance between two adjacent electrodes focused microbeads and human leukemia HL-60 cells to regions 10-15 µm in diameter at 15 V peak to peak at a frequency of 10 kHz (Fig. 6.1(d)).

6.2.2 Acoustic Focusing

Acoustic waves can generate pressure gradients in a fluid transporting suspended particles either to the pressure nodes (minimum pressure amplitude) or the antinodes (maximum pressure amplitude). Particles can be trapped or focused with resonating transducers generating the acoustic wave field (confocal or planar fields). When a standing wave is generated in a medium, the acoustic pressure at x can be described by

$$\Delta p(x) = p_0 \sin(kx) \cos(\omega t) \tag{6.2}$$

where p_0 is the acoustic pressure amplitude, k is the wave number of ultrasonic radiation ($k = 2\pi/\lambda$, λ is the wavelength), x is the distance from the nodal position in the medium, ω is the angular frequency, and t is time. An acoustic radiation force on a particle can be represented as,

$$F_{ac} = -\frac{4}{3}\pi R^3 k E_{ac} A \sin(2kx) \tag{6.3}$$

where R is the particle radius, E_{ac} is the averaged acoustic energy density, and A is the constant given by density, compressibility and the sound velocity in the medium and particle. When A is positive, the particles move to the nodal position of the acoustic standing wave.[15]

Goddard *et al.* attached a piezoceramic crystal to the external surface of a glass capillary tube with an outer diameter of 3.92 mm and an internal diameter of 1.9 mm to generate an acoustic standing wave (Fig. 6.2(a)).[16] Particles of 10 µm diameter were focused to the central axis of the tube of 35 µm width. Due to the planar nature of the standing wave field, the focusing direction of the particles is only one-dimensional. In addition to this localized transducer, acoustic excitation of the entire structure was introduced to achieve three-dimensional focusing. The resulting diameter of the focused stream was ~40µm. The large acoustic energy transfer through this whole cylindrical structure eliminates the need for accurate alignment of a transducer.

To form standing bulk acoustic waves, the channel material must have excellent acoustic reflection properties. Shi *et al.* used standing surface acoustic waves (SSAW) integrated with a soft polymer material, poly(dimethylsiloxane) (PDMS) having poor reflection properties.[17] A PDMS-based microfluidic channel was positioned between the two interdigitated transducers (IDTs) deposited on a piezoelectric substrate ($LiNbO_3$) (Fig. 6.2(b)). When two coherent AC signals were subsequently applied to both IDTs, the constructive interference of the two SAW

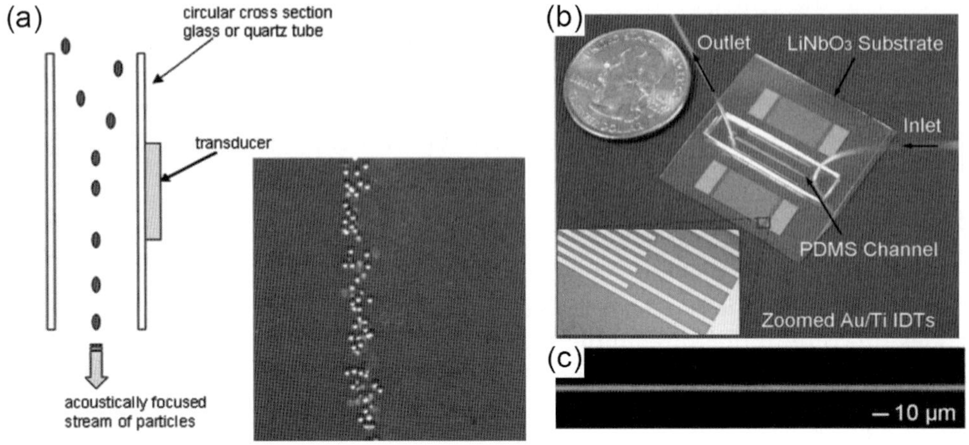

Figure 6.2. (a) Schematic illustration of an acoustic focusing flow cell described by Goddard *et al.* (2006).[16] [G. Goddard, J. C. Martin, S. W. Graves and G. Kaduchak, Ultrasonic particle concentration for sheathless focusing of particles for analysis in a flow cytometer. *Cytometry A* **69A**, 66–74 (2006). Copyright John Wiley & Sons. Reproduced with permission.] (b) Photograph of the bonded SSAW focusing device consisting of a $LiNbO_3$ substrate with two parallel IDTs and a PDMS channel with an inset of enlarged photograph of IDTs and (c) the recorded fluorescent image of focused particles by Shi *et al.* (2008).[17] [J. Shi, X. Mao, D. Ahmed, A. Colletti and T. J. Huang, *Lab Chip* **8**, 221–223 (2008). Reproduced by permission of The Royal Society of Chemistry.]

propagating in opposite directions resulted in the periodic distribution of the pressure nodes and anti-nodes on the substrate. Because suspended particles inside the channel could be forced toward either the pressure nodes or antinodes, the channel width of 50 μm covered only one pressure node. The focused width of particles (1.9 μm in diameter) was 5 μm with an acoustic wavelength of 100 μm and 10 μm with a wavelength of 200 μm (Fig. 6.2(c)).

6.2.3 Optical Focusing

Focusing of particles for microflow cytometers has not been demonstrated with optical forces. However, the scattering force in the direction of laser beam propagation can be used to deviate particles from their streamline for focusing. Kim *et al.* have utilized this lateral force from laser light to separate microparticles and distribute them across the microchannel according to their sizes.[18] The spatially averaged constant scattering force for a laser beam with a Gaussian intensity profile is expressed as follows.

$$F^* = \frac{n_0 P}{4c}\left(\frac{d_p}{\omega_0}\right)^2 Q^* \sqrt{\frac{\pi}{2}}\,\mathrm{erf}\left(\sqrt{2}\right) \tag{6.4}$$

where n_0 is the refractive index of the medium, P is the power of the laser beam, c is the speed of light in a vacuum, d_p is the particle diameter, ω_0 is the width of

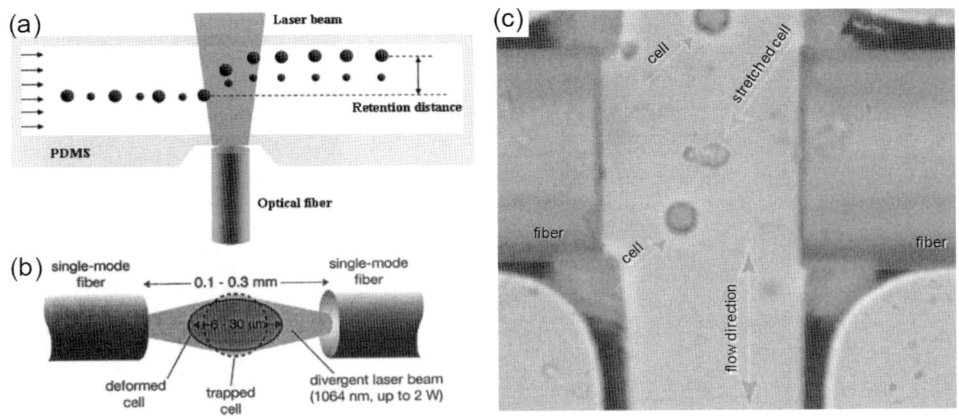

Figure 6.3. (a) Schematics of the optical particle separation by Kim *et al.* (2008). [Reprinted with permission from S. B. Kim, S. Y. Yoon, H. J. Sung and S. S. Kim, Cross-type optical particle separation in a microchannel. *Anal. Chem.* **80**, 2628–2630 (2008). Copyright 2008 American Chemical Society.] (b) An optical stretcher where cell is stably trapped between two opposing divergent laser beams and (c) photograph showing one cell trapped and stretched out while two others are being transported in flow by Lincoln *et al.* (2004).[19] [B. Lincoln, H. M. Erickson, S. Schinkinger, F.Wottawah, D.Mitchell, S. Ulvick, C. Bilby and J. Guck, Deformability-based flow cytometry. *Cytometry A* **59A**, 203–209 (2004). Copyright John Wiley & Sons. Reproduced with permission.]

the laser beam, and Q^* is a constant that depends on the refractive indices of the particle and medium. The optical force is proportional to the square of a particle diameter, so the retention distance increases with particle size (Fig. 6.3(a)). When designing the optical method for particle focusing, an important factor is the size of the particle compared to the wavelength of light, so consideration of electromagnetic theory should be included.

Lincoln *et al.* proposed a microfluidic optical stretcher, which utilizes the optical forces to induce cell deformation.[19] The deformability can be used as an alternative to fluorescence tagging for distinguishing individual cells having different viscoelastic properties. Two optical fibers are aligned to form divergent counter-propagating light beams perpendicular to a flow channel and trap cells at the midpoint between the two sources (Figs. 6.3(b), (c)). This scheme of creating the intense center with Gaussian beam profile to drive particles towards the midpoint of the axis can also be applicable in particle focusing technology.

6.2.4 Hydrodynamic Focusing

Fluidic devices can be microfabricated with accurate channel dimensions at micro/nano-scale. In microfabricated devices, a detailed understanding of fluid/particle transport has contributed to the development of new techniques for self-ordering of biological particles without external forces. The steric hindrance mechanism between a particle and a channel wall allows the self-focusing

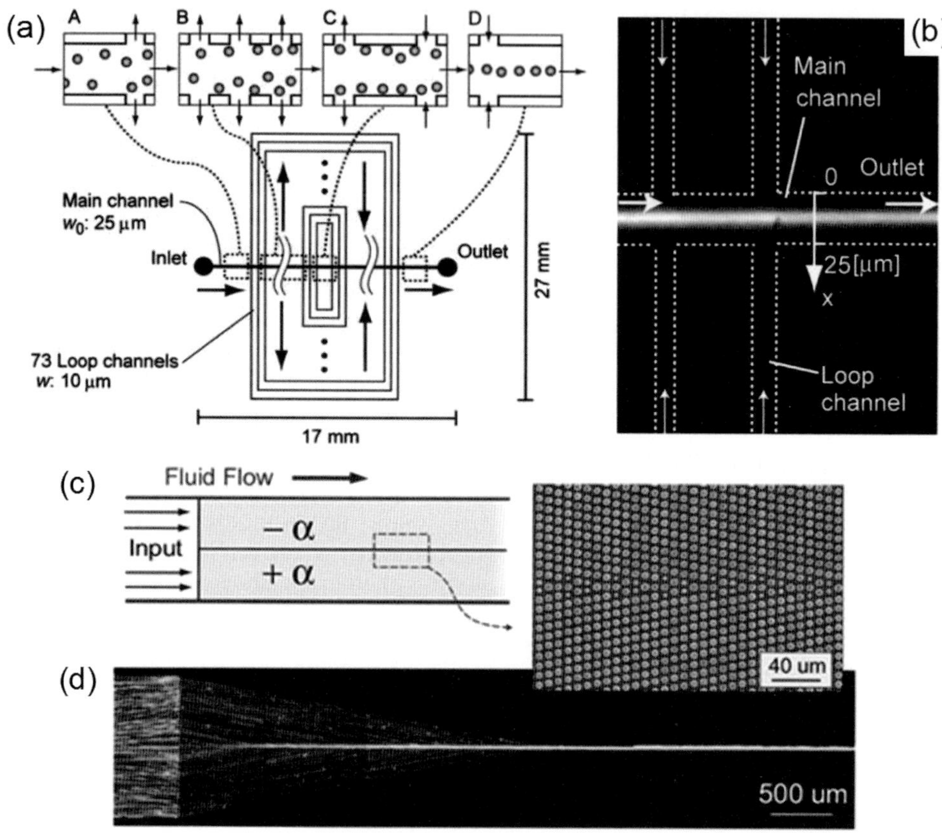

Figure 6.4. (a) Schematic illustration of sheathless particle focusing by hydrodynamic filtration and (b) the recorded fluorescent image of focused particles by Aoki *et al.* (2008).[20] [R. Aoki,M. Yamada,M. Yasuda andM. Seki, In-channel focusing of flowing microparticles utilizing hydrodynamic filtration. *Microfluid. Nanofluid.* **6,** 571-576 (2009). Reproduced with kind permission of Springer Science+Business Media.] (c) The microfluidic focusing device consisting of micropillar arrays tilted toward the channel center and (d) thereby focused 2.7 μm diameter fluorescent beads by Morton *et al.* (2008).[24] [Reprinted with permission from K. J. Morton, K. Loutherback, D. W. Inglis, O. K. Tsui, J. C. Sturm, S. Y. Chou and R. H. Austin, Hydrodynamic metamaterials: microfabricated arrays to steer, refract, and focus streams of biomaterials. *Proc. Natl. Acad. Sci. USA* **105,** 7434-7438 (2008). Copyright (2008) National Academy of Sciences, USA.]

of micron and submicron particles without sheath flows. Seki and coworkers have employed splitting and recombining microchannel networks for sheathless focusing of microparticles using a hydrodynamic principle called hydrodynamic filtration (Fig. 6.4(a)).[20] In their earlier work, the similar channel networks were used to sort microparticles and blood cells by size.[21] While small particles readily flowed out through the branch channels, larger particles formed thicker layers along a sidewall, and this steric barrier forced other particles flow though the main channel, not to the branch channels. The flow volume draining into the branch channels determined whether a particle flowed through the branch channel or not. For sheathless focusing, the authors adjusted the drain volumetric flow rate so that particles larger than 3.5 μm in diameter could not flow out through the branch channels.[20] When the particle-free flows were collected enough to be used for sheath flows, the drained flows joined the main stream, simultaneously focusing the main stream. The authors reported the sheathless focusing of 5 μm polystyrene particles within the standard deviation of 0.7 μm in 25 μm-wide channels (Fig. 6.(b)). They also confirmed that their focusing principle was not affected by flow rates ranging from 1 to 4 μL/min.

The deterministic lateral displacement (DLD) array, pioneered by Austin and colleagues has seen widespread use in sheathless particle focusing as well as high-resolution size separation.[22–24] The DLD array utilizes a kind of steric hindrance mechanism in which particles interact with a large array of pillars, repeating similar hindrance processes with hydrodynamic filtration. The larger the particles are, the farther is their distance from the pillar. The repeated physical alignment with the tilted pillar array allows large particles to migrate across a microchannel. For sheathless focusing, the authors adjusted the tilting direction of the pillar array toward the channel center and thereby focused 2.7 μm particles to the channel center (Fig. 4(c),(d)).[24] The repeated sieving processes in the DLD array also have been used for continuous, high-resolution size separation of microbeads, DNA, and blood cells.[22,23]

Convective flows by anisotropic grooves patterned on a microchannel have been investigated as a means of simply deflecting particles into a certain equilibrium position. Choi *et al.* described a unique particle ordering principle called hydrophoresis that refers to the movement of suspended particles under the influence of a microstructure-induced pressure field.[25,26] The hydrophoretic ordering principle is governed by anisotropic obstacles, a kind of the physical (steric) barrier. Upon application of a fluid flow into the channel, the anisotropic fluidic resistance of the V-shaped obstacles generates rotational fluid streams (Fig. 6.5(a)).[27] These streams force particles to migrate laterally and into the center of the channel. The streamlines starting at the center move upward or downward in the z-direction along with the particles, and their motions are determined by steric hindrance mechanism. The steric hindrance occurs when the obstacles prevent rotational flows of large particles that are observed in relatively smaller particles. In short, the particle-obstacle interaction deflects the large particles from their streamline and leads to particular equilibrium flow paths; this is called hydrophoretic ordering. A particle with a diameter that is similar to the obstacle gap will steer its position toward the center of the z-axis due to the particle-wall interaction. Therefore, the particle can be focused to the channel center and remain in its focused position. The authors reported the sheathless focusing of 10 μm and 15 μm polystyrene particles within the standard deviation of 22 μm and 18 μm in 1 mm-wide channels, respectively (Fig. 6.4(b)).[27] They confirmed that their focusing principle was not affected by the flow rate ranged from 2 to 9 μL/min. The equilibrium position in hydrophoretic self-ordering also

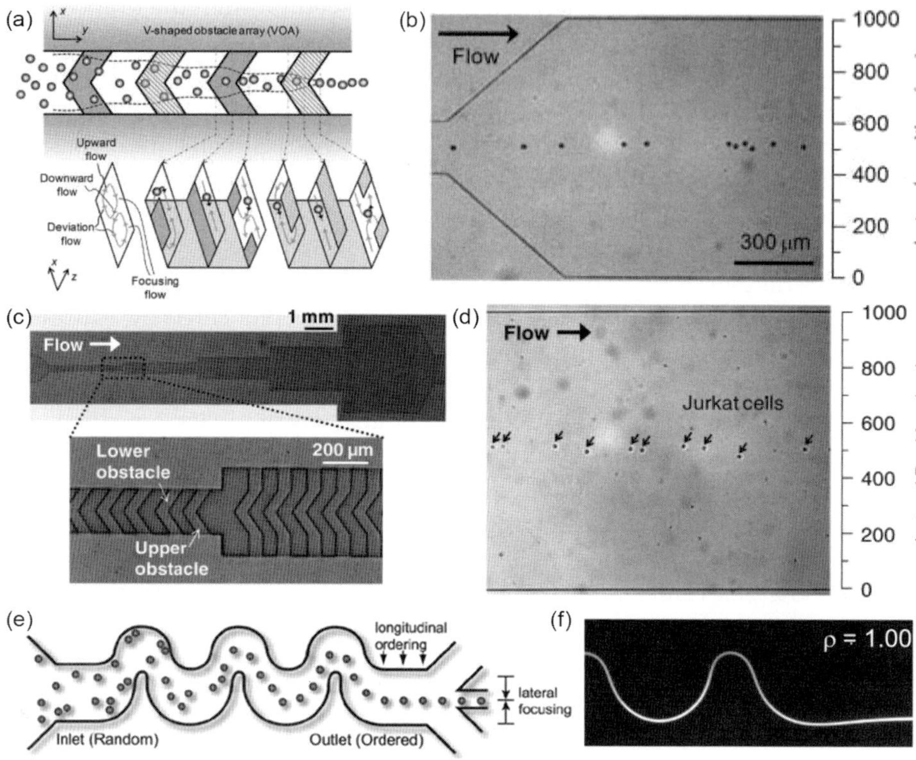

Figure 6.5. (a) Schematic illustration of sheathless particle focusing by hydrophoresis and (b) the recorded image of focused 15 µm beads by Choi *et al.* (2008).[27] [S. Choi, S. Song, C. Choi and J. K. Park, Sheathless focusing of microbeads and blood cells based on hydrophoresis, Small 4, 634 (2008). Copyright Wiley-VCH Verlag GmbH & Co. KGaA. Reproduced with permission.] (c) The microfluidic device consisting of exponentially increasing obstacle arrays and (d) thereby focused Jurkat cells by Choi *et al.* (2008). Reproduced with permission from.[28] [Reprinted with permission from S. Choi and J.-K. Park, Sheathless hydrophoretic particle focusing in a microchannel with exponentially increasing obstacle arrays. *Anal. Chem.* **80**, 3035–3039 (2008). Copyright 2008 American Chemical Society.] (e) Schematic drawing of the inertial self-focusing and (f) the recorded fluorescent image of focused 10 µm beads by Carlo *et al.* (2007).[30] [Reprinted with permission from D. D. Carlo, D. Irimia, R. G. Tompkins and M. Toner, Continuous inertial focusing, ordering, and separation of particles in microchannels. *Proc. Natl. Acad. Sci. USA* **104**, 18892–18897 (2007). Copyright (2007) National Academy of Sciences, USA.]

depends on the particle size. The smaller particles were found to have larger focusing variation. In more recent work, the authors described a modification to this approach (Fig. 6.4(c)).[28] In this case, a microfluidic device with exponentially increasing obstacle arrays was proposed to reduce the size dependence of hydrophoretic ordering. The anisotropic fluidic resistance of the V-shaped obstacles generates transverse flows, along which particles are focused to the channel center. In the channel with exponentially increasing widths, the bent obstacles extended from the V-shaped obstacles increase the focusing efficiency of the particles. By using this method, the sheathless focusing of Jurkat cells within the standard deviation of 8.7 μm in 1 mm-wide channels was achieved at a flow rate of 4 μL/min (Fig. 6.4(d)).[28]

Flowing particles suspended in a fluid are subjected to inertial lift forces such as shear-gradient and wall-induced lift forces.[29] These forces acting on the particles are negligible at relatively low particle Reynolds number (Re_p). The particle Reynolds number is defined as the ratio of the particle inertia to the viscous force;

$$Re_p = \frac{\rho D^2 U}{\mu D_h} \tag{6.5}$$

where ρ is the fluid density, D is the particle diameter, U is the maximum fluid velocity, and μ is the dynamic fluid viscosity. D_h is the hydraulic diameter, defined as

$$D_h = \frac{2wh}{w+h} \tag{6.6}$$

where w and h are the width and height of the channel, respectively. At higher Re_p of order 1, the inertial lift forces dominate particle behaviors even in microscaled channels. When balancing the inertial forces with a drag force, the drift velocity induced by the inertial forces is proportional to the volume of the particle. This inertial method can be classified into the field-based methods, in that it uses physical forces acting on particles. However, due to its hydrodynamic nature, we include this inertial focusing method in flow-assisted/hydrodynamic methods. Carlo *et al.* have employed curving microchannels to demonstrate the inertial ordering and focusing of microparticles.[30] By the interaction of the inertial forces and a convective flow such as Dean flow, micron-sized particles migrate toward an equilibrium position across the channels. The authors report the sheathless focusing of 10 μm polystyrene particles within the standard deviation of 80 nm. Below the Re_p value of 0.15, the level of the inertial forces is insufficient to focus the particles and thereby widens the focused stream. Above the critical Re_p value, the particle focusing is perturbed by mixing due to Dean flow. The inertial forces acting on relatively smaller particles are insufficient to make them accurately positioned to the equilibrium position. This size dependence of the inertial particle ordering causes the size separation. In the similar curving channels, platelets, the smallest cell type of blood cells were isolated from the other types of blood cells with an enrichment ratio of 100-fold.[31] Bhagat *et al.* recently proposed a spiral microchannel for the complete separation between two particle sizes and demonstrated the inertial separation of 7.32 μm and 1.9 μm particles.[32]

6.3 CHALLENGES OF SHEATHLESS FOCUSING METHODS

Recent advances in microfluidics have overcome many disadvantages of flow cytometry by replacing conventional flow chambers with sheathless focusing devices and eliminating a large volume of sheath liquid. However, there are several practical considerations for real cytometric applications of the focusing devices. The first limitation is two-dimensional focusing that focus particles only in the lateral direction, and not in the vertical direction. Most sheathless focusing devices have demonstrated only for two-dimensional focusing. Even some devices for three-dimensional focusing show the high-dependence of flow rate and particle size that prevents a practical implementation of the focusing method. The major drawback of two-dimensional particle focusing is a broad detection volume in the vertical direction, and subsequent impairment of illumination and signal collection efficiency. If particles are allowed to flow through the broad detection volume, any measures to obtain quantitative information will be confounded by variations in their flow speed of a parabolic velocity profile and illumination. Recent efforts have begun to address this problem by covering a laterally focused sample flow with convective, rotational flows. Howell *et al.* reported three-dimensional hydrodynamic focusing with a single sheath flow by employing a set of slant grooves that fully wraps the sheath solution around the sample solution.[33] The authors also demonstrated the separation of the sample and sheath fluid (called 'unsheathing') by using another set of grooves mirroring the first. Using the similar convective approach, Sato *et al.* developed a symmetrically grooved microchannel that induces local rotational streams and thereby fully surrounds a sample flow with a single sheath flow.[34] These sheath flow designs further can be incorporated into the sheathless devices to achieve three-dimensional focusing.

Sheathless focusing devices have inherent strengths and weaknesses. The focusing efficiency of the devices largely depends on particle size. In the field-based methods utilizing electric, acoustic, and optical fields, the force acting on particles is typically proportional to their volume. Flow-assisted methods utilize steric hindrance mechanisms that rely on particle size. Thus, these methods have been used for size separation of micro and submicron particles. For many cytometric applications, this size-dependence is a weakness for stable focusing of particles variable in size, but can be a new, useful technique for simpler microflow cytometers that utilize fewer parameters than conventional flow cytometers. For instance, enumerating separately focused blood cells according to their sizes may be able to provide rapid diagnostic information for microflow hemocytometers.

The final consideration for real cytometric applications of the sheathless focusing devices arises from the flow rate dependence of the profile of focused particles. During the sheathless focusing processes, particles experience the repeated exposure of external fields such as electric, acoustic, optical, and inertial fields, or the repeated interaction with microstructures for steric hindrance effects. For field-based methods, the migration distance of particles relies on the exposure times to the fields at a given field strength. At over a critical flow rate, the focusing profile can be broadened due to the insufficient migration of particles into a targeted focusing position. For flow-assisted methods, the effect of flow rates on the focusing efficiency is negligible, but inertial lift forces can make a focused stream unstable at a high flow rate over $Re_p = 1$. The flow-rate limitation can be a hurdle for the development of high-throughput microflow cytometry. Conventional flow cytometers are capable of processing from thousands to tens-of-thousands of particles per second. However, current microfluidic focusing devices have been limited in throughput rate of tens of particles per second. Thus, continued efforts should be addressed to solve this problem by parallelizing multiple focusing channels in a single device.

6.4 OUTLOOK FOR THE FUTURE

Microfluidics holds the potential to enhance the way in which conventional flow cytometry have been conducted significantly. One important area for miniaturizing flow cytometry is sheathless particle focusing. While significant improvement has been shown in this field, there are still many more challenges for future development. The real integration of sheathless focusing devices into flow cytometry has yet to be achieved. However, once three-dimentional directional manipulation is completely realized for sheathless focusing, sheathless particle focusing devices will offer new opportunities for low-cost flow cytometry and personalized diagnostics with advantages of low sample and reagent volumes and short analysis time.

ACKNOWLEDGMENTS

This work was supported by the Korea Science and Engineering Foundation (KOSEF) NRL Program grant funded by the Korea government (MEST) (R0A-2008-000-20109-0), and by the Nano/Bio Science and Technology Program (2005-01291) of the MEST, Korea. The authors also thank the Chung Moon Soul Center for BioInformation and BioElectronics, KAIST.

References

[1] M. Rieseberg, C. Kasper, K. F. Reardon and T. Scheper, Flow cytometry in biotechnology. *Appl. Microbiol. Biotechnol.* **56**, 350–360 (2001).

[2] J. W. Tung, K. Heydari, R. Tirouvanziam, B. Sahaf, D. R. Parks, L. A. Herzenberg and L. A. Herzenberg, Modern flow cytometry: A practical approach. *Clin. Lab. Med.* **27**, 453–468 (2007).

[3] P. J. Crosland-Taylor, A device for counting small particles suspended in a fluid through a tube. *Nature* **171**, 37–38 (1953).

[4] J. V. Watson, The early fluidic and optical physics of cytometry. *Cytometry* **38**, 2–14 (1999).

[5] D. A. Ateya, J. S. Erickson, P. B. Howell Jr, L. R. Hilliard J. P. Golden and F. S. Ligler, The good, the bad, and the tiny: A review of microflow cytometry. *Anal. Bioanal. Chem.* **391**, 1485–1498 (2008).

[6] D. Huh, W. Gu, Y. Kamotani, J. B. Grotberg and S. Takayama, Microfluidics for flow cytometric analysis of cells and particles. *Physiol. Meas.* **26**, R73–R98 (2005).

[7] T. D. Chung and H. C. Kim, Recent advances in miniaturized microfluidic flow cytometry for clinical use. *Electrophoresis* **28**, 4511–4520 (2007).

[8] A. Y. Fu, C. Spence, A. Scherer, F. H. Arnold and S. R. Quake, A microfabricated fluorescence-activated cell sorter. *Nat. Biotechnol.* **17**, 1109–1111 (1999).

[9] C. Simonnet and A. Groisman, Two-dimensional hydrodynamic focusing in a simple microfluidic device. *Appl. Phys. Lett.* **87**, 114104 (2005).

[10] C. Simonnet and A. Groisman, High-throughput and high-resolution flow cytometry in molded microfluidic devices. *Anal. Chem.* **78**, 5653–5663 (2006).

[11] N. Sundararajan, M. S. Pio, L. P. Lee and A. A. Berlin, Three-dimensional hydrodynamic focusing in polydimethylsiloxane (PDMS) microchannels. *J. Microelectromech. Syst.* **13**, 559–567 (2004).

[12] H. Morgan, D. Holmes and N. G. Green, 3D focusing of nanoparticles in microfluidic channels. *IEE Proc.-Nanobiotechnol.* **150**, 76–81 (2003).

[13] D. Holmes, H. Morgan and N. G. Green, High throughput particle analysis: Combining dielectrophoretic particle focussing with confocal optical detection. *Biosens. Bioelectron.* **21**, 1621–1630 (2006).

[14] C. Yu, J. Vykoukal, D. M. Vykoukal, J. A. Schwartz, L. Shi and P. R. C. Gascoyne, A three-dimensional dielectrophoretic particle focusing channel for microcytometry applications. *J. Microelectromech. Syst.* **14**, 480–487 (2005).

[15] T. Masudo and T. Okada, Ultrasonic radiation — Novel principle for microparticle separation. *Anal. Sci.* **17**, i1341–i1344 (2001).

[16] G. Goddard, J. C. Martin, S. W. Graves and G. Kaduchak, Ultrasonic particle-concentration for sheathless focusing of particles for analysis in a flow cytometer. *Cytometry A* **69A**, 66–74 (2006).

[17] J. Shi, X. Mao, D. Ahmed, A. Colletti and T. J. Huang, Focusing microparticles in a microfluidic channel with standing surface acoustic waves (SSAW). *Lab Chip* **8**, 221–223 (2008).

[18] S. B. Kim, S. Y. Yoon, H. J. Sung and S. S. Kim, Cross-type optical particle separation in a microchannel. *Anal. Chem.* **80**, 2628–2630 (2008).

[19] B. Lincoln, H. M. Erickson, S. Schinkinger, F. Wottawah, D. Mitchell, S. Ulvick, C. Bilby and J. Guck, Deformability-based flow cytometry. *Cytometry A* **59A**, 203–209 (2004).

[20] R. Aoki, M. Yamada, M. Yasuda and M. Seki, In-channel focusing of flowing microparticles utilizing hydrodynamic filtration. *Microfluid. Nanofluid.* **6**, 571-576 (2009).

[21] M. Yamada and M. Seki, Hydrodynamic filtration for on-chip particle concentration and classification utilizing microfluidics. *Lab Chip* **5**, 1233–1239 (2005).

[22] L. R. Huang, E. C. Cox, R. H. Austin and J. C. Sturm, Continuous particle separation through deterministic lateral displacement. *Science* **304**, 987–990 (2004).

[23] J. A. Davis, D. W. Inglis, K. J. Morton, D. A. Lawrence, L. R. Huang, S. Y. Chou, J. C. Sturm and R. H. Austin, Deterministic hydrodynamics: taking blood apart. *Proc. Natl. Acad. Sci. USA* **103**, 14779–14784 (2006).

[24] K. J. Morton, K. Loutherback, D. W. Inglis, O. K. Tsui, J. C. Sturm, S. Y. Chou and R. H. Austin, Hydrodynamic metamaterials: microfabricated arrays to steer, refract, and focus streams of biomaterials. *Proc. Natl. Acad. Sci. USA* **105**, 7434–7438 (2008).

[25] S. Choi and J.-K. Park, Continuous hydrophoretic separation and sizing of microparticles using slanted obstacles in a microchannel. *Lab Chip* **7**, 890–897 (2007).

[26] S. Choi, S. Song C. Choi and J.-K. Park, Continuous blood cell separation by hydrophoretic filtration. *Lab Chip* **7**, 1532-1538 (2007).

[27] S. Choi, S. Song, C. Choi and J.-K. Park, Sheathless focusing of microbeads and blood cells based on hydrophoresis. *Small* **4**, 634–641 (2008).

[28] S. Choi and J.-K. Park, Sheathless hydrophoretic particle focusing in a microchannel with exponentially increasing obstacle arrays. *Anal. Chem.* **80**, 3035–3039 (2008).

[29] E. S. Asmolov, The inertial lift on a spherical particle in a plane Poiseuille flow at large channel Reynolds number. *J. Fluid. Mech.* **381**, 63–87 (1999).

[30] D. D. Carlo, D. Irimia, R. G. Tompkins and M. Toner, Continuous inertial focusing, ordering, and separation of particles in microchannels. *Proc. Natl. Acad. Sci. USA* **104**, 18892–18897 (2007).

[31] D. D. Carlo, J. F. Edd, D. Irimia, R. G. Tompkins and M. Toner, Equilibrium separation and filtration of particles using differential inertial focusing. *Anal. Chem.* **80**, 2204–2211. (2008).

[32] A. A. S. Bhagat, S. S. Kuntaegowdanahalli and I. Papautsky, Continuous particle separation in spiral microchannels using dean flows and differential migration. *Lab Chip* **8**, 1906–1914 (2008).

[33] P. B. Howell Jr, J. P. Golden, L. R. Hilliard, J. S. Erickson, D. R. Mott and F. S. Ligler, Two simple and rugged designs for creating microfluidic sheath flow. *Lab Chip* **8**, 1097–1103 (2008).

[34] H. Sato, Y. Sasamoto, D. Yagyu, T. Sekiguchi and S. Shoji, 3D sheath flow using hydrodynamic position control of the sample flow. *J. Micromech. Microeng.* **17**, 2211–2216 (2007).

Chapter Seven

Two-Dimensional Particle Focusing: Sheath Flow on Two Sides

Jaeho Shin and Michael Ladisch*

Department of Agricultural and Biological Engineering, Laboratory of Renewable Resources Engineering, Weldon School of Biomedical Engineering, Purdue University, West Lafayette, IN 47907, USA

*ladisch@purdue.edu

The ability to obtain precise information from the particles traveling through a cytometer requires adequate focus of the sample stream. One approach to obtaining a focused stream is the induction of sheathed flow and utilization of the hydrodynamic characteristics of 10 to 100 micron-wide channels to obtain a flow inequality that focuses particles into a narrow band, i.e. sheathed flow. While there have been many examples of microcytometry in the literature, only a few have succeeded in completely sheathing the stream. This chapter reviews several recent approaches to achieving focused sample introduction in a manner that may be suitable for the microflow channels associated with flow cytometers. The fabrication of these channels shares fabrication techniques based on two-dimensional networks in microelectronics. Appropriate design, characterization of surface properties, and optimization of channel geometry that enhances stable sheath flow is discussed.

7.1 IMPORTANCE OF MICROFLUIDIC FLOW TO FLOW CYTOMETRY

The earliest flow cytometry systems consisted of a small capillary tube to deliver particles or cells past a modified microscope with electronic detectors for counting and measuring sizes. The problem associated with this technique was that large cells often clogged small-diameter tubes. On the other hand, large-diameter capillary tubes caused significant challenges in focusing the target cells and detecting them in a wide path. Further the probability of two or more cells passing though the interrogation region at the same time increased. Conventional cytometry solved this challenge by injection of particles into the center of a sheath fluid.[1] Thus in the large benchtop laboratory systems, the combined flow enables fluid

The Microflow Cytometer by J S Kim & F S Ligler

www.panstanford.com
978-981-4267-41-0

dynamics to cause the smaller diameter particles to be focused in the center of the larger bore stream.

In these systems, sheathed flow is generated by placing a glass capillary inside a large diameter tube. However, this kind of structure requires an expensive fabrication technique to make the flow cells. For an inexpensive and portable microflow cytometer, microfabricated particle focusing components have been explored by many research groups. The benefits and challenges of sheathless particle focusing solutions have been reviewed in the preceding chapter. Microfabricated designs for three-dimensional particle focusing will be discussed in the following chapter. Though three-dimensional hydrodynamic flow focusing is often favored for particle focusing and reduction of clogging, two-dimensional particle focusing has benefits for a microchip flow cytometer and is discussed here in the context of reducing both cost and volume of the sample analysis.[2] In this section, the focusing of particles or cells that pass between streams of sheath fluid on two sides of the microfluidic channels will be discussed.

7.2 CHARACTERISTICS OF 2D MICROFLUIDIC HYDRODYNAMIC FOCUSING

When the sample stream is injected into middle of a sheath stream, cells in sample stream align into a relatively smaller cross-section of the sheath stream. This is called hydrodynamic focusing. Focusing of the sample stream is important for flow cytometers to reduce adherence of cells to channel walls that could lead to capillary blockage, reduction of flow, and fouling. Hydrodynamic focusing also enhances the effectiveness of an optical detector by directing flow of the particles within the optical path. For effective hydrodynamic focusing and interrogation, cells should present themselves in a single file in a laminar flow of liquid through the channel. Sample fluids containing the cells will not mix with the sheath fluid because of the laminar flow in the focused stream at low Reynolds numbers. In most microfluidic channels, the Reynolds number (*Re*) can be calculated based on hydraulic diameter D_h:

$$D_h = \frac{4A}{P} \tag{7.1}$$

$$Re = \frac{\rho D_h V_{avg}}{\mu} \tag{7.2}$$

where A is the cross sectional area of the channel, P is the wetted perimeter of the channel, ρ is the fluid density, V_{avg} is the average flow velocity, and μ is the dynamic viscosity of the fluid. Under typical dimensions of microfluidic channels and a low rate of fluid flow, the Reynolds number is generally very small. For example, when the hydrodynamic diameter is about 20 μm (corresponds to a height of 10 μm in a wide rectangular channel) and average flow velocities are approximately 1 cm/s, the Reynolds numbers are less than 1 in this microfluidic channel, and are in the laminar flow regime. Turbulent flow occurs when Reynolds numbers are over 2000. Based on pressure-driven flow, a viscous drag at the wall results in higher velocity in the center, thus flow through microchannels has a parabolic velocity profile. The various flow velocities of the sample stream due to contact with channel walls would be reduced by introducing sheathed flows. Thus, the cells can be transported in a predictable manner through micro-fluidic channels by sheathed flow in the laminar flow regime.

In order to achieve stable flow profiles, pumping action that minimizes pulsation is needed such as through hydraulic amplification or syringe pumps. In this type of pump a pulseless flow is achieved through the action of a syringe or other hydraulic cylinder that pushes the fluid through the system using a continuous mechanical or hydraulic action. This avoids pulsation as long as the stroke of the pump is long enough so that there is ample liquid volume for a run. Since the volume of microfluidic devices is small this criterion may be achieved with pumps that hold as little as 10 mL. We have used syringe pumps for this purpose.

7.2.1 Review of Progress in Microfluidic Flow Methods

The general concept of two-dimensional hydrodynamic focusing in microfluidic channels is described in Fig. 7.1.[3] The sample flow is restricted laterally within the center of the channel by two sheathed flows. As sample flow rate increases, the width of the sample stream should also increase. Likewise, the decrease of the sample flow rate relative to the sheath flow rate causes the reduction in the width of the sample stream, resulting in highly focusing state (see Fig. 7.2)[3] Flow rates in the sheathed flow channels and the sample channel determine the width of a sample stream which can be as small as 1–2 μm.

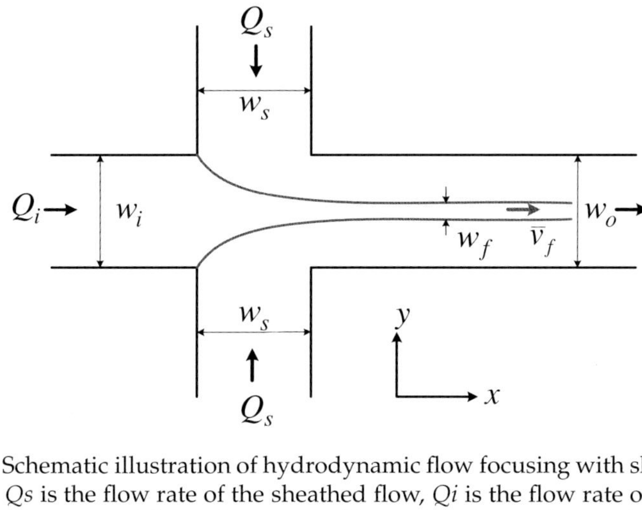

Figure 7.1. Schematic illustration of hydrodynamic flow focusing with sheath flow on two sides, where *Qs* is the flow rate of the sheathed flow, *Qi* is the flow rate of the sample flow, and *Qo* is the flow rate of the outlet stream. [Reproduced from Lee *et al.* *J. Micromech. Microeng.* **16**, 1024–1032 (2006) with permission from Institute of Physics and IOP Publishing.]

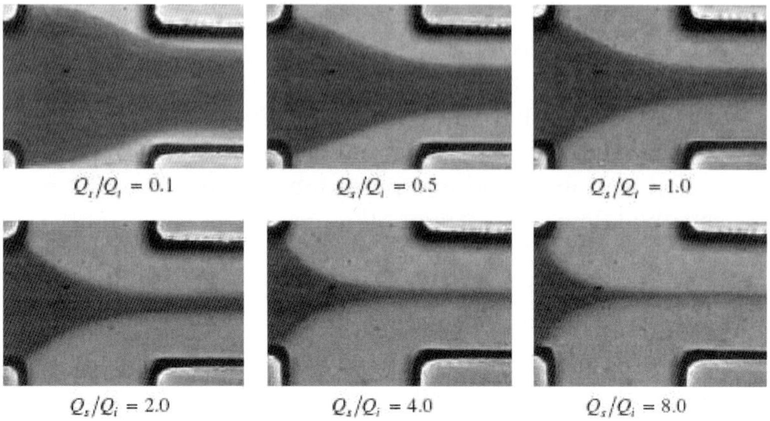

Figure 7.2. Experimental images of the hydrodynamic focusing effect in microchannels with sheath flow on two sides (top view), outlet channel width = 250 μm and height = 445μm, *Qs* = sheath flow rate on one side, *Qi* = sample flow rate. [Reproduced from Lee *et al. J. Micromech. Microeng.* **16**, 1024–1032 (2006) with permission from Institute of Physics and IOP Publishing Limited.]

In general, a narrow width of sample stream, ideally the same size as the cells to be analyzed, is required for analysis using the flow cytometer. However, the flow rate of sample stream must also address the needs of specific targeted applications and configurations of the microfluidic cytometer to perform various functions. For example, higher sample stream flow rates are required in the qualitative measurements such as immunophenotyping.[4,5] Microfluidic mixers[6–8] that utilize hydrodynamic focusing may achieve complete mixing within the channel length by decreasing the width of the focused mixing stream from the sheathed stream. The fluids are difficult to be mixed with each other in the laminar regime of flow in the microchannels, and diffusion dominates in the mixing condition. Thus, relatively slow flow rates of the mixing fluids are required in order to reduce the width of mixing stream, which results in short diffusion time.

Hydrodynamic focusing is one the most popular tools for any microfluidic cytometer due to its ability to align particles for interrogation in a simple and adaptable way. However, liquid-based hydrodynamic focusing requires a large volume of aqueous sheath liquid, approximately 250∼1000 ml per 1 ml of biological sample. As an alternative, the use of ambient air to provide focusing not only avoids the need for continuous pumping of sheath liquid but also results in a significant reduction in the size of the overall system.[9] At the micron scale, surface forces of channel walls play an important role in the formation of air-liquid interfaces. Huh *et al.* introduced air-liquid two-phase flow in hydrophobic microfluidic channels to produce a focused sample stream.[10] Air sheath flows from the two side inlets focus the sample liquid in the focusing chamber as described in Fig. 7.3. The size of the focusing chamber is gradually narrowed in order to attain a high speed sample stream. At a sample flow rate of 5 ml/hr, the sample stream becomes unstable and breaks into droplets.

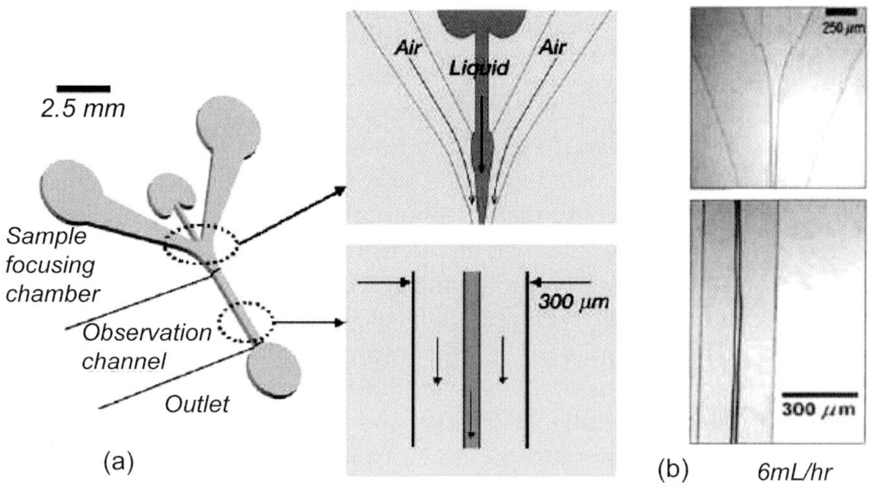

Figure 7.3. (a) Schematic diagram of basic mechanism of focusing by air-liquid two phase (b) Images of focusing chamber by air-water two-phase. [Reprinted from Huh *et al. Biomedical Microdevices* **4**, 141–149 (2002) with permission from Kluwer Academic Publishers.]

Another technique uses surface chemistry in order to focus a stream in a channel introduced by using the press-fit method. By sandwiching a hydrophilic glass fiber between two hydrophobic surfaces such as PDMS and OTS wafer, the unique geometry of surface forces form a fluid layer next to the fiber. This configuration was applied to hydrodynamic focusing to count cells in a boundary flow. The liquid flow is focused along the fiber within 20 μm in width as shown in Fig. 7. 4.[11]

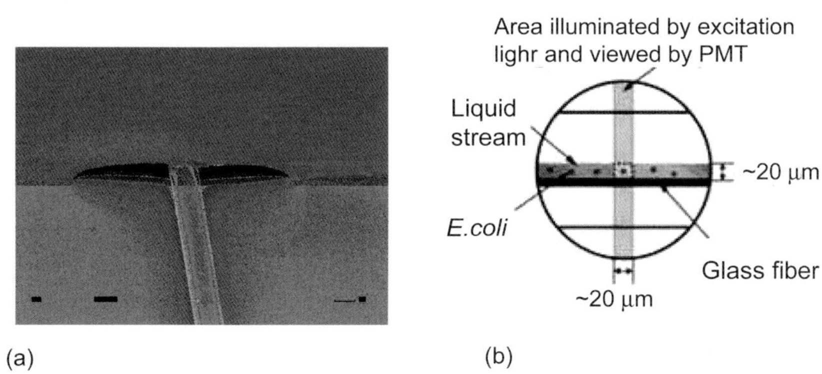

Figure 7.4. (a) SEM of a press-fit microdevice. Glass fiber is sandwiched between PDMS and a glass substrate (b) Press-fit microdevice as a cell counting and detection tool. Liquid stream is controlled in 20 μm by making underlying silicon wafer hydrophobic. Shaded area represents the region viewed by PMT detector with flowing fluorescent *E. coli.* [Reprinted with permission from Huang *et al.* Microfiber-directed boundary flow in press-fit microdevices fabricated from self-adhesive hydrophobic surfaces. *Anal. Chem.* **77** 3671–3675 (2005). Copyright 2005 American Chemical Society.]

7.3 MICROFLUIDIC CHANNELS AND FABRICATION

Microfluidic systems will expand the potential applications of flow cytometry in terms of high throughput, volume efficiency, and low cost, and will offer the opportunity to develop new analytical methods. Most of the early development on microfluidic systems evolved from the silicon chip industry and concentrated on microelectronic technology, such as photolithography on silicon and glass.[12] More recently, microfluidic channels made using polymers have received attention due to their flexibility and reproducibility. Polymers are easily molded and no corrosive etching is needed.[13] Poly(dimethysiloxane) (PDMS), which is the most popular prototyping elastomer in soft lithography, has a combination of properties that makes it attractive for assembling microfluidic devices. PDMS is mechanically sturdy, with a Young's modulus of 750 kPa, and is also optically transparent. This quality makes it easy to monitor the flow of products that have distinct optical properties, thus allowing various detection schemes in the flow cytometry system.

Other properties include: being relatively impermeable to most liquids so that PDMS-molded channels permit transport of liquid analytes; curability at comparably low temperature; reversible sealing to itself or a wide range of materials by van der Waals force;[14] and irreversible sealing with oxygen plasma treatment. Its surface chemistry is easily controlled by chemical modification.[15,16] Therefore, PDMS is widely used in today's microfluidic prototyping systems. Typical fabricated methods for microfluidic channels by PDMS and the SEM image of microfluidic channels to two-dimensional particle focusing are shown in Figs. 7.5(a) and 7.5(b), respectively.[17]

The fabrication process for the PDMS channel can be described by the following four main steps:

- Patterning of the Channel: Photoresist is spun on Si wafer, exposed to UV light through a mask, cured, and processed to provide a mold.
- Casting PDMS: PDMS resin solutions are mixed with curing agents, poured over the mold, and cured.
- Sealing the Channel: After peeling the PDMS off the mold, a glass cover is bonded on it.
- Generating the Fluid Inlet/Outlet Ports: Capillary tubes or needle inserts on glass or PDMS sides are added to make the fluid inlets and outlet.

Once the PDMS-based microchannels are fabricated, pressure-driven flow can be controlled by pump or vacuum. However, pressure drops are much higher in microchannels compared to mm sized tubing and channels due to the micron sized dimensions. When a positive pressure is applied, microchannels need to be sealed in order to enable higher pressure operation. Although PDMS can contact, conform to, and form reversible seals with a variety of materials, it can only withstand a pressure of 5 psi or less.[14] With irreversible sealing, however, PDMS can withstand up to 30~40 psi. The most common method of irreversible sealing is through plasma oxygen treatment. This process generates silanol groups that

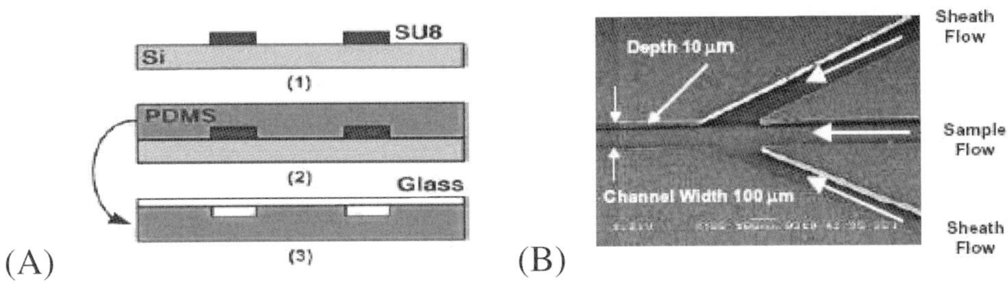

Figure 7.5. (a) A typical fabricated method to make microfluidic channels in PDMS (b) The SEM image of the microfluidic channels used for 2-D particle focusing. [Both figures are reproduced from Lee *et al. Sensors, Proceedings of IEEE* **1**, 308–311 (2004) with permission from IEEE.]

enable sealing of PDMS to itself or to a wide range of materials.[14] Although oxygen plasma use is commonplace, it suffers from hydrophobicity recovery, which is a process where PDMS regains hydrophobicity within 10 minutes. This process is highly influenced by applied partial electrical discharge, humidity and ambient temperature.[18,19]

Using conventional microfabrication methods, two-dimensional sheathed flow can be introduced into the focusing channel either vertically or with a Y-junction shape. The horizontal hydrodynamic focusing channel using a microfabrication technique described in Fig. 7.6 was studied by Lee *et al.*[20] The microfluidic channels were made from poly(methylmethacrylate) (PMMA) using a hot embossing method. The small inner nozzle is located in the middle of the outer nozzle with sheath flows generated around it. The combined flow reduces a sample stream to a width that is smaller than 10 μm.

For design of a flow channel to transport cells, conventional microfluidic fabrication can readily create thedesired patterns and dimensions of the microchannel.

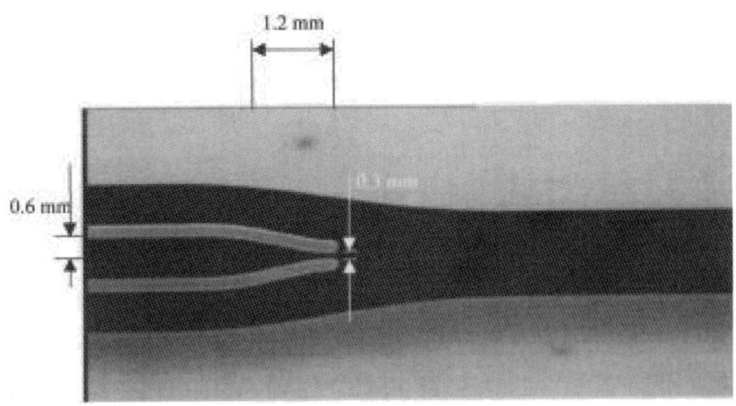

Figure 7.6. Horizontal focusing channel made of PMMA using hot embossing, [Reprinted from Lee *et al. Trans. ASME I* **123**, 672–679 (2001) with permission from ASME.]

However, certain channel shapes are difficult to construct by typical microfabrication methods. The wet chemical etching method is favorable for fabricating a deeper channel which is difficult to achieve by the photolithography. A simple V-groove microchannel was achieved by a combined approach of conventional photolithography and wet chemical etching, resulting in a lateral width of 20 ∼ 25 μm and a depth of 50 μm. This device was used for differential blood cell counts without sheath flow.[21]

The press-fit technique reported by Huang *et al.* utilized a flat PDMS film to cover a glass fiber resting on a glass slide, described in Fig. 7.7.[11] A microflow chamber was fabricated in a single step by sandwiching a glass fiber between the PDMS and the OTS wafer. The flexible PDMS deformed over the glass fiber, forming a reversible seal with the glass side to obtain flow channels on both sides of the glass fiber.

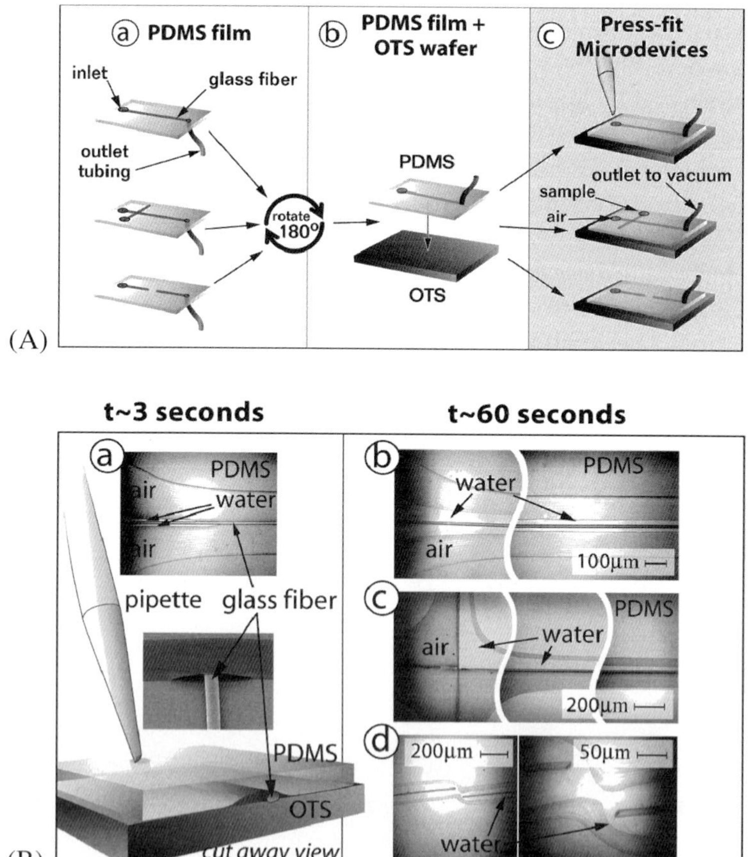

Figure 7.7. (a) The fabrication of a press-fit microchannel[11] (b) Microfiberdirected flows.[11] [Reprinted with permission from Huang *et al.* Microfiber-directed boundary flow in press-fit microdevices fabricated from self-adhesive hydrophobic surfaces. *Anal. Chem.* **77** 3671–3675 (2005). Copyright 2005 American Chemical Society.]

Channels produced by the press-fit technique have three walls, unlike the four walls found in photolithographic channels. The actual geometry of the device is derived from the cylindrical shape of the glass fiber and the elastic deformation of the PDMS. This gives rise to a cylindrical surface along the fiber and an elliptical surface at the PDMS wall. In order to generate equal width along channel length, the capillary force is applied in the press-fit microchannel. For irregular shapes and geometries, it is desirable to predict the solutions to a capillary filling problem. A hydraulic diameter of irregular shapes is a convenient way to predict a good solution for the press-fit microchannel. Huang *et al.* described initial wetting behavior in hydrophilic and hydrophobic press-fit microchannel.[22] After initial wetting, flow can be induced by capillary force. The advantage of using a press-fit device lies not only in its rapid assembly, but also easy functionalization. Each fiber can be modified with different chemical or biological entities such that each microchannel can have its own identity. Thus, press-fitting microfibers with PDMS enable microchannels that are controlled by geometry, surface wettability and surface chemistries, and which are particularly useful for prototyping new devices and systems for various applications in microfluidics, and the channel of modified surfaces can be rapidly constructed on top of a substrate, resulting in a channel that directs the fluid and particles that it contains into a band suitable for optical monitoring.

7.4 CRITICAL ISSUES AND FUTURE OUTLOOK

The rapid growth of the microfluidic device technology has pushed the cytometer forward in terms of both miniaturization and reduction of manufacturing costs. Microfluidic systems were the first developed using microelectronic technologies and silicon substrates. Although use of silicon wafers and glass covers to fabricate microchannels in cytometers would make mass production by the current miocroelectronic techniques feasible, they require special techniques to connect components such as valves and pumps.[23] Silicon may be inappropriate for use in a microfabricated cytometer due to the opacity to both visible and ultraviolet light.

Furthermore, the surface properties of silicon, including hydrophilicity and oxidation, may result in surface fouling and a lack of biocompatibility with respect to biological samples.[24] Thus, microchannels made from silicon wafers are often difficult to use in a continuous manner while obtaining consistent results. In the past decade, polymer-based materials such as PDMS and PMMA have therefore become popular for forming microfluidic channels for microflow cytometry. However, soft lithography techniques are not always readily adaptable to mass production as is the case for photolithographic techniques. Imprinting or hot embossing has the potential to provide a cost-effective solution to overcome drawbacks of soft lithography while utilizing polymer substrate in microfluidic structures.

Two-dimensional hydrodynamic focusing can be a feasible and attractive solution for creating flow channels for microflow cytometers, if a balance can be achieved between adequate particle focusing and limiting complexity/cost. The main challenge appears to be the wide sample streams that result in low detection efficiencies. It will be important to continue to develop two-dimensional flow channels for microfluidic cytometry applications.

The extensive effort devoted to three-dimensional hydrodynamic focusing.[2,25–27] for microflow cytometry has created methods for tight focusing of the sample stream along the vertical direction in the channel dimensions, but continues to suffer from problems associated with preparing complex device designs and utilizing difficult fabrication techniques.[28] Microfluidic channels in a compact and portable flow cytometer chamber require methods to achieve various channel shapes inexpensively and simply, for instance, numerous inlets and outlets, complex flow paths, multi-phase sheath flow pattern and optimized channel dimensions. High sensitivity optical detectors[29] or other non-optical detection techniques such as impedance spectroscopy and high-bandwidth radio frequency counter require focused samples and thereby may expand applications of two-dimensional focusing in a cytometer. In addition, surface chemistry needs to be considered in two-dimensional hydrodynamic focusing because the sample can directly make contact with channel walls, causing adsorption or sample deformation. Well-controlled surface chemistries along the microchannel walls may prevent this adsorption.

Considering nontraditional sheathing ideas to enable the sample stream to interact with diverse interfaces have utility in two-dimensional microfluidic channels. A press-fit microfluidic device with a hydrophilic channel wall to guide liquid flow against air-phased hydrophobic walls has a two-dimensional solid/liquid/air interface. Other example combinations are air/liquid/air.[10] as well as various liquid/liquid/liquid interfaces. Huh *et al.*[30] presented three liquid streams separated by two air flows, merging five different channels to one channel.

The critical issues that must be addressed are (1) cost-effective fabrication of microfluidic devices in a manner that makes them compatible with the detectors with which they will be integrated, and (2) characterization of surface and channel properties that promote stable sheath flow.

ACKNOWLEDGMENTS

The material in this work was supported by Agricultural Research Service of USDA (Grant # 1935-42000-049-00D). We thank Dr. Eduardo Ximenes and Dr. Young-mi Kim for helpful comments and review of this chapter.

References

[1] P. J. Crosland-Taylor, A device for counting small particles suspended in a fluid through a tube. *Nature* **171**, 37–38 (1953).

[2] S. Chung, S. J. Park, J. K. C. Kim, C. D. C. Han and J. K. Chang, Plastic microchip flow cytometer based on 2- and 3-dimensional hydrodynamic flow focusing. *Microsystem Technologies* **9**, 525–533 (2003).

[3] G.-G. Lee, C.-C. Chang, S.-B. Huang and R.-J. Yang, The hydrodynamic focusing effect inside rectangular microchannels. *J. Micromech. Microeng.* **16**, 1024–1032 (2006).

[4] C. H. Dunphy, Contribution of flow cytometric immunophenotyping to the evaluation of tissues with suspected lymphoma. *Cytometry* **42**, 296–306 (2000).

[5] A. T. Moriarty, L. Wiersema, W. Snyder, P. K. Kotylo and D. W. McCloskey, Immunophenotyping of cytologic specimens by flow cytometry. *Diagnostic Cytopathology* **9**, 252–258 (2006).

[6] J. Voldman, M. L. Gray and M. A. Schmidt, An integrated liquid mixer/valve. *Journal of Microelectromechanical Systems* **9**, 295–302 (2000).

[7] Z. Zhang, P. Zhao and G. Xiao, Focusing-enhanced mixing in microfluidic channels. *Biomicrofluidics* **2**, 014101, 1–9 (2008).

[8] D. E Hertzog, X. Michalet, M. Jager, X. Kong, J. G. Santiago, S. Weiss and O. Bakajin, Femtomole mixer for microsecond kinetic studies of protein folding. *Anal. Chem.* **76**, 7169–7178 (2004).

[9] D. Huh, W. Gu, Y. Kamotani, J. B. Grotberg, and S. Takayama, Microfluidics for flow cytometric analysis of cells and particles. *Physiol. Meas.* **26**, R73–R98 (2005).

[10] D. Huh, Y. C. Tung, H. H. Wei, J. B. Grotberg, S. J. Skerlos, K. Kurabayashi and S. Takayama, Use of air-liquid two-phase flow in hydrophobic microfluidic channels for disposable flow cytometers. *Biomedical Microdevices* **4**, 141–149 (2002).

[11] T. T. Huang, D. G. Taylor, M. Sedlak, N. S. Mosier and M. R. Ladisch, Microfiber-directed boundary flow in press-fit microdevices fabricated from self-adhesive Hydrophobic Surfaces. *Anal. Chem.* **77** 3671–3675 (2005).

[12] A. Manz, N. Graber and H. M. Widmer, Miniaturized total chemical analysis systems: A novel concept for chemical sensing. *Sensors & Actuator B* **1**, 244–248 (1990).

[13] J. Godin, C. H. Chen, S. H. Cho, W. Qiao, F. Tsai and Y. H. Lo,, Microfluidics and photonics for Bio-System-on-a-Chip: A review of advancements in technology towards a microfluidic flow cytometry chip. *J. Biophoton.* **1**, 355–376 (2008).

[14] J. McDonald and G. Whitesides, Poly(dimethylsiloxane) as a material for fabricating microfluidic devices. *Acounts of Chemical Research* **35**, 491–499 (2002).

[15] H. Makamba, J. H. Kim, K. Lim, N. Park and J. H. Hahn, Surface modification of poly(dimethylsiloxane) microchannels. *Electrophoresis* **24**, 3607–3619 (2003).

[16] A. Papra, A. Bernard, D. Juncker, N. B. Larsen, B. Michel and E. Delamarche, Microfluidic networks made of poly(dimethylsiloxane), Si and Au coated with polyethylene glycol for patterning proteins onto surfaces. *Langmuir* **17**, 4090–4095 (2001).

[17] S.-S. Lee, S.-I. Izuo and K.-I. Inatiomi, A CAD study on micro flow cytometer and its application to bacteria detection. *Sensors, Proceedings of IEEE* **1**, 308–311 (2004).

[18] M. Arnao, M. Acosta and J. DelRio, Garcia-Canovas, F., Inactivation of peroxidase by hydrogen peroxide and its protection by a reductant agent. *Biochim. Biophys. Acta* **1038**, 85–89 (1990).

[19] J. Kim, M. Chaudhury and M. Owen, Hydrophobic recovery of polydimethysiloxane elastomer exposed to partial electrical discharge. *Journal of Colloid Interface Sciences* **226**, 231–236 (2000).

[20] G. B. Lee, C. I. Hung, B. J. Ke G. R. Huang, B. H. Hwei, H. F. Lai, Hydrodynamic focusing for a micromachined flow cytometer. *Trans. ASME I* **123**, 672–679 (2001).

[21] E. Altendorf, D. Zebert, M. Holl and P. Yager, Differential blood cell counts obtained using a microchannel based flow cytometer. *Transducers* 531–534 (1997).

[22] T. T. Huang, D. G. Taylor, K.-S. Lim, M. Sedlak, R. Bashir, N. S. Mosier and M. R. Ladisch, Surface-directed boundary flow in microfluidic channels. *Langmuir* **22**(14), 6429–6437 (2006).

[23] G. M. Whitesides, The Origins and the future of microfluidics. *Nature* **442**, 368–373 (2006).

[24] L.-K. Chau, T. Osborn, C.-C. Wu and P. Yager, Microfabricated silicon flow-cell for optical monitoring of biological fluids. *Analytical Sciences* **15**, 721–724 (1999).

[25] C. H. Lin, G. B. Lee and B. H. Hwei, A novel micro flow cytometer with 3-dimensional focusing utilizing dielectrophoretic and hydrodynamic forces. *Micro Electro Mechanical Systems* **19**, 439–442 (2003).

[26] C. Yu, J. Vykoukal, D. M. Vykoukal, J. A. Schwartz, L. Shi and P. C. Gascoyne, A Three-Dimensional Dielectrophoretic Particle Focusing Channel for Microcytometry Applications. *Journal of Microelectromechanical Systems* **14**, 480–487 (2005).

[27] X. Mao, J. R. Waldeisen and T. J. Huang, Microfluidic Drifting - Implementing three-dimensional hydrodynamic focusing with a single-layer planar microfluidic device. *Lab Chip* **7**, 1260–1262 (2007).

[28] T.-D. Chung and H.-C. Kim, Recent advances in miniaturized microfluidic flow cytometry for clinical use. *Electrophoresis* **28**, 4511–4520 (2007).

[29] K. Singh, X. Su, C. Liu, C. Capjack, W. Rozmus and C. J. Backhouse, A miniaturized wide-angle 2D cytometer. *Cytometry Part A* **69A**, 307–315 (2006).

[30] D. Huh, A. H. H. TKaczyk, B. J. Chang, Y. Wei, H. H. Grotberg, J. B. Kim, C. J. K. Kurabayashi and S. Takayama, Reversible switching of high-speed air-liquid two-phase flows using electrowetting-assisted flow-pattern change. *J. Am. Chem. Soc.* **125**, 14678–14679 (2003).

Chapter Eight

Three-Dimensional Particle Focusing

Peter B. Howell

Center for Bio/Molecular Science and Engineering, Naval Research Lab, 4555 Overlook Ave. SW, Washington DC, USA.
peter.howell@nrl.navy.mil

The ability to obtain precise information from the particles traveling through a cytometer is greatly hindered if the sample stream is not adequately focused. While there have been many examples of microcytometry in the literature, only a few have succeeded in completely sheathing the stream. This typically involves the addition of two or more sheath inlets to perform the vertical focusing. Recently, grooves in the channel walls have been used to move the sheath fluid completely around the sample stream, providing a simpler approach to 3-dimensional focusing.

8.1 INTRODUCTION

While there have been many examples of microfluidic cytometers in the literature, almost all have used inherently 2-dimensional (2-D) structures to perform the sheathing. The sheathing that they perform is therefore commonly referred to as 2-D, although in practice the sample stream is only sheathed in the horizontal dimension.[1–8] This is a simplification of the true 3-D sheathing that occurs in traditional bench-scale cytometers and has significant drawbacks. However, the classic annulus is extremely difficult to manufacture using standard microfabrication technologies.

When viewed from above, the 2-D designs show the classic cytometer behavior, with the sample stream being compressed and focused into a narrow band. Some reports have shown a reproducible nominal width as narrow as 50 nm,[9] althoughthe actual lower limit is set by the diffusion of the species of interest. Unfortunately, the sample stream still spans the depth of the channel between the top and bottom surfaces, which can lead to several potential problems when deployed in real-world situations. The most obvious problem is the danger of fouling of the channel surface by the cells and particles or by solutes in the sample.

The Microflow Cytometer by J S Kim & F S Ligler

www.panstanford.com
978-981-4267-41-0

In addition to the loss of potentially valuable cells to the walls, such fouling in the vicinity of the optical interrogation region it can render the chip unusable.

There is a balance of effects that must be considered when engineering the interrogation region. This volume of space is defined by the intersection of the illumination beam and the volume from which the light collection optics can gather light. Higher numerical aperture (NA) optics are preferable, because they can both more effectively focus the illumination and gather emitted light from a wider solid angle. One effect of using higher NA optics is that the interrogation region is made smaller. This has the added benefit of reducing the background of the measurement because a smaller volume of the fluid surrounding each particle is co-interrogated. This is particularly important when performing sandwich assays, because the labeling fluorophore is often present in solution with the particles. Ideally, the interrogation volume should be situated away from the channel walls to avoid light scattered from the wall surface or the native fluorescence of the wall material. Unfortunately, none of this is possible when the sample stream is in contact with the top and bottom of the channel. In this situation, the interrogation region must necessarily encompass the entire depth of the channel or risk missing some of the particles entirely.

When the precision of the measurements is considered, the situation is even worse. The strength of the signal received from a particle depends upon its location within the interrogation volume. In order to avoid introducing significant variation in the data, the sample stream should be substantially smaller than the interrogation volume. By far the best way of doing this is to focus the stream both horizontally and vertically. Perhaps the best demonstration of this can be seen in Lin *et al.*,[10] where hydrodynamic focusing was used to focus the particles laterally, and the dielectrophoretic forces were used to focus them vertically. This allowed them to turn the vertical focusing on and off independent of the horizontal focusing. They found that the resulting signal was both more precise and more intense when the vertical focusing was turned on.

8.2 HYDRODYNAMIC FOCUSING

To date, only a fraction of the microcytometer articles have presented structures that fully sheath the sample stream. This speaks to the difficulty producing structures to do so using existing microfabrication technologies. Several additional alignment steps are often required. For operation, some of the designs require as many as 6 or 7 sheath inlets to fully ensheath the sample stream. This can place tight constraints on the quality of the pumps providing the sheath streams. Careful control of the relative flow rates of multiple sheath streams is essential. If there is any variation, the sample stream will move out of alignment with the interrogation optics. Although this requires complex support plumbing in the form of multiple, precisely matched of pumps, far more sensitive and precise measurements can me made. In many cases, the complexity can be at least partially alleviated by bifurcating structures that split a single input to the chip into multiple channels to provide the sheath streams. In those cases, the relative flow rates of the different sheath streams are set by the relative resistances of the channels providing them. However, the clogging, fouling or temperature gradients that may occur in real-world situations must be avoided; otherwise the flow ratios will shift and change the position of the sample stream.

The most straight-forward means of moving beyond 2-dimensional sheathing was put forward by Chung *et al.*[11] They used a cross intersection similar to many of the previous designs, but with the simple modification of making the two sheath channels substantially taller than the sample channel. In addition to laterally squeezing the sample stream, the two sheath streams could travel over the sample, forcing it downward. The vertical confinement did not appear to be completely successful, as a small amount of sample fluid formed a narrow band reaching to the top of the channel. Nonetheless, an improvement in the distribution of fluorescent intensity was seen when the vertical focusing was present. Unfortunately, this system still puts the sample stream in contact with the surface of the channel. Fouling is still a danger, and scatter from the surface cannot be avoided. One application where this design is well suited is in microfluidic Coulter counters because electrodes can be placed on the bottom surface where they will be in direct contact with the sample stream. Scott *et al.* presented such a design, in which the sample stream was sheathed on three sides by a low conductivity solution, creating an adjustable "virtual aperture".[12]

Two papers from the Vellekoop lab have presented an alternate design that provided similar results.[13,14] The sample fluid was introduced into the center of an 800 μm wide channel through an 80 μm via. Like the Chung design, the sample stream flowed along the center bottom of the channel. The channel was then tapered down to 50 μm and two more sheath streams were introduced at a cross to further focus the sample. The sample stream still had a tendency to be taller than wide, but at the highest flow ratios, the height could be brought down to less than 10 μm, and the width to 2.5 μm.

For optical interrogation, it is preferable to be able to sheath the sample on all sides, fully isolating it from the walls of the channel. The earliest reports of microcytometers that could achieve this was put forward by Miyake *et al.*[15,16] Their system consisted of three 100 μm thick photo-etched metal plates. Two glass plates formed the top and the bottom, allowing optical interrogation of the sample stream. They demonstrated that this kind of design has a considerable reduction in pressure drop compared to a conventional cytometer. The reduced pressure drop is essential to miniaturization, because it removes the need for large, high pressure pumps.

Sundarajan *et al.* presented a microcytometer manufactured in five layers of polydimethylsiloxane (PDMS).[17] The center of the design was a cylindrical chamber, set perpendicular to the direction of flow, that spanned all the levels. The design required careful alignment in order to function properly. The central level had the sample inlet bounded by two sheath inlets and opposing the outlet, much like the 2-D microcytometers. Two more sheath streams entered the cylindrical chamber at the top and bottom layers to provide vertical sheathing. Their initial design had the top and bottom inlets coming from one side, but it was discovered that this asymmetry translated into an asymmetry in the resulting sample stream. The asymmetry was eliminated by either adding a second, opposing inlet to the top and bottom layers (bringing the total number of sheath inlets to six), or by placing the top and bottom sheath inlets directly above and below the sample inlet so that the system was symmetrical.

There are more recent designs that have similar topologies to the Miyake design. Munsen *et al.* presented a design with an aim toward preventing adsorption of species to the channel walls during T-sensor analysis.[18] The device was made from seven layers of laser-cut Mylar bonded together with pressure-sensitive adhesive. When fully assembled, the sample stream was injected from a nozzle that was cantilevered into the middle of the flow channel. A variation of this system has been commercialized by Micronics.[19]

Yang *et al.* produced a cytometer lithographically that was a convincing attempt to reproduce the annular nozzle found in commercial cytometers, including a central nozzle and a tapered channel.[20] The open nozzle was produced in photoresist by two separate illuminations at oblique angles. A third oblique illumination was used to create the tapered channel. In all, three masks and four exposures were necessary to produce this quite complex structure, but the result was remarkably close to a miniaturized version of the nozzle found in commercial cytometers.

Simonnet and Groisman have presented two similar designs, made by soft lithography (Fig. 8.1).[21,22] Only two lithographic levels were necessary, and these could be produced in a single monolithic stack for molding of PDMS. The vertical focusing was accomplished via a high channel intersecting several much shallower channels at a series of cross intersections. The deep channel was first filled with sheath fluid, then the sample was introduced to either side from a pair of shallower channels. This put the sample stream at the bottom of the deep channel. Then a second set of shallow channels introduced another sheath stream, which went under the sample stream and lifted it to the middle of the channel. A standard cross intersection was then used to provide the horizontal focusing. While this system requires that as many as seven sheath lines be carefully controlled, the size,

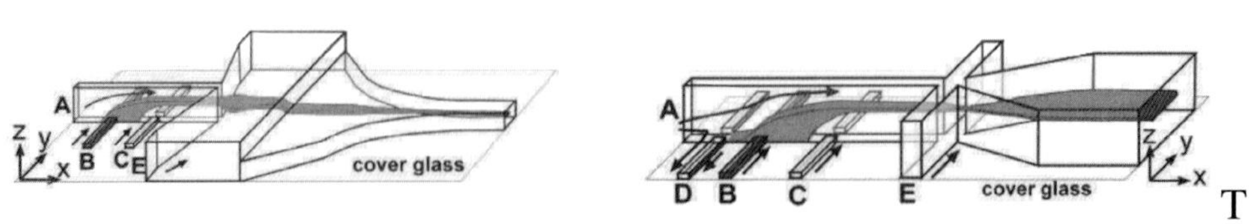

Figure 8.1. The two Simonnet and Groisman designs. The design on the left is intended to be used as a typical flow cytometer. The design on the right focuses the sample stream into a flat sheet so that it can be imaged with a microscope. Color reference – pg. 341.

aspect ratio, and position of the sample stream can be controlled. One application that they used to increase the throughput was to form the sample stream into a wide, shallow sheet for imaging with a CCD camera. Chang *et al.* presented a comprehensive numerical analysis of the flow behavior of this design.[23]

There are a few 3-D hydrodynamic focusing designs that require only a single sheath inlet. One such design was presented by Sato *et al.*[24] The sample stream was first introduced through a small via into the top of the channel in a manner similar to that seen in the Vellekoop design.[13,14] Then a series of chevron-shaped grooves brought fluid from the side of the channel to a point above the sample stream, pushing it downward. With enough chevrons the sample stream was pushed into the middle of the channel, but there was significant distortion. Using inclined, backside exposure, they were able to create a series of diagonal grooves running up the sides of the channel. These served to bring fluid up from the bottom of the channel and prevented deformation of the sample stream while it was being pushed downward.

Howell *et al.* presented two microcytometer designs that required only one or two sheath inlets to achieve full 3-D sheathing.[25] The designs were an outgrowth of the Naval Research Laboratory (NRL) Fluidic Toolbox project, in which fluid streams were controllably rearranged by a series of grooves cut in the channel surfaces.[26,27] In the simplest example, the sample and sheath streams were introduced to the channel at a T intersection, so that the two streams flowed side-by-side. A set of grooves then wrapped the sheath stream completely around the sample. As with any cytometer design, the cross-sectional size of the sample stream could be controlled by the relative ratio of the two flow rates, but unlike many of the previous designs, the change was not symmetric with respect to the channel. Instead, one side of the stream remained stationary while the other expanded and contracted. There are situations where this behavior may be undesirable. For those cases, a second design was developed in which the sample stream was initially focused horizontally at a cross intersection, much like many of the 2-D designs. Then a set of chevrons cut into the top and bottom of the channel created the vertical focusing. Because this design was vertically symmetric, it did not suffer from the distortion seen in the Sato design, and grooves were not needed on the side of the channel.

So far, all of the designs mentioned took advantage of the flow characteristics in the Stokes regime, where inertial effects could be assumed to be negligible. In practice, this is not always a safe assumption, but many of the designs will continue to function even at moderate Reynolds numbers.[28] There is one microcytometer design, presented by Mao *et al.* that depends on inertial effects for its operation.[29] It takes advantage of the formation of a microfluidic Dean vortex to provide the vertical sheathing.[30] When a pressure driven fluid stream is forced around a sufficiently tight bend, the centripetal acceleration experienced by the faster moving fluid in the middle of the channel causes a secondary flow toward the outside of the curve. A matching flow toward the inside of the curve is experienced by the fluid at the top and bottom of the channel. They exploited this phenomenon by placing the sample stream on the inside wall of the bend, and sheath fluid on the outside. As the fluids moved around a 90° bend, the sheath fluid moved inward along the top and bottom of the channel until it reached the inner wall, thereby vertically focusing the sample stream. The channel straightened after the bend and two more sheath inlets on either side focused the sample stream horizontally to place it at the center of the channel. Unlike previous designs, the proper operation of this microcytometer required a specific flow rate for the sample and first sheath inlets so that the appropriate centripetal forces could be achieved in the bend. This complication was offset by the fact that only one level was required to manufacture the device.

8.3 DIELECTROPHORETIC FOCUSING

A somewhat less popular, but quite powerful technique for 3-dimensional hydrodynamic focusing is the use of dielectrophoresis.[31–38] Its slow adoption may in part be due to the added complexity involved in manufacturing chips containing integrated electrodes. This initial complexity is offset, however, by eliminating the need for any sheath streams.

Typically, an AC electric field is applied to the solution. This eliminates any influence of electrophoretic mobility and also decreases the risk of bubble formation on the electrodes. Unlike electrophoresis, which depends on the native charge on the surface of a particle, dielectrophoresis depends on its polarizability. The particle does not have to be charged. Instead the factors affecting the dielectrophoretic force are complex, depending on the size, shape and material of the particle.[39] A full quantitative description is beyond the scope of this chapter, but a brief introduction to some of the proportionalities would be instructive.

Firstly, the force experienced is a monotonic function of the difference in polarizability between the particle and the surrounding solution. Particles which are more polarizable than the medium will tend to be pulled into regions with high field gradients. This situation is referred to as positive dielectrophoresis. Particles with polarizabilities below that of the medium will be repelled from high field gradients in a situation known as negative dielectrophoresis. The polarizability of cells can be strongly dependent on the frequency of the AC field, which raises the possibility of doing separation or sorting.

Secondly, the force is also proportional to the first derivative of the square of the electric field, which means that it is particularly strong near the edges of the electrodes and drops off quickly as one moves away. The forces typically become insignificant more that 300 μm away from the electrode surface.[39] The electrodes must be integrated into the microfluidic chip in order for the field gradient to have a significant effect. The most common configuration is to place lithographically patterned electrodes on the substrate that forms the top and bottom of the channel. To produce focusing, conditions are always configured to produce negative dielectrophoresis so that particles are pushed into the middle of the channel.

This also prevents cells from entering the high-field regions near the surface of the electrode, where they may be damaged or destroyed.[40] Generating negative dielectrophoresis is trivial when dealing with polystyrene beads, but can require adjusting the frequency of the field or the conductivity of the solution when dealing with biological cells.

The most comman production technique is to pattern the two glass slides, which are then brought together with an intermediate polymer layer to define the fluid channels. Electrodes are typically arranged in top/bottom pairs (Fig. 8.2). The opposing electrodes of a pair are typically driven 180° out of phase. For focusing, the most common configuration is to run an electrode pair diagonally into the channel from either side to create a funnel. Particles passing down the channel under pressure-driven flow are repelled from the electrodes and driven to the center of the channel. The funnel shape causes the lateral focusing, while the vertical focusing is created by the natural repulsion of the particles from the high-field regions near the electrodes on the top and bottom surfaces.

One feature of this technique is that particles are being focused laterally without being spread out longitudinally, effectively concentrating them. This increases the throughput, particularly when dealing with relatively dilute suspensions. There is the increased probability of multiple particles passing through the interrogation volume at the same time, but most cytometric assays can be performed at sufficiently dilute concentrations to prevent this problem.

Fiedler *et al.* were among the first to use a set of electrodes to push particles into a single stream.[32] Their chips consisted of platinum/titanium electrodes on the bottom surface and indium tin oxide on the top for visibility. In addition to focusing, they demonstrated a set of quadrapole and octapole cages for trapping particles, and a switch for sorting. The relative ease with which sorting can be integrated is a definite advantage of dielectrophoretic systems.[31–33,41]

One potential downside of the application of dielectrophoresis to real-world samples is that the sample solution is still allowed to be in contact with walls of the channel. Dissolved species will not be focused, and may foul the sides of the channel. Also, depending on the conditions, some of the particulate components may experience positive dielectrophoresis, and be pulled into the electrodes,

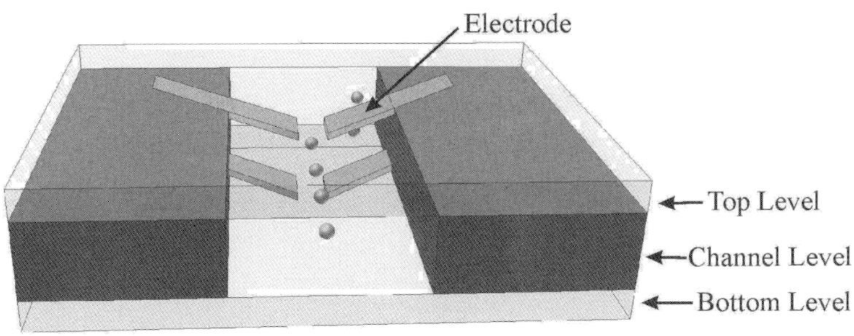

Figure 8.2. Typical configuration of electrodes for dielectrophoretic focusing.

where they would build up until a clog occurred. Of course, periodically turning off the electric field would help to remove the build-up.

8.4 HYDROPHORETIC FOCUSING

Recently, a new technique, known as hydrophoresis, has been presented as a means of focusing particles with no need for sheath inlets or electrodes.[42–44] Instead, a series of grooves or ridges move the particles to a desired location within the channel. Understanding the mechanisms behind hydrophoresis requires careful examination of the behavior of fluids flowing over a groove. Figure 8.3 is an isometric view of a length of channel with a chevron-shaped groove cut into the bottom. An example of a flow path is also plotted from a point A on the upstream end to a point B on the downstream end. Fluid streams flowing along the bottom center of the channel will become entrained in the groove and move toward the sides of the channel. In the bulk of the channel volume, there will be a net flow in the opposite direction to compensate.

Figure 8.4 shows a map of the displacement created by the groove in Fig. 8.3, as predicted by the NRL Fluidic Toolbox.[45] The figure shows the channel in cross-section. There are two symmetric, triangular regions on either side of the center point of the bottom of the channel with a characteristic height, h. The action of the chevron is to move the fluid in these regions to the similar regions near the edge of the channel.

Outside those regions, the net movement of fluid is toward the center of the bottom of the channel. To a first approximation, a particle's center follows the fluid

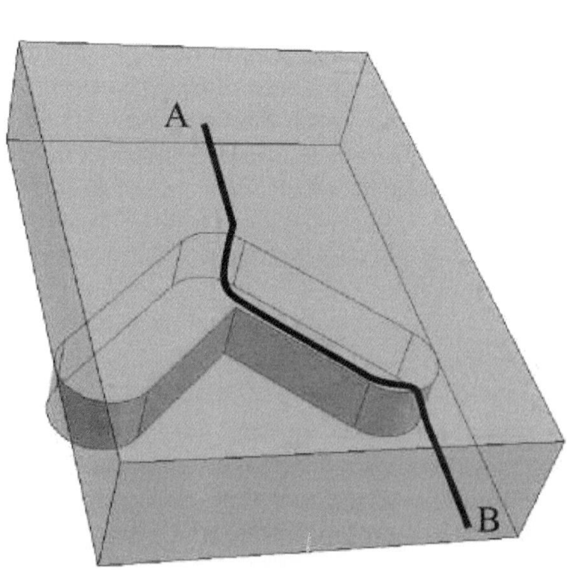

Figure 8.3. Isometric view of chevron-shaped groove in a channel.

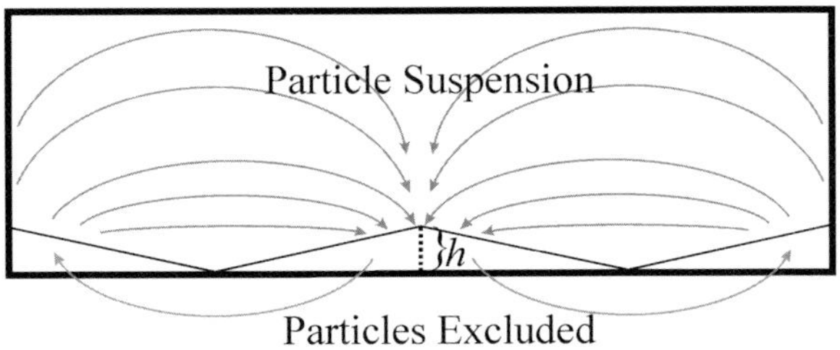

Figure 8.4. Secondary flows created by a chevron-shaped groove.

stream that it originally occupies. To use Fig. 8.3 as an example, a spherical particle with its center at A will follow the stream line until its center is at B. However, spherical particles with radius greater than h are physically excluded from having their centers fall within the triangles. They are therefore excluded from any flow paths that follow the grooves toward the outside of the channel and instead must necessarily be carried to the center. Unfortunately, the particles are brought to a location at the bottom of the channel, rather than in the middle. As already mentioned, this has implications for the sensitivity and precision of any optical measurements taken from the particles.

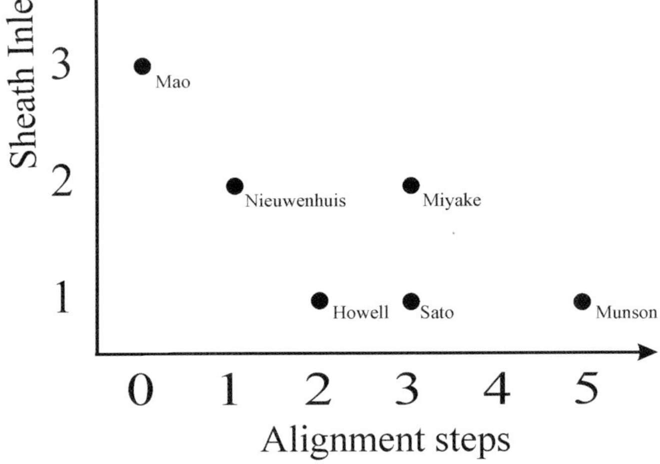

Figure 8.5. Various hydrodynamic focusing designs arranged in order of difficulty of fabrication (horizontal axis) and integration (vertical axis).

Like dielectrophoresis, hydrophoresis focuses the particles without diluting them, but it also suffers from some similar downsides. Small contaminating species will not be focused, and will still be in contact with the entire surface of the channel. Fouling of the surface is therefore a significant danger.

8.5 OTHER MEANS OF FOCUSING

Lastly, Leshansky *et al.* have demonstrated focusing of particles using a viscoelastic solution of high molecular weight polymer.[46] The shear occurring near the wall of the channel creates a net force on the particle away from the wall. The system works best when the particle diameter is a sizable fraction of the channel height or width. In a 45 μm deep channel, they were able to demonstrate effective focusing of 8 μm particles, but 5 μm particles were not fully focused.

8.6 CONCLUSIONS

As can be seen, there are a variety of ways of approaching the need to produce 3-D focusing for a microcytometer. Hydrophoresis can focus particles without the need of a sheath stream, but unfortunately, cannot focus them to the center of the channel, which is desired for optical interrogation. Dielectrophoresis requires added manufacturing capabilities, and can be dependent on the matrix conditions, but shows great promise for particle focusing, trapping and sorting.

To date, the most common technique has been hydrodynamic focusing. Within the technique, there have been many approaches to produce complete isolation of the sample stream. The difficulties in manufacturing and using the various designs vary widely. Figure 8.5 compares the current approaches, plotting each of the microcytometer designs according to an estimate of their complexity. The horizontal axis is a rudimentary measure of the difficulty of fabrication, quantified as the number of alignment steps required to produce the given design. The vertical axis is a measure of the complexity required to integrate such a design into a microcytometer, as indicated by the number of sheath inlets required for operation. As already mentioned, it is possible to have one pump feed multiple sheath inlets, but the value provides a basis for comparison. The obvious ideal would be a microcytometer located as close to the origin as possible. In terms of what is available, one may wish to sacrifice simplicity on one axis in favor of the other depending on his or her application, production capabilities and packaging

ACKNOWLEDGEMENTS

The work presented here was performed under NIH grant UO1 AI075489 and ONR/NRL 6.2 work unit 6336. The views presented here are those of the authors and do not represent the opinion or policy of the US Navy or Department of Defense.

References

[1] G. Blankenstein and U. D. Larsen, Modular concept of a laboratory on a chip for chemical and biochemical analysis. *Biosensors & Bioelectronics* **13**, (3–4), 427–438 (1998).

[2] D. P. Schrum, C. T. Culbertson, S. C. Jacobson and J. M. Ramsey, Microchip flow cytometry using electrokinetic focusing. *Analytical Chemistry* **71**(19), 4173–4177 (1999).

[3] J. Kruger, K. Singh, A. O'Neill, C. Jackson, A. Morrison and P. O'Brien, Development of a microfluidic device for fluorescence activated cell sorting. *Journal of Micromechanics and Microengineering* **12**(4), 486–494 (2002).

[4] N. Pamme, R. Koyama and A. Manz, Counting and sizing of particles and particle agglomerates in a microfluidic device using laser light scattering: Application to a particle-enhanced immunoassay. *Lab Chip*, **3**(3), 187–192 (2003).

[5] L. M. Fu, R. J. Yang, C. H. Lin, Y. J. Pan and G. B. Lee, Electrokinetically driven micro flow cytometers with integrated fiber optics for on-line cell/particle detection. *Analytica Chimica Acta* **507**(1) 163–169 (2004).

[6] B. H. Kunst, A. Schots and A. Visser, Design of a confocal microfluidic particle sorter using fluorescent photon burst detection. *Review of Scientific Instruments* **75**(9), 2892–2898 (2004).

[7] A. J. de Mello and J. B. Edel, Hydrodynamic focusing in microstructures: Improved detection efficiencies in subfemtoliter probe volumes. *Journal of Applied Physics* **101**(8) (2007).

[8] D. A. Ateya, J. S. Erickson, P. B. Howell, L. R. Hilliard, J. P. Golden and F. S. Ligler, The good, the bad, and the tiny: a review of microflow cytometry. *Analytical and Bioanalytical Chemistry* **391**(5), 1485–1498 (2008).

[9] J. B. Knight, A. Vishwanath, J. P. Brody and R. H. Austin, Hydrodynamic focusing on a silicon chip: Mixing nanoliters in microseconds. *Physical Review Letters* **80**(17), 3863–3866 (1998).

[10] C. H. Lin, G. B. Lee, L. M. Fu and B. H. Hwey, Vertical focusing device utilizing dielectrophoretic force and its application on microflow cytometer. *Journal of Microelectromechanical Systems* **13**(6), 923–932 (2004).

[11] S. Chung, S. J. Park, J. K. Kim, C. Chung, D. C. Han and J. K. Chang, Plastic microchip flow cytometer based on 2 and 3-dimensional hydrodynamic flow focusing. *Microsystem Technologies-Micro-and Nanosystems-Information Storage and Processing Systems* **9**(8), 525–533 (2003).

[12] R. Scott, P. Sethu and C. K. Harnett, Three-dimensional hydrodynamic focusing in a microfluidic Coulter counter. *Review of Scientific Instruments* **79**(4) (2008).

[13] J. H. Nieuwenhuis, J. Bastemeijer, P. M. Sarro and M. J. Vellekoop, Integrated flow-cells for novel adjustable sheath flows. *Lab Chip* **3**(2), 56–61 (2003).

[14] G. Hairer, G. S. Parr, P. Svasek, A. Jachimowicz and M. J. Vellekoop, Investigations of micrometer sample stream profiles in a three-dimensional hydrodynamic focusing device. *Sensors and Actuators B-Chemical* **132**(2), 518–524 (2008).

[15] R. Ohki, H. Yamazaki, I. and T. Takagi, Flow cytometric analysis by using micromachined flow chamber. *JSME International Journal Series B-Fluids and Thermal Engineering* **43**(2) 219–224 (2000).

[16] R. Miyake, H. Ohki, I. Yamazaki and T. Takagi, Investigation of sheath flow chambers for flow cytometers — (Micro machined flow chamber with low pressure loss). *JSME International Journal Series B-Fluids and Thermal Engineering* **40**(1), 106–113 (1997).

[17] N. Sundararajan, M. S. Pio, L. P. Lee and A. A. Berlin, Three-dimensional hydrodynamic focusing in polydimethylsiloxane (PDMS) microchannels. *Journal of Microelectromechanical Systems* **13**(4), 559–567 (2004).

[18] M. S. Munson, M. S. Hasenbank, E. Fu and P. Yager, Suppression of non-specific adsorption using sheath flow. *Lab on a Chip* **4**(5), 438–445 (2004).

[19] C. Lancaster, A. Kokoris, M. Nabavi, J. Clemmens, P. Maloney, J. Capadanno, J. Gerdes and C. F. Battrell, Rare cancer cell analyzer for whole blood applications: Microcytometer cell counting and sorting subcircuits. *Methods* **37**(1), 120–127 (2005).

[20] R. Yang, D. L. Feeback and W. J. Wang, Microfabrication and test of a three-dimensional polymer hydro-focusing unit for flow cytometry applications. *Sensors and Actuators a-Physical* **118**(2), 259–267 (2005).

[21] C. Simonnet and A. Groisman, High-throughput and high-resolution flow cytometry in molded microfluidic devices. *Analytical Chemistry* **78**(16), 5653–5663 (2006).

[22] C. Simonnet and A. Groisman, Two-dimensional hydrodynamic focusing in a simple microfluidic device. *Applied Physics Letters* **87**(11), 114104 (2005).

[23] C. C. Chang, Z. X. Huang and R. J. Yang, Three-dimensional hydrodynamic focusing in two-layer polydimethylsiloxane (PDMS) microchannels. *Journal of Micromechanics and Microengineering* **17**(8), 1479–1486 (2007).

[24] H. Sato, Y. Sasamoto, D. Yagyu, T. Sekiguchi and S. Shoji, 3D sheath flow using hydrodynamic position control of the sample flow. *Journal of Micromechanics and Microengineering* **17**(11), 2211–2216, (2007).

[25] P. B. Howell, J. P. Golden, L. R. Hilliard, J. S. Erickson, D. R. Mott and F. S. Ligler, Two simple and rugged designs for creating microfluidic sheath flow. *Lab Chip*, **8**, 1097–1103 (2008).

[26] D. R. Mott, P. B. Howell, J. P. Golden, C. R. Kaplan, F. S. Ligler and E. S. Oran, Toolbox for the design of optimized microfluidic components. *Lab Chip* **6**(4), 540–549 (2006).

[27] D. R. Mott, K. Obonschain, P. B. Howell and E. S. Oran, In *The Numerical Toolbox: An Approach for Modeling and Optimizing Microfluidic Components*, Mechanical Research Communications, Reno, Nevada, 8/1, Reno, Nevada, (2008).

[28] P. B. Howell, D. R. Mott, F. S. Ligler, J. P. Golden, C. R. Kaplan and E. S. Oran, A combinatorial approach to microfluidic mixing. *Journal of Micromechanics and Microengineering* **18**(11), 115019 (2008).

[29] X. L. Mao, J. R. Waldeisen and T. J. Huang, "Microfluidic drifting" — Implementing three-dimensional hydrodynamic focusing with a single-layer planar microfluidic device. *Lab Chip* **7**(10), 1260–1262 (2007).

[30] P. B. Howell, D. R. Mott, J. P. Golden and F. S. Ligler, Design and evaluation of a Dean vortex-based micromixer. *Lab Chip* **4**(6), 663–669 (2004).

[31] D. Holmes, M. E. Sandison, N. G. Green and H. Morgan, On-chip high-speed sorting of micron-sized particles for high-throughput analysis. *IEE Proceedings-Nanobiotechnology*, **152**(4), 129–135 (2005).

[32] S. Fiedler, S. G. Shirley, T. Schnelle and G. Fuhr, Dielectrophoretic sorting of particles and cells in a microsystem. *Analytical Chemistry* **70**(9), 1909–1915 (1998).

[33] T. Muller, G. Gradl, S. Howitz, S. Shirley, T. Schnelle and G. Fuhr, A 3-D microelectrode system for handling and caging single cells and particles. *Biosensors & Bioelectronics* **14**(3), 247–256 (1999).

[34] D. Holmes N. G. Green, and H. Morgan, Microdevices for dielectrophoretic flow-through cell separation. *IEEE Engineering in Medicine and Biology Magazine* **22**(6), 85–90 (2003).

[35] H. Morgan, D. Holmes and N. G. Green, 3D focusing of nanoparticles in microfluidic channels. *IEE Proceedings - Nanobiotechnology* **150**(2), 76–81 (2003).

[36] S. K. Ravula, D. W. Branch, C. D. James, R. J. Townsend, M. Hill, G. Kaduchak, M. Ward and I. Brener, A microfluidic system combining acoustic and dielectrophoretic particle preconcentration and focusing. *Sensors and Actuators B-Chemical* **130**(2), 645–652 (2008).

[37] D. Holmes, H. Morgan and N. G. Green, High throughput particle analysis: Combining dielectrophoretic particle focussing with confocal optical detection. *Biosensors & Bioelectronics* **21**(8), 1621–1630 (2006).

[38] C. H. Yu, J. Vykoukal, D. M. Vykoukal, J. A. Schwartz, L. Shi and P. R. C. Gascoyne, A three-dimensional dielectrophoretic particle focusing channel for microcytometry applications. *Journal of Microelectromechanical Systems* **14**(3), 480–487 (2005).

[39] R. Pethig, Dielectrophoresis: Using inhomogeneous AC electrical fields to separate and manipulate cells. *Critical Reviews in Biotechnology* **16**(4), 331–348 (1996).

[40] A. Menachery and R. Pethig, Controlling cell destruction using dielectrophoretic forces. *IEE Proceedings-Nanobiotechnology* **152**(4), 145–149 (2005).

[41] J. Voldman, M. L. Gray, M. Toner and M. A. Schmidt, A microfabrication-based dynamic array cytometer. *Analytical Chemistry* **74**(16), 3984–3990 (2002).

[42] S. Choi and J. K. Park, Continuous hydrophoretic separation and sizing of microparticles using slanted obstacles in a microchannel. *Lab Chip* **7**(7), 890–897 (2007).

[43] S. Choi and J. K. Park, Sheathless hydrophoretic particle focusing in a microchannel with exponentially increasing obstacle arrays. *Analytical Chemistry* **80**(8), 3035–3039 (2008).

[44] S. Choi, S. Song, C. Choi and J. K. Park, Sheathless focusing of microbeads and blood cells based on hydrophoresis. *Small* **4**(5), 634–641 (2008).

[45] D. R. Mott, P. B. Howell, J. P. Golden, C. R. Kaplan, F. S. Ligler and E. S. Oran, In *A Lagrangian Advection Routine Applied to Microfluidic Component Design*, 44th AIAA Aerospace Sciences Meeting and Exhibit, Reno, Nevada, 9-12 January 2006, 2006 American Institute of Aeronautics and Astronautics: Reno, Nevada, p 10 (2006).

[46] A. M. Leshansky, A. Bransky, N. Korin and U. Dinnar, Tunable nonlinear viscoelastic "focusing" in a microfluidic device. *Physical Review Letters* **98**(23) (2007).

Chapter Nine

Fluidic Control: Pumps and Valves

Siyang Zheng,[1] Kashan Shaikh[2,*] and Jun Xie[2,†]

[1]*Department of Bioengineering, Pennsylvania State University 224 Hallowell Building, University Park, PA 16802, USA. E-mail:siyang@psu.edu*

[2]*General Electric Global Research Center, One Research Circle, Niskayuna, NY 12309, USA*

[*]*shaikh@ge.com*

[†]*E-mail:xiej@ge.com*

In this chapter, current state of the art micropump and microvalve technologies are surveyed. Several micropumps and microvalves microflow cytometers are highlighted. The importance and challenge of fluidic system miniaturization for microflow cytometry is also discussed.

9.1 INTRODUCTION: THE IMPORTANCE OF FLOW CONTROL IN FLOW CYTOMETRY

Precise control of fluid flow is of utmost importance for microflow cytometers.[1] Before particles or cells enter the interrogation region, they normally need to be in a single-file stream so that each particle/cell can be analyzed individually. The fluidic control system of a typical flow cytometer utilizes a sheath fluid surrounding the sample stream to focus the particles/cells into the single-file stream. This stream must be stable without appreciable pulsation and should be well aligned with the detection system for optimal, reproducible sensing of each particle/cell. The shape, width, and height of the detected signal pulse of a particle/cell passing through the interrogation region are closely related to its position and speed in the flow stream, so any flow perturbation will translate into variations in the detected signal. Fluidic system designers must therefore be careful to select pumps and valves that minimize unwanted fluidic noise introduced into the system. In order to shrink the size of the fluidic system, micropumps and microvalves can be integrated onto the microflow cytometer device. Micropumps and microvalves have been extensively reviewed by other authors.[2–6] This chapter gives a brief review of

The Microflow Cytometer by J S Kim & F S Ligler

www.panstanford.com
978-981-4267-41-0

various micropumps and microvalves that are particularly amenable to microflow cytometers.

9.2 METHOD OF PUMPING

Despite continued research in micropumps over the past twenty years, most microflow cytometry studies still rely on external components for fluidic control. External pressure-driven pumping systems, including syringe pumps.[7–10], compressed gas sources,[11] and vacuum suction[12] or vacuum pumps,[13] have been popular choices. Pumping from inlets with hydrodynamic focusing normally requires at least two syringe pumps or pressure sources, while vacuum suction needs only one vacuum source at the outlet with proper designed flow resistances for various channels and limited adjustment during operation.

Micropumps can be classified based on the mechanisms they use to produce flow and pressure.[2] We will discuss two primary categories: displacement micropumps and dynamic micropumps.

9.2.1 Displacement Micropumps

Displacement pumps cause fluid to move by exerting pressure on the working fluid using one or more moving boundaries.

9.2.1.1 Periodic Displacement Micropumps

Reciprocating displacement pumps make up the majority of those used in the microscopic domain, although rotary displacement pumps are popular in the macroscopic domain. In reciprocating displacement pumps, oscillatory movements of mechanical parts are used to exert pressure forces on the working fluid to displace fluid. Diaphragm (or membrane) pumps dominate reciprocating displacement micropumps. In general, a diaphragm pump is composed of a pump chamber filled with working fluid and bounded on one side by a diaphragm, a driver for actuation, and two passive check valves (one at the inlet and one at the outlet). To pump fluids, the driver actuates the diaphragm and alternatively increases and decreases the pump chamber volume. Working fluid is drawn into the pump chamber during the suction stroke and pushed out of the pump chamber during the discharge stroke.

The pump chamber can be one chamber or multiple chambers in series or in parallel to eliminate the use of valves or to improve the performance.[14,15] Peristaltic micropumps are micropumps with multiple chambers and without valves that operate in peristalsis mode similar to those in the macroscale.[16] Silicon, glass, brass, titanium, and polymers (including silicone rubber, polyimide, polyxylylene, and polycarbonate) have been reported as the diaphragm material.

Several actuation mechanisms have been explored, including piezoelectric (in both lateral-strain[16–18] and axial-strain[19,20] configurations), thermopneumatic,[21,22] electrostatic,[23–25] and external pneumatic.[26] Piezoelectric diaphragm pumps have piezoelectric material mounted on one side of the diaphragm. Applied electric field generates strain either in the lateral or in the axial direction of the piezoelectric material and thus deflects the diaphragm. Due to the nature of the piezoelectric materials, piezoelectric drivers typically require high actuation voltage and generate large force but limited deflection. Thermopneumatic diaphragm pumps have a secondary chamber on the other side of the diaphragm and hold the secondary working fluid. Volume expansion of the secondary fluid due to heating deflects the diaphragm and thus pumps out the primary fluid. A variation of the thermopneumatic diaphragm pump is the 'bubble' pump.[27] Instead of using a secondary fluid, the expansion of the primary fluid volume is generated by its phase change is regulated by two diffuser structures; the primary working fluid is pumped directly. Electrostatic diaphragm micropumps typically have patterned electrodes in a parallel plate fashion on top of the diaphragm. The electrostatic force between the parallel plate electrodes bows the diaphragm. Electrostatic diaphragm micropumps can provide appreciable force at moderate voltages and the force increases as the electrode on the deflected diaphragm moves closer to the electrode on the opposing surface. External pneumatic micropumps are the most popular micropumps so far for lab-on-chip applications. These pumps have a typical configuration of three pump chambers in series and operate peristaltically. The pumps can be batch fabricated efficiently with soft lithography in PDMS and reproducible fabrication of over a thousand of pumps on a single chip has been demonstrated.[28] These pumps do not have fully integrated actuators and require an external pneumatic source as well as high speed valve connections.

The differential pressure of micropumps, which is an important parameter for pump performance, is roughly proportional to the pump compression ratio (the ratio of stroke volume and dead volume) and inversely proportional to the compressibility of the working fluid.[2] Therefore, reducing dead volume and the compressibility of the working fluid can improve pumping performance. Any bubbles present in the working fluid can increase the compressibility. Miniaturization of the pump diaphragm makes the micropump more susceptible to bubbles in working fluid, which is a significant challenge for micro diaphragm pump design.

9.2.1.2 Aperiodic Displacement Micropumps

In aperiodic displacement pumps, the pressurized moving boundary does not move in a reciprocating or generally periodic fashion. Without recharging the pressure source, these pumps are suitable only to pump a finite volume of fluid. Pneumatic aperiodic displacement pumps have been implemented at the microscale.[29,30] These pumps are in general low power and robust, but require closed-loop flow control, active valves for bidirectional flow, and recharging for long-term use. Thermal aperiodic displacement pumps are based on local heating and boiling of the liquid in a closed-end microchannel.[31,32] Electrolysis pumps.[33,34] use microelectrodes to induce an electrochemical reaction of water to generate hydrogen and oxygen gas to move fluid. The required voltage and power is much smaller compared with electrostatic pumps and the pressure generated can be very large in a short time.

9.2.2 Dynamic Micropumps

Dynamic (kinetic) micropumps require continuous addition of energy (mechanical or otherwise) to the working fluid. They convert the energy either into pressure directly or first to kinetic energy then to pressure in order move liquid continuously. In general, these pumps have no mechanical moving parts and are simpler to fabricate than the displacement micropumps. Due to the fact that there is no reciprocating mechanical movement, these pumps move fluid continuously without pulsations.

9.2.2.1 Electrokinetic (EK) Micropumps

EK pumps are based on either electroosmotic flow or electrophoretic flow. Electroosmotic (EO) micropumps have been the most popular micropumps in capillary electrophoresis and its on-chip implementation.[35] EO pumps require immobiled charges on the channel wall. These charges can come from ionizable materials (e.g. glass device wall, silica particles packed-bed[36,37]) or strongly adsorbed ions from liquid. When an electrolyte is in contact with the surface, an electric double layer (EDL) is formed with counter ions migrating to the surface to provide charge neutrality. If an electric field is applied between the two ends of the channel, the counter ions in the Debye layer will experience the Coulomb force and move under the influence of the applied longitudinal electric field. The fluid inside the channel moves because of the drag of moving counter ions. The electroosmotic flow has a characteristic 'plug' flow profile, as opposed to the parabolic profile of pressure-driven flow. EO micropumps normally require high electric field in the range of 50-5,000 V/cm across the channel length. The pumping performance is sensitive to the surface properties and charge. Large bubbles inside the channel can block the current and reduce the flow rate significantly.

Electrophoretic (EP) micropumps are based on the movement of charged particles in the fluid when an electric field is applied to both ends of the channel. The overall flow is a combination of electroosmotic flow and electrophoretic flow, and normally electroosmotic flow dominates. So to achieve only electrophoretic flow, the electroosmotic flow needs to be inhibited.

9.2.2.2 Electrohydrodynamic (EHD) Micropumps

Electrohydrodynamic micropumps use the electrostatic forces on ions in dielectric fluids to move the fluids directly. The forces can be Coulomb force, polarization force,[38] and force due to the inhomogeneity of the fluid permittivity. Operation of the pumps using Coulomb force requires space charges inside the dielectric fluid. Space charges can be generated by a direct charge-injection electrochemical reaction on the electrode surface (injection EHD pumps), disassociation of ionizable groups at the electrode/fluid interface (conduction EHD pumps), or induction of charges in an inhomogeneous working fluid (induction EHD pumps). Injection EHD pumps typically need metal electrodes with sharp features and application of very high electric field over 100 kV/cm.[39–41] Conduction EHD pumps require applying a much weaker electric field to the electrodes inside a dielectric fluid. The Coulomb forces exerted on the ions caused by the dissociation of ionizable groups at the electrode/fluid interface drag the fluid to induce flow. A common implementation for induction EHD pumps is to induce charges by applying traveling waves of electric potential across the fluid.[42,43] The major limitation for EHD micropumps to be used in microflow cytometry is that the working fluid has to be dielectric fluid with low electric conductivity (10^{-12}–10^{-6} S/cm), at least 3 orders of magnitude lower than the conductivity of blood.

9.2.2.3 Gravity

Gravity-driven flow is a method that has been widely used in conventional analysis. Water-based columns connected to the inlets and outlets are the common choice. The operation and construction of these systems are simple. The fluid inside can be electrically conductive or dielectric, aqueous or organic. On the other hand, gravity-driven flow systems are sensitive to the viscosity of the fluid and the change of surface effects, such as air bubbles. They are also bulky compared with fully integrated microfluidic systems.

9.2.2.4 Centrifugal

Microfluidic devices can be fabricated on a CD-shaped substrate. These lab-on-a-CD devices use centrifugal force to drive fluid flow by simply rotating the devices on a platform and exploiting capillary force to control flow inside the microchannels. The devices can be fabricated economically by hot embossing of plastic materials. Several companies and institutes, including Gyros,[44] SpinX, and HSG-IMIT/IMTEK,[45] have developed lab-on-a-disc systems mainly for *in vitro* biological assays. Multiple tests can be performed on a single CD simultaneously and this batch process increases system throughput significantly.

Disk pump is another type of pump based on centrifugal force and fluidic drag. In a disk micropump implementation, the disk is formed by a ferromagnetic bar.[46] Due to the adhesion between the liquid and the rotating disc surface, the liquid enters the machine through the axis and is pumped out due to centrifugal force.

9.2.2.5 Other Pumping Mechanisms

An on-chip syringe pump based on an electrostatically controlled linear inchworm actuator has been integrated for precise flow rate of 19-27 pL/s.[47]

Evaporation has been studied as an effective pumping mechanism. In one evaporation-based pump, liquid propulsion is achieved by controlled evaporation of a liquid through a membrane into a gas space containing a sorption agent. Under the condition that the vapor partial pressure inside the sorption chamber is below saturation, fluid will continuously evaporate from the membrane and be replaced by capillary force. A constant flow rate of 35 nL/s over six days of continuous pumping has been reported. Another evaporation pump has been made by fixing a filter paper plug with a vent tube at the outlet port.[48] It requires no peripheral equipment and provides steady flow in the μL/min range.

9.3 MICROVALVES

According to Oh and Ahn, microvalves can be categorized into two large groups: *active valves* and *passive valves*.[5] Active microvalves use mechanical, non-mechanical, or external mechanisms to actuate moving parts, while passive microvalves utilize flow conditions or fluid properties to control flow. Valving time, burst pressure, leak rate, flow regulation, and power are the important performance characteristics of microvalves. Normally open (NO) and normally closed (NC) are the two common microvalve modes of operation.

Active valves employ active actuation to improve the valve performance. The fabrication and operation of these active valves are generally more complicated than passive valves. A variety of mechanical actuation mechanisms including magnetic,[49] electrictrostatic,[50,51] piezoelectric,[19,52] thermopneumatic,[53–56] and bimetallic.[57] have been reported. These operating mechanisms are similar to those of the periodic displacement micropumps described previously. Non-mechanical actuation mechanisms using electrochemical, phase change, and rheological methods have been attempted. Electrochemical valves normally use oxygen and/or hydrogen gases generated by electrolysis to generate pressure exerted on a deflectable membrane.[58,59] Phase-change actuation involves change in structure or shape of a material as a result of a change in phase of that material due to applied stimuli (e.g., temperature, pH). This change in structure or shape allows the material to either block fluids or allow them to pass. One example is the stimulus-responsive hydrogel or sol-gel materials, which have the ability to undergo abrupt volume changes or liquid-solid transition in response to the surrounding environment without the requirement of an external power source. The stimuli can be physical (e.g. temperature, electrical field, light) or chemical (e.g. pH, glucose, carbohydrate, and antigen). D. J. Beebe *et al.* proposed a hydrogel-based microvalve regulated by pH,[60] and later a drug delivery application was demonstrated. Other examples of phase-change valves include the thermopnuematic manipulation of wax.[61,62] and low melting temperature metal alloys.[63] External actuation can be either modular or pneumatic. The external actuation is attractive mainly due to low/no leak flow under high input pressure, but suffers from difficulty in miniaturization. A modular rotary micro-switching valve, which can dispense liquid sample from one inlet to ten individual outlets, used built-in micro-solenoid for actuation and a steel ball for automatically positioning the outlets.[64]

External pneumatic microvalves have by far been the most popular type. A classic example is the PDMS valve first pioneered by Stephen Quake's lab at Caltech,[65] which uses gas pressure in a control channel to open/close the microfluidic channel underneath that is separated by a thin PDMS membrane. Prefabricated screw, pneumatic, and solenoid valves have been embedded in the bulk device structure recently to achieve more reliable and reproducible performance.[66] A three-way pneumatic valve was made for μTAS applications,[67] with disposable Si channel part, and silicone rubber on Si actuator part. A low-melting-temperature metal alloy has also been used as a working fluid for external pneumatic actuation and latching capabilities (thermopneumatic phase change operation).[63]

Check valves have diode-like behavior, allowing flow in only one direction. Most check valves are incorporated at the inlet and outlet of micropumps. The backpressure and pumping rate are negatively affected by the valve leakage. Most check valves are normally closed, passive flap valves. The flap structure can be circular diaphragm,[17,21,68] cantilever,[23,25,69] or tethered-plate,[14,19,70] although non-flap structures like ball-type check valves.[71] have been reported. Recently, floating-disk check valves for self-pressure-regulating flow controls were demonstrated with a biocompatible polymer parylene and used in a retinal prosthesis application.[72] One design achieved near ideal diode operation with zero forward-cracking pressure and zero reverse leakage. The other design demonstrated a pressure-bandpass check valve with 0-100 mmHg pressure regulation range.

The flow rectification can also be achieved by using fixed geometry channel structures with direction-dependant flow resistance. These passive diffuser-type check valves are also called 'valveless' pumps. They do not have moving parts, instead they adopt structures like nozzle-diffuser[18,20,22,73–75] or Tesla structures.[76] The absence of moving parts makes them advantageous for biological samples since the cells and other particles in these samples are less likely to become clogged in the valve or be damaged.

Passive capillary valves utilize the dependence of capillary force on geometry[77] or surface properties[78] to achieve liquid flow regulation. Various geometries have been designed for passive capillary stop valves and trigger valves on hydrophilic polymer coated Au/Ti Si substrates with dry-etched channels for flow control.[77] Passive capillary stop valves use a geometry change in a microfluidic channel to regulate flow.[79] The liquid from one stop valve will wait for the liquid from the other stop valve at a Y-junction to avoid trapping of air bubbles when the two liquids meet.[80]

9.4 MICROPUMPS AND MICROVALVES IN MICROFLOW CYTOMETRY

Honeywell Corporation developed a pneumatic aperiodic displacement pump as a low-power and low-cost flow controller for a portable flow cytometer using hydrodynamic focusing in the DARPA BioFlip program.[11] A manually pressurized

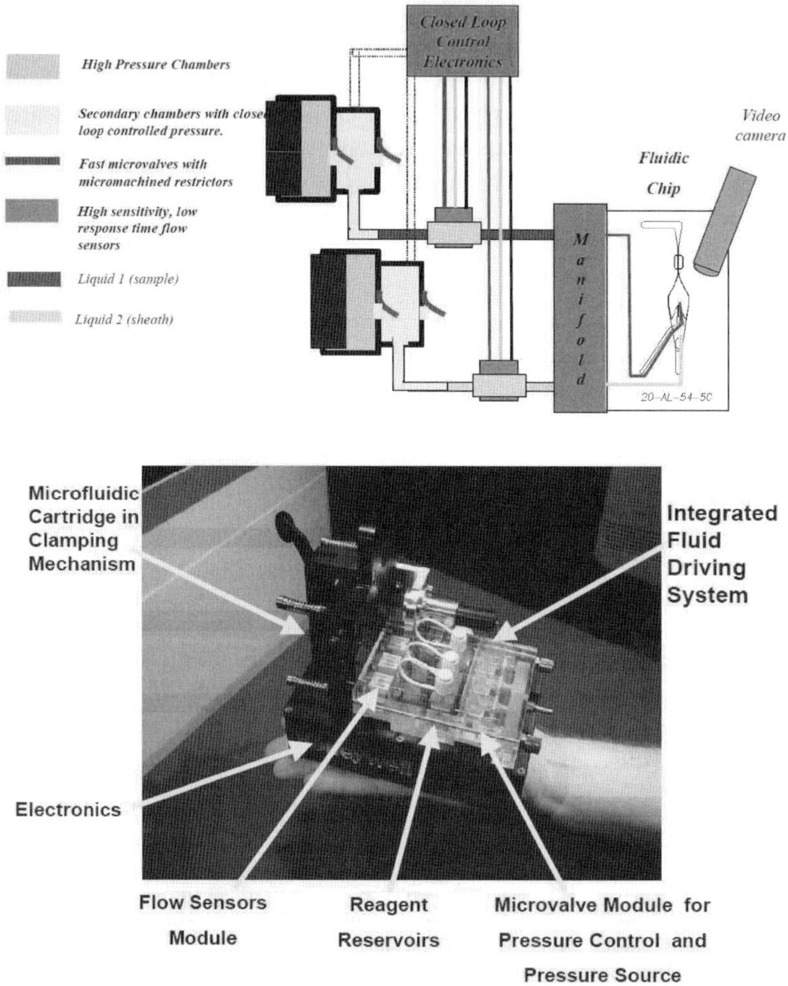

Figure 9.1. Honeywell's pneumatic aperiodic displacement pump and its use in microflow cytometer. Top: Schematic diagram of a 2-channel micro flow controller based on MEMS-based microvalves and flow sensors. Bottom: Integrated version of the 3-channel micro flow controller showing its various components. [Reprinted with permission from Cabuz, E., Schwichtenberg, J., DeMers, B., Satren, E., Padmanabhan, A. and Cabuz, C., in *Solid- State Sensors, Actuators and Microsystems Workshop* 2002, Hilton Head, SC, 2002. Images courtesy of Honeywell International Inc.] Color reference – pg. 341.

chamber was used as the pressure source. It produced pressure-driven, pulse-free liquid flows in the nL/sec to mL/sec range. Closed loop control, two high speed active microvalves, and a sensitive liquid flow sensor were also integrated.

EK pumps have been used for flow focusing and sample switching in microflow cytometry.[81] The EK pumps can be easily fabricated and the particle movement inside the channel can be controlled instantly and finely adjusted.

J. M. Ramsey's group from Oak Ridge National Lab used electrophoretic flow without electroosmotic flow to focus bacterium *Escherichia coli* and subsequently detect light scattering and fluorescence.[82] The device surface was covalently coated with Poly(dimethylacrylamide) (PDMA) to reduce cell adsorption and inhibit electroosmotic flow. The overall negative charges of the bacteria cells allow electrophoretic transport. Optimal focusing was achieved by applying 150 and 2000V to the sample and waste reservoirs respectively. The high voltage was believed to induce electroporation and cell death for majority of the *Escherichia coli* cells.

S. R. Quake's group demonstrated a disposable micro fluorescence-activated cell sorter (μFACS) chip fabricated with PDMS by multilayer soft lithography.[83] EO pumps were implemented with platinum electrodes inserted at three ports of the T-shape channel. Both forward and reverse sorting algorithms were implemented with the latter allowing rare-event capture at a rate independent of the switching speed of the device. Throughput of the device was reported to be up to 20 cells/s. Sorting living *Escherichia coli* cells achieved recovery of 20% viable cells in electric field up to ~100 V/cm.

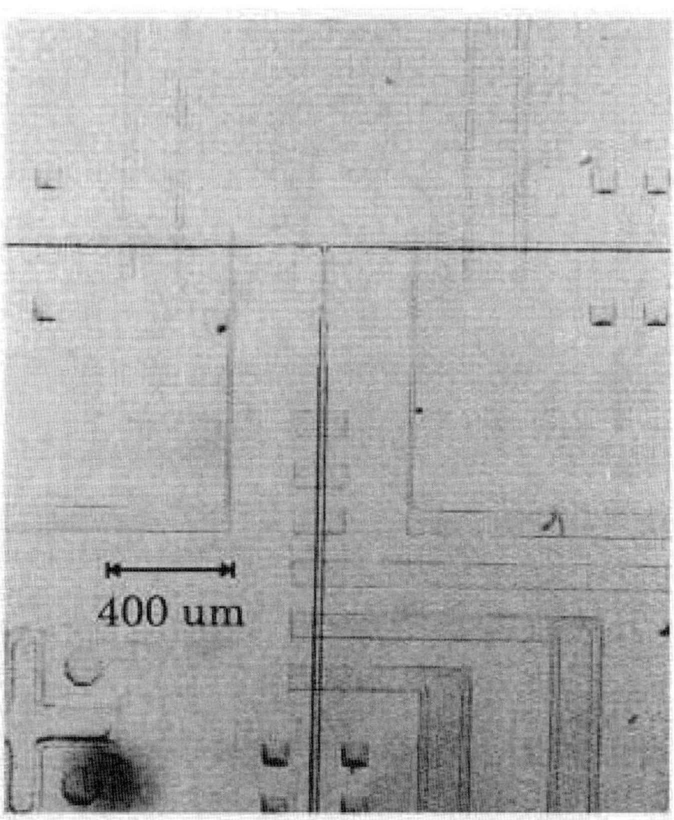

Figure 9.2. A disposable micro fluorescence activated cell sorter from S. R. Quakes group. (Reprinted with permission from Fu, A. Y., Chou, H.-P., Spence, C., Arnold, F. H., Quake, S. R., *Anal. Chem.* **74**, 2451-2457, 2002.)

One significant problem for EK micropumps in microflow cytometer systems is that the particles are exposed to high electric field. In most of the cases, mammalian cells will be electroporated irreversibly and lose viability. One proposed solution is electroosmotically induced hydraulic pumping. In one device, electric potential is applied to one inlet and one outlet of a T-shape channel. Viscous polymer coating in the outlet inhibits electroosmotic flow. Thus, pressure driven flow is generated in the other outlet with no potential applied.[84] In another device, an electrode pair separated by 50 μm in the middle of a straight microfluidic channel is used to generate electroosmotic flow. The fluid in the remainder of the channel moves with field-free pressure driven flow.[85] Another interesting implementation is field-free electroosmotic micropumps.[86] The surfaces of the top two arms of the Y-shape channel are positively and negatively charged by multiple layers of polyelectrolytes. During operation, and applied electric field between the ports of the two upper arms induces electroosmotic flow. The fluid flow in the lower vertical channel is field free and the direction of the flow is determined by the coated charges and the polarity of the applied field. A flow rate of 262.4 nL/min under 1.0 kV/cm electric field and a throughput of 120 particles/min have been demonstrated.

Using laser induced heating of a thermoreversible gelation polymer as an active valve for cell sorting was reported by T. Funatsu's group.[87] The response time of the sol-gel transformation of the block copolymer was 3 ms and flow switching of 120 ms was achieved. The flow was driven by a syringe pump. Sorting of microbeads and fluorescently labeled *Escherichia coli* cells was demonstrated.

G.-B. Lee's group presented a complete microflow cytometry system using microfluidic devices.[88] The micropumps were a modified form of Quake's PDMS pump. Instead of using three separate electromagnetic valves for pneumatic control of the single valve, a serpentine-shaped control gas channel is employed on a straight liquid channel. The time-phased deflection of successive membranes along the liquid channel caused by the traverse of compressed air generates peristaltic liquid flow. The sorting was accomplished using three PDMS valves at the three outlets.

Tung *et al.*[89] showed a novel peristaltic pumping mechanism using external braille-display actuators that is similar to soft lithography pumps and valves. To alleviate pulsatile flow inherent in the pin actuator-based peristaltic pumping, diffusers were integrated in the channels. A funnel was designed for efficient loading of samples containing small number of cells and was also positioned on the chip to prevent physical damage to the samples by the squeezing action of Braille pins during actuation. Gravity driven flow has been used with DEP sorting for microflow cytometry systems.[90] Groisman *et al.* demonstrated using gravity-driven flow for 3D flow focusing and achieved a high throughput (17,000 particles/s) microflow cytometer, which is equivalent to the speed of conventional flow cytometers.[91] The pressures were generated hydrostatically by using long vertical rails with precise rulers and sliding stages. The liquid height difference between the two individual syringes could be controlled with a precision of 0.1 mm (or 1 Pa).

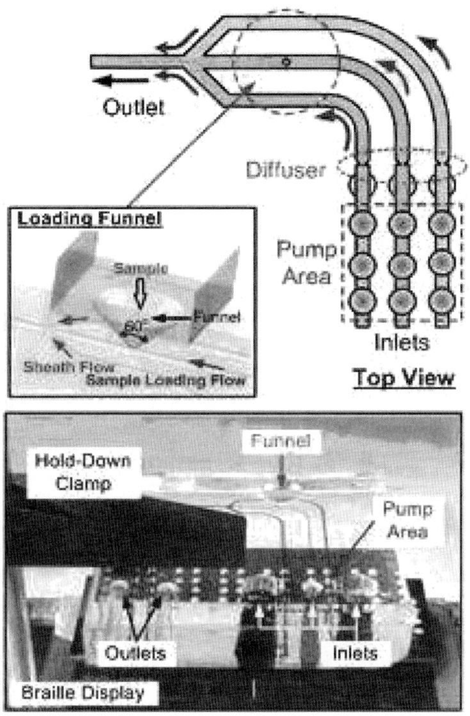

Figure 9.3. A micro pumping system in a micro flow cytometer is based Braille pin actuator to achieve peristaltic pumping for hydrodynamic focusing.[89] [Y.-C. Tung, Y.-S. Torisawa, N. Futai and S. Takayama, *Lab Chip* **7**, 1497 (2007). Reproduced by permission of The Royal Society of Chemistry.] Color reference – pg. 342.

Y. -M. Wang *et al.* presented a microflow cytometry system using gravity force for pumping and a combination of gravity force and electrophoretic force for cell sorting.[92] During operation, the microchip has to be in an upright position. Each cell falls down under gravity in the microchannels, passes through the fluorescence detector, and is then was sorted by the combined effects of gravity and electric field. Fluorescently labeled HeLa cells have been sorted with the device.

Hydrogel-based flow control systems have been demonstrated in microflow cytometry. T. Funatsu *et al.* presented a system in which flow switching was performed by the sol–gel transfer of the thermosensitive hydrogel generated by focused infrared (IR) laser irradiation inside microfluidic channel.[93,94] High-speed sorting around 10 ms (improved from previously 120 ms) without sorting errors makes the system a viable sorting solution. Although sorting was performed in simple PDMS-glass microchannels without any electrical stimulation and mechanical valve structures, a two-phase flow of aqueous solution and sol-gel needed to be maintained.

One key requirement for the majority of flow cytometers is precise hydrodynamic focusing. Many researcher have explored alternative methods for

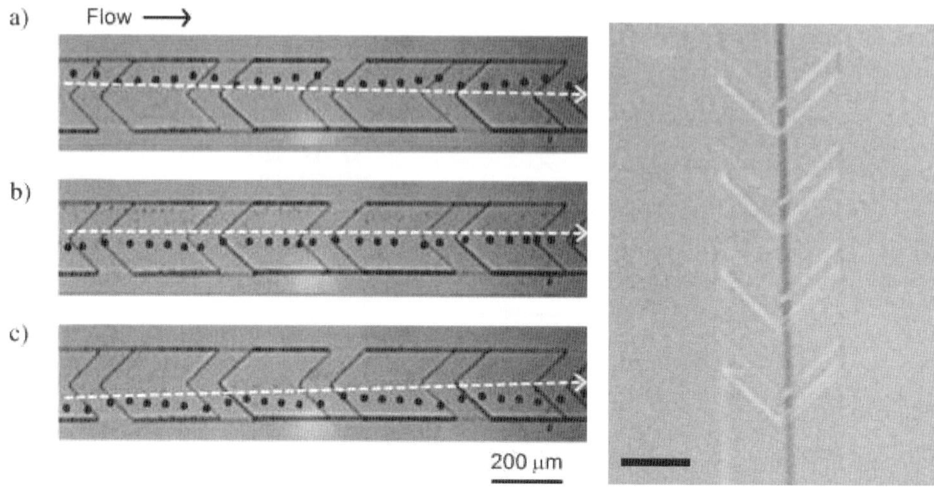

Figure 9.4. Two examples of using groove structures as alternative way to achieve hydrodynamic focusing.[95,96] [P. B. H. Jr, J. P. Golden, L. R. Hilliard, J. S. Erickson, D. R.Mott and F. S. Ligler, **Lab Chip** 8, 1097 (2008). Reproduced by permission of The Royal Society of Chemistry, S. Choi, S. Song, C. Choi and J. K. Park, Sheathless focusing of microbeads and blood cells based on hydrophoresis, *Small* 4, 634 (2008). Copyright Wiley-VCH Verlag GmbH & Co. KGaA. Reproduced with permission.]

hydrodynamic focusing so that complexity of the pumping system can be reduced. Choi *et al.*[95] used groove structures in the micro channel to achieve sheathless particle focusing based on hydrophoresis. Howell *et al.*[96] demonstrated using groove structures in the micro channel to achieve 3D hydrodynamic focusing while the sheath and sample streams merged in a typical 2D hydrodynamic focusing configuration. Holmes *et al.*[97] demonstrated using negative dielectrophoresis to ensure that all the particles pass through the detection region with a constant velocity and at a reproducible height.

9.5 CONCLUSION AND OUTLOOK

In order for the microflow cytometer to become portable or even a handheld instrument, the critical pumps and valves will first have to be miniaturized. While a large amount of promising research has demonstrated working micropumps and valves for microfluidic devices, the integration of these pumps and valves into microflow cytomers has been limited. One major reason is that it is still difficult to create pulsation-free, high flow-rate micropumps that do not adversely affect the contents of the sample (*i.e.*, cells). Non-mechanical types of pumps, such as electro kinetic pumps, are good candidates for pulsation-free pumping, but are limited by solvent properties, and may affect cells. Mechanical pumps based on pneumatic or piezoelectric actuation are widely used for other microfluidic systems due to their flow rate and pressure handling capability, but satisfactory stable pulsation-free pumping is hard to achieve. In conventional flow cytometers, various damping mechanisms have been employed to smooth the flow, but those designs are not very straightforward to scale down. Novel methods of achieving hydrodynamic focusing could lead to simplified microfluidic design and ease the requirements on pumps and valves. As the microfluidic field progresses forward, new technology in micropumps and micro valves amenable to microflow cytometery will hopefully emerge to enable it to migrate from a gold standard in the laboratory to a field-deployable tool that has many applications in life sciences, biodefense, and clinical diagnostics.

References

[1] D. A. Ateya, J. S. Erickson, P. B. Howell, L. R. Hilliard, J. P. Golden and F. S. Ligler, *Analytical and Bioanalytical Chemistry* **391**, 1485 (2008).

[2] D. J. Laser and J. G. Santiago, *Journal of Micromechanics and Microengineering* **14**, R35 (2004).

[3] N. T. Nguyen, X. Y. Huang and T. K. Chuan, *Journal of Fluids Engineering-Transactions of the ASME* **124**, 384 (2002).

[4] L. Chen, S. Lee, J. Choo and E. K. Lee, *Journal of Micromechanics and Microengineering* **18**, 2008).

[5] K. W. Oh and C. H. Ahn, *Journal of Micromechanics and Microengineering* **16**, R13 (2006).

[6] P. Woias, *Sensors and Actuators B-Chemical* **105**, 28 (2005).

[7] M. A. McClain, C. T. Culbertson, S. C. Jacobson, N. L. Allbritton, Sims, C. E. and J. M. Ramsey, *Anal. Chem.* **75**, 5646 (2003).

[8] Z. Wang, J. El-Ali, M. Engelund, T. Gotsaed, I. R. Perch-Nielsen, K. B. Mogensen, D. Snakenborg, J. P. Kutter and A. Wolff, *Lab on a Chip*, **4**, 372 (2004).

[9] C., Lancaster, M. Kokoris, M. Nabavi, J. Clemmens, P. Maloney, J. Capadanno, J. Gerdes and C. F. Battrell, *Methods* **37**, 120 (2005).

[10] R. Bernini, E. De Nuccio, F. Brescia, A. Minardo, L. Zeni, P. Sarro, R. Palumbo and M. Scarfi, *Analytical and Bioanalytical Chemistry* **386**, 1267 (2006).

[11] E. Cabuz, J. Schwichtenberg, B. DeMers, E. Satren, A. Padmanabhan and C. Cabuz, In *Solid-State Sensors, Actuators and Microsystems Workshop 2002*, Hilton Head, SC, (2002).

[12] S. Chung, S. J. Park, J. K. Kim, C. Chung, D. C. Han and J. K. Chang, *Microsystem Technologies* **9**, 525 (2003).

[13] T. Stiles, R. Fallon, T. Vestad, J. Oakey, D. W. M. Marr, J. Squier and R. Jimenez, *Microfluidics and Nanofluidics* **1**, 280 (2005).

[14] S. Shoji, S. Nakagawa and M. Esashi, *Sensors and Actuators A, Physical* **21**, 189 (1990).

[15] K.-S. Yun, I.-J. Cho, J.-U. Bu, C.-J. Kim and E. Yoon, *Journal of Microelectromechanical Systems* **11**, 454 (2002).

[16] J. Smits, *Sensors and Actuators A, Physical* **21**, 203 (1990).

[17] H. T. G. van Lintel, F. C. M. van de Pol and S. Bouwstra, *Sensors and Actuators* **15**, 153 (1988).

[18] M. Koch, N. Harris, R. Maas, A. G. R. Evans, N. M. White and A. Brunnschweiler, *Measurement Science and Technology* **8**, 49 (1997).

[19] M. Esashi, S. Shoji and A. Nakano, *Sensors and Actuators* **20**, 163 (1989).

[20] A. Olsson, P. Enoksson, G. Stemme and E. Stemme, *Journal of Microelectromechanical Systems* **6**, 161 (1997).

[21] F. C. M. van de Pol, H. T. G. van Lintel, M. Elwenspoek and J. H. J. Fluitman, *Sensors and Actuators A* **21**, 198 (1990).

[22] O. C. Jeong and S. S. Yang, *Sensors and Actuators A: Physical* **83**, 249 (2000).

[23] M. Richter, R. Linnemann and P. Woias, *Sensors and Actuators A: Physical* **68**, 480 (1998).

[24] J. Xie, J. Shih, Q. A. Lin, B. Z. Yang and Y. C. Tai, *Lab Chip*, **4**, 495 (2004).

[25] R. Zengerle, J. Ulrich, S. Kluge, M. Richter and A. Richter, *Sensors and Actuators A: Physical* **50**, 81 (1995).

[26] M. A. Unger, H.-P. Chou, T. Thorsen, A. Scherer and S.R. Quake, *Science* **288** 113, (2000).

[27] J. H. Tsai and L. Lin, *Journal of Microelectromechanical Systems* **11** (2002).

[28] T. Thorsen, S. J. Maerkl and S. R. Quake, *Science* **298**, 580 (2002).

[29] N. R. Tas, J. W. Berenschot, T. S. J. Lammerink, M. Elwenspoek and A. van den Berg, *Anal. Chem.* **74** 2224, (2002).

[30] C. P. Jen and Y. C. Lin, *Journal of Micromechanics and Microengineering* **12**, 115 (2002).

[31] K. Handique, D. T. Burke, C. H. Mastrangelo and M. A. Burns *Anal. Chem.* **73**, 1831 (2001).

[32] K. J. Thomas and C. J. K. Chang-Jin, *Journal of Applied Physics* **83**, 5658 (1998).

[33] S. Bohm, B. Timmer, W. Olthuis and P. Bergveld, *Journal of Micromechanics and Microengineering* **10**, 498 (2000).

[34] J. Xie, Y. N. Miao, J. Shih, Q. He, J. Liu, Y. C. Tai and T. D. Lee, *Analytical Chemistry* **76**, 3756 (2004).

[35] R. S. Ramsey and J. M. Ramsey, *Anal. Chem.* **69**, 1174 (1997).

[36] S. Yao, D. E. Hertzog, S. Zeng, J. C. Mikkelsen and J. G. Santiago, *Journal of Colloid and Interface Science* **268**, 143 (2003).

[37] S. Zeng, C.-H. Chen, J. C. Mikkelsen and J. G. Santiago, *Sensors and Actuators B: Chemical* **79**, 107 (2001).

[38] J. Darabi, M. M. Ohadi and D. DeVoe, *Journal of Microelectromechanical Systems* **10**, 98 (2001).

[39] A. Richter, A. Plettner, K. A. Hofmann and H. Sandmaier, *Sensors and Actuators A: Physical* **29**, 159 (1991).

[40] S.-H. Ahn and Y.-K. Kim, *Sensors and Actuators A: Physical* **70**, 1 (1998).

[41] J. Darabi, M. Rada, M. Ohadi and J. Lawler, *Journal of Microelectromechanical Systems* **11**, 684 (2002).

[42] S. F. Bart, L. S. Tavrow, M. Mehregany and J. H. Lang, *Sensors and Actuators A: Physical* **21**, 193 (1990).

[43] G. Fuhr, T. Schnelle and B. Wagner, *Journal of Micromechanics and Microengineering* **4**, (217 (1994).

[44] P. Andersson, G. Jesson, G. Kylberg, G. Ekstrand and G. Thorsen, *Analytical Chemistry* **79**, 4022 (2007).

[45] J. Ducree, S. Haeberle, S. Lutz, S. Pausch, F. V. Stetten and R. Zengerle, *Journal of Micromechanics and Microengineering* **17**, S103 (2007).

[46] J. Atencia and D. J. Beebe, *Lab Chip*, **6**, 567 (2006).

[47] R. Yokokawa, T. Saika, T. Nakayama, H. Fujita and S. Konishi, *Lab Chip*, **6**, 1062 (2006).

[48] Z.-R. Xu, C.-H. Zhong, Y.-X. Guan, X.-W. Chen, J.-H. Wang and Z.-L. Fang, *Lab Chip* **8**, 1658 (2008).

[49] A. Gaspar, M. E. Piyasena, L. Daroczi and F. A. Gomez, *Microfluidics and Nanofluidics* **4**, 525 (2008).

[50] L. Yobas, M. A. Huff, F. J. Lisy and D. M. Durand, *Journal of Microelectromechanical Systems* **10**, 187 (2001).

[51] N. Vandelli, D. Wroblewski, M. Velonis and T. Bifano, *Journal of Microelectromechanical Systems* **7**, 395 (1998).

[52] I. Chakraborty, W. C. Tang, D. P. Bame and T. K. Tang, *Sensors and Actuators A: Physical* **83**, 188 (2000).

[53] D. Baechi, R. Buser and J. Dual, *Sensors and Actuators A: Physical* **95**, 77 (2002).

[54] J. Fahrenberg, W. Bier, D. Maas, W. Menz, R. Ruprecht and W. K. Schomburg, *Journal of Micromechanics and Microengineering* **5**, 169 (1995).

[55] C. Goll, W. Bacher, B. Stgens, D. Maas, W. Menz and W. K. Schomburg, *Journal of Micromechanics and Microengineering* **6**, 77 (1996).

[56] W. K. Schomburg and C. Goll, *Sensors and Actuators A: Physical* **64**, 259, (1998).

[57] H. Jerman, *Journal of Micromechanics and Microengineering* **4**, 210 (1994).

[58] H. Suzuki and R. Yoneyama, *Sensors and Actuators B. Chemical* **96**, 38 (2003).

[59] D. E. Lee, S. Soper and W. J. Wang, Microsystem Technologies-Micro-and Nanosystems-Information Storage and Processing Systems, **14**, 1751 (2008).

[60] D. J. Beebe, J. S. Moore, J. M. Bauer, Yu, Q., Liu, R. H., Devadoss, C. and B.-H Jo, *Nature* **404**, 588 (2000).

[61] R. Pal, M. Yang, Johnson, B. N., Burke, D. T. and Burns, M. A., *Analytical Chemistry* **76**, 3740 (2004).

[62] R. H. Liu, J. Bonanno, J. Yang, R. Lenigk and P. Grodzinski, *Sensors and Actuators B: Chemical* **98**, 328 (2004).

[63] K. A. Shaikh, S. F. Li and C. Liu, *Journal of Microelectromechanical Systems* **17**, 1195 (2008).

[64] T. Hasegawa, K. Nakashima, F. Omatsu and K. Ikuta, *Sensors and Actuators A: Physical* **143**, 390 (2008).

[65] S. R. Quake and A. Scherer, *Science* **290**, 1536 (2000).

[66] S. E. Hulme, S. S. Shevkoplyas and G. M. Whitesides, *Lab Chip* **9**, 79 (2009).

[67] T. Ohori, S. Shoji, K. Miura and A. Yotsumoto, *Sensors and Actuators A: Physical* **64**, 57 (1998).

[68] S. Bohm, W. Olthuis and P. Bergveld, *Sensors and Actuators A: Physical* **77**, 223 (1999).

[69] J. Voldman, M. L. Gray and M. A. Schmidt, *J. Microelectromech. Syst.* **9**, 295 (2000).

[70] W. L. Benard, H. Kahn, A. H. Heuer and A, H. M., *J. Microelectromech. Syst.* **7**, 245 (1998).

[71] M. C. Carrozza, N. Croce, B. Magnani and P. Dario, *Journal of Micromechanics and Microengineering* **5**, 177 (1995).

[72] P. J. Chen, D. C. Rodger, M. S. Humayun and Y. C. Tai, J. Microelectromech. Syst. **17**(6), 1352 (2008).

[73] E. Stemme and G. Stemme, *Sensors and Actuators A: Physical* **39**, 159 (1993).

[74] H. Andersson, W. van der Wijngaart, P. Nilsson, P. Enoksson and G. Stemme, *Sensors and Actuators B: Chemical* **72**, 259 (2001).

[75] X. N. Jiang, Z. Y. Zhou, X. Y. Huang, Y. Li, Y. Yang and C. Y. Liu, *Sensors and Actuators A: Physical* **70**, 81 (1998).

[76] F. Forster, R. Bardell, M. Afromowitz, N. Sharma and A. Blanchard, In *ASME Int. Mechanical Engineering Congress and Exposition*, San Francisco, CA (1995).

[77] M. Zimmermann, P. Hunziker and E. Delamarche, *Microfluidics and Nanofluidics* **5**, 395 (2008).

[78] G. Londe, A. Chunder, A. Wesser, L. Zhai and H. J. Cho, *Sensors and Actuators B: Chemical* **132**, 431 (2008).

[79] C. H. Ahn, J. W. Choi, G. Beaucage, J. H. Nevin, J. B. Lee, A. Puntambekar and J. Y. Lee, *Proceedings of the IEEE* **92**, 154 (2004).

[80] J. Melin, N. Roxhed, G. Gimenez, P. Griss, W. van der Wijngaart and G. Stemme, *Sensors and Actuators B: Chemical* **100**, 463 (2004).

[81] L.-M. Fu, R.-J. Yang, C.-H. Lin, Y.-J. Pan and G.-B. Lee, *Analytica Chimica Acta*, **507**, 163 (2004).

[82] M. A. McClain, C. T. Culbertson, S. C. Jacobson and J. M. Ramsey, *Anal. Chem.* **73**, 5334 (2001).

[83] A. Y. Fu, C. Spence, A. Scherer, F. H. Arnold and S. R. Quake, *Nature Biotechnology* **17**, 1109 (1999).

[84] C. T. Culbertson, R. S. Ramsey and J. M. Ramsey, *Anal. Chem.* **72**, 2285 (2000).

[85] T. E. McKnight, C. T. Culbertson, S. C. Jacobson and J. M. Ramsey, *Anal. Chem.* **73**, 4045 (2001).

[86] S. Joo, T. D. Chung and H. C. Kim, *Sensors and Actuators B: Chemical* **123**, 1161 (2007).

[87] Y. Shirasaki, J. Tanaka, H. Makazu, K. Tashiro, S. Shoji, S. Tsukita and T. Funatsu, *Anal. Chem.* **78**, 695 (2006).

[88] S. Y. Yang, S. K. Hsiung, Y. C. Hung, C. M. Chang, T. L. Liao and G. B. Lee, *Measurement Science & Technology* **17**, 2001 (2006).

[89] Y.-C. Tung, Y.-S. Torisawa, N. Futai and S. Takayama, *Lab Chip* **7**, 1497 (2007).

[90] S. Fiedler, S. G. Shirley, T. Schnelle and G. Fuhr, *Anal. Chem.* **70**, 1909 (1998).

[91] C. Simonnet and A. Groisman, *Anal. Chem.* **78**, 5653 (2006).

[92] B. Yao, G. A. Luo, X. Feng, W. Wang, L. X. Chen and Y. M. Wang, *Lab Chip* **4** (603 (2004).

[93] T. Arakawa, Y. Shirasaki, T. Izumi, T. Aoki, H. Sugino, T. Funatsu and S. Shoji, *Measurement Science & Technology* **17**, 3141 (2006).

[94] Y. Shirasaki, J. Tanaka, H. Makazu, K. Tashiro, S. Shoji, S. Tsukita and T. Funatsu, *Analytical Chemistry* **78**, 695 (2006).

[95] S. Choi, S. Song, C. Choi and J. K. Park, *Small* **4**, 634 (2008).

[96] P. B. H. Jr, J. P. Golden, L. R. Hilliard, J. S. Erickson, D. R. Mott and F. S. Ligler, *Lab Chip* **8**, 1097 (2008).

[97] D. Holmes, J. K. She, P. L. Roach and H. Morgan, *Lab Chip* **7**, 1048 (2007).

Chapter Ten

Integrated Optics

Yahya Hosseini and Karan V. I. S. Kaler*

Department of Electrical and Computer Engineering, Schulich School of Engineering, University of Calgary 2500 University Dr., NW, Calgary, Canada
**kaler@ucalgary.ca*

10.1 INTRODUCTION

The previous chapters were devoted to and discussed various approaches in establishing sheath flow in microfluidic channels for cytometric applications. The thrust of this chapter is to examine and review current integrated optics solutions for chip-based optical detection methods and approaches employed to monitor the contents of the microchannel fluidic-chips.

This chapter first examines the key features of the conventional optical components utilized in optical detection systems and then reviews the current research direction and approaches concerning the development of miniaturized optical components (light sources, waveguides, microlenses, and optical detectors) and their integration onto or fabrication within a microfluidic chip.

10.2 CONVENTIONAL DETECTION SYSTEMS IN MICROFLOW CYTOMETERS

Fluorescence-based optical detection methods are widely used in microflow cytometers due to their superior selectivity and sensitivity compared to the conventional light scattering and absorbance detection methods. Traditionally fluorescence-based detection systems, utilizing laser-induced fluorescence (LIF) sources, are widely used as they can precisely focus a high power, monochromatic beam on samples in a micron-sized area of the microfluidic channel, resulting in a high irradiation of particles. Less expensive approaches utilize mercury and/or xenon lamps which can be mounted on an epifluorescence microscope.

Several optical lenses and filters have to be used to focus and collimate the incident light beam into the interrogation area. Upon irradiation of the samples,

The Microflow Cytometer by J S Kim & F S Ligler

www.panstanford.com
978-981-4267-41-0

the resulting fluorescence, absorbance or scattered signal is subsequently collected by focusing lenses and/or spectroscopic filters.

Photomultiplier tubes (PMTs), avalanche photodiodes (APDs), and charged-coupled devices (CCDs) are commonly used as detectors in many such cases. PMTs and APDs are very sensitive optical detectors, and are capable of counting even single photons under suitable conditions. PMTs are usually used to amplify optical signals, due to their very high internal gain (up to 160 dB), while APDs are much smaller in size compared to PMTs, but their internal gain is significantly lower–typically around 40∼60 dB. Numerous researchers have reported on the use of epifluorescence microscopes combined with PMTs and APDs to detect various types of biochemical samples.[1–4]

Unlike the single-point detectors such as PMTs, CCD cameras can also be utilized for image-based data analysis. CCD cameras are used for conserving spatial resolution between different regions of microfluidic channels. Several examples of monitoring the cell sorting process using CCD cameras have been reported.[5–7]

In one of the earliest works, Fu *et al.* microfabricated a disposable fluorescence activated cell sorting (FACS) device in polydimethylsiloxane (PDMS), using soft lithography.[1] This approach and similar advancements have enabled the replacement of the mechanically complex fluidic components of the conventional FACS instrument by a microfluidic chip. This PDMS-based microfluidic device is fabricated with three channels (see Fig. 10.1) joined at a T-junction. The cells in the channel can be manipulated by electro-osmotic flow[8] using platinum electrodes in each well.

A fluorescence-based optical detection system is placed ahead of the T-junction and along the inlet channel so that cells or particles tagged with fluorochromes

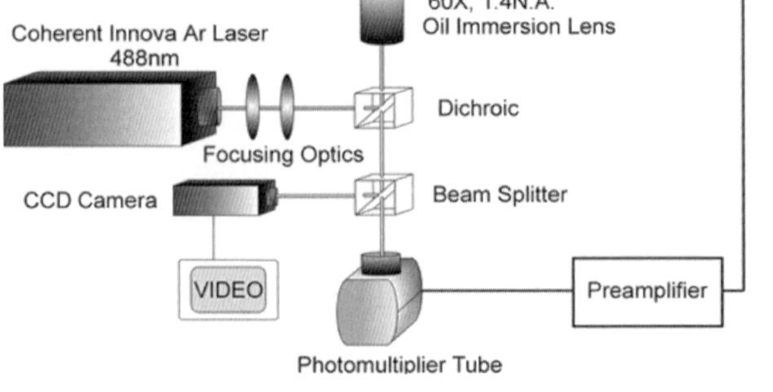

Figure 10.1. Detailed schematic diagram of the micro-FACS.

are excited with the laser beam near the T-junction. The fluorescence emission is detected by the photomultiplier tube (PMT), and digitized signal is utilized to control the electro-osmotic flow utilized to affect cell manipulation. Utilizing this device, *Escherichia coli* cells (HB101, 1 μm diameter) of labelled with fluorochromes with emission at different wavelengths were handled and sorted, at rates of up to 33,000 cells per hour.

The fluidic system in this micro-FACS device has been miniaturized utilizing microfluidics, however, many of the optical components (laser source, PMT detector, and optical filters) and electronics needed for fluorescence detection and data analysis are located off-chip, and remain very much the same as in conventional bench top cytometers.

10.3 ON-CHIP INTEGRATION OF OPTICAL COMPONENT

The 'on-chip' integration and interfacing of optical components to the microfluidic domain is key to the development of compact, stand-alone microflow cytometers. The integration of the light sources and detectors onto the microfluidic chip will greatly assist efforts in system miniaturization. Further technological improvements are anticipated by leveraging parallel fabrication and/or integration of miniaturized detectors to the microfluidic devices, facilitating the monitoring of fluidic channels at various points. Furthermore, the integration of waveguides and/or optical fibers, and microlenses to the microfluidic chip may further help to improve signal to noise ratio and the sensitivity of detection.

10.3.1 On-chip Integration of Waveguides

To circumvent the need for a number of optical filters and to reduce the complexity of the detection system, waveguides may be employed and incorporated inside the microchannels.[4,5,9–14] Waveguides are usefully employed to control the light transmission path where the illumination is confined and propagated along the waveguide structure and thus illuminate only specific locations in the microfluidic channels. These waveguides are generally used for optical coupling of an external light source (laser) with microfluidic channels carrying analyte of interest and a set of photodetectors. One method of interfacing waveguides to the microfluidic device is to insert the tip of commercially available optical fibers into pre-fabricated passively aligned elements.[4,5,9] Optical waveguides can be furthermore be fabricated and interfaced to microfluidic devices using a variety of different materials including SiO_2,[10,11] SU-8,[12,13] and PDMS.[14]

Figure 10.2 shows a schematic diagram of the microflow cytometer, developed by Fu *et al.*, and fabricated utilizing soda lime glass, incorporating optical fibers (9 μm core and 125 μm cladding) that optically interface with the microchannels.[5] One end of the optical fiber is connected to the light source and the detector, and the other end is inserted into the microchannels. The fluorescent emission was detected using an APD, and subsequently analyzed using a digital microcomputer,

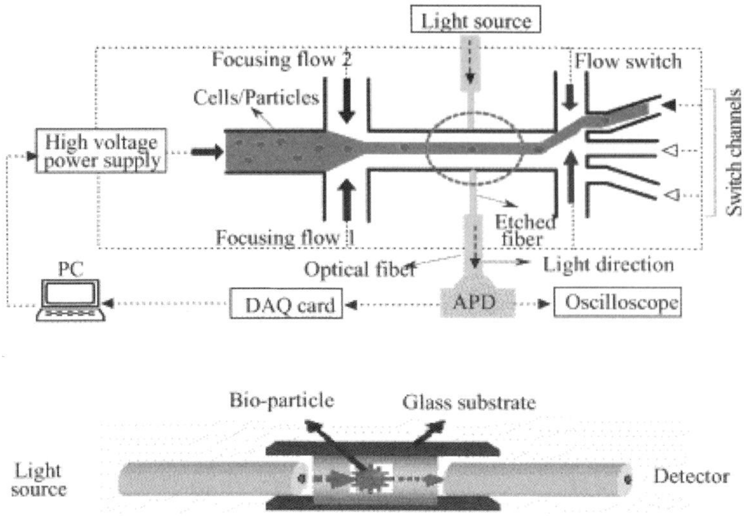

Figure 10.2. (a) Schematic representation of experimental setup for the microflow cytometer; (b) detection light propagates in the waveguide structure and its intensity changes as particles pass through the detection region. [Reprinted from *Analytica Chimica Acta* **507**, L.-M. Fu, R.-J. Yang, C.-H. Lin, Y.-J Pan and G.-B. Lee, Electrokinetically driven microflow cytometers with integrated fiber optics for on-line cell/particle detection, 163–169. Copyright (2004), with permission from Elsevier.] Color reference – pg. 342.

which eventually controls the electrokinetic flow switches of microfluidic chip upon detection of different types of cells and particles to achieve sorting. The experimental finding reported suggested that this device is capable of sorting microscopic sized particles (~10 μm in diameter) with a variation of 11.2% and red blood cell (6~8 μm) with a variation of 38.5%.

It is important to note that such integration and interfacing of optical fibers to the microchannels in the above microflow cytometer is not straightforward. The integration is particularly problematic in systems that incorporate multiple light sources and detectors sensitive to different optical wavelengths.

Ruano *et al.* developed SiO_2 waveguides, which split the incident beam into 16 beams by a series of Y-branches, to excite distinct regions of a glass microfluidic channel for parallel fluorescence and absorbance detection of biological analytes.[11] After detection, the fluorescence emission was measured using a CCD camera, and the results were subsequently analyzed by image processing software. The fluorescence measurement of 24 μm Cy5 dye solution showed a signal-to-noise (SNR) ratio of 9.14, with the limit of detection (LOD) in an individual assay (within the array) of 320×10^{-18} moles of fluorophore.

Thus, optical waveguides are an effective means of optical coupling sources and detectors to microfluidic channels provided due care is taken to minimize scattering at fluid-waveguide interface. Furthermore, waveguides, if used over extended lengths, may result in signal loss, hence the waveguide fabrication processes should be carefully implemented and controlled to minimize attenuation of the optical signal and keep it at an acceptable level.

10.3.2 On-chip Integration of Optical Detectors

The "on-chip" integration of custom CMOS sensors with the microfluidic chips or the fabrication of photodiodes inside the microfluidic device has led to miniaturization of optical detection systems.[12,15–22] Recently, a custom near-field CMOS optical sensor chip, consisting of a two-photodiode strip sensor and a 2×20 photodiode array, was housed underneath a glass microfluidic chip and used to determine both the shape and size of microscopic particles.[15] Utilizing this device, quasi-static measurements of particles (27 – 67 μm), traversing the interrogation region at the velocity of ~2 cm/sec, was obtained.

The above near-field interface of the CMOS sensor to the microfluidic device has circumvented the need for optical filters and/or lenses. The principle drawback of this system is that a vast amount of off-chip signal processing is required to analyze the output of 40 photodiodes in the array of the sensor and thus limits its applications requiring real-time on-chip detection.

A CMOS silicon detector arrays and replica-molded elastomeric microchannels were combined to enable absorption and fluorescent spectroscopy in the visible and near-UV wavelength range.[16] Prior to the fabrication of the fluidic channels on the detector arrays, Si_3Ni_4/SiO_2 filters can be deposited on the detector array. In order to achieve desired wavelength response, the filter layers may be fabricated on each detector array with an appropriate thickness. The device detection sensitivity was experimentally determined using a variety of dye molecules, including fluorescein, bromophenol blue, and orange G. The LOD of this device was not specified.

A hydrogenated amorphous silicon (a-Si:H) PIN photodiode with a hole at the center of detector, was directly fabricated on a glass substrate of microfluidic device for fluorescence detection.[17] As shown in Fig. 10.3, the photodetector was positioned in such a way that the excitation beam passed through the hole and the emitted light was collimated by microlenses on to the photodetector surface. In this device, the distance between the center of microchannel and the PIN photodiode was ~ 3.6 mm. The LOD of this device for fluorescein was 17 nM, sufficient to enable DNA fragment sizing and chiral analysis of glutamic acid.

Hartley *et al.* reported an integrated digital CMOS sensor with a glass microfluidic chip capable of detecting and counting microscopic particles, as shown in Fig. 10.4.[18] In this microfluidic chip, the embedded dielectrophoresis (DEP) microelectrodes enable sorting and separation of particles and cells according to their type and morphology using a non-uniform electric field. The CMOS sensors were directly coupled to the microfluidic channels utilizing a flip-chip bonding technique, enabling the near-field detection of microscopic particles (the distance between the photodiodes and the microchannel is ~100 μm). The CMOS optical sensor benefits from a linear array of sixteen active pixel sensors (APSs). Following

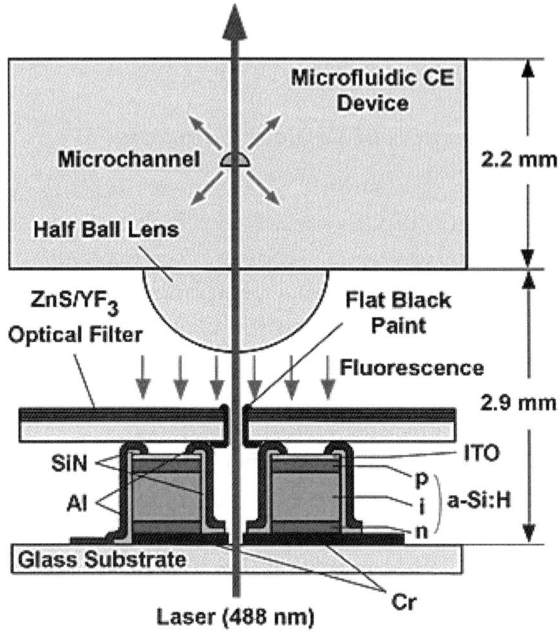

Figure 10.3. Schematic cross-sectional view of the hybrid integrated a-Si:H fluorescence detector with a microfluidic electrophoresis device. [Reprinted with permission from T. Kamei, B. M. Paegel, J. R. Scherer, A. M. Skelley, R. A Street and R. A. Mathies, Integrated hydrogenated amorphous Si photodiode detector formicrofluidic bioanalytical devices. *Anal. Chem.* **75**, 5300–5305 (2003). Copyright 2003 American Chemical Society.] Color reference – pg. 343.

the separation chamber, as different types of particles pass over the photodiode array of the CMOS sensor, the incident light intensity falling on the photodiodes is modulated. This modulation of the light intensity is manifested as a perturbation on the sensor's output voltage, and results in the detection of particles. The CMOS sensor benefits from the spatial filter topology and digital block which perform analog to digital conversion function and data serialization. Using this scheme, a vast amount of data processing has been performed on chip. The digital output of the CMOS sensor can be accessed over a simple interface by an embedded microcontroller. This device was used for real-time detection of 16 μm polystyrene microspheres.

Using the same hybrid microfluidic chip, Hosseini *et al.* have reworked the CMOS optical sensor, replacing the original APS array with a parallel arrangement two-APS-array, spaced 30 um apart.[19] This arrangement of the dual APS arrays has facilitated both velocity and size measurements in addition to the counting of different sized particles. The capabilities of this device were demonstrated by detecting and categorizing 14.6 μm, 10 μm, and 6 μm polystyrene microspheres based on their size. The device, under laminar flow conditions, was able to follow and quantify particle velocities up to be ~500 μm/sec.

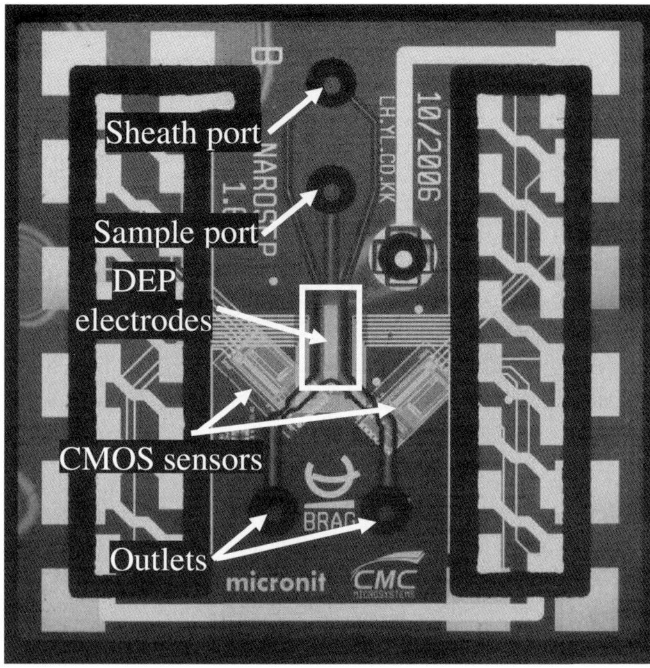

Figure 10.4. The photograph of the CMOS optical sensors integrated to the microfluidic glass chip with DEP microelectrodes. Color reference – pg. 343.

10.3.3 On-chip Integration of Light Sources

On-chip integration of light sources in microflow cytometers has been demonstrated utilizing light-emitting diodes (LEDs) and dye laser diodes housed on the microfluidic device.[12,20–23] Novak *et al.* reported on the development of a micro flow cytometer, combining an on chip LED, optical filters, dichroic mirrors, a photodiode, and an on-board amplification device.[20] The LOD of this device based on fluorescein measurement was determined to be 1.96 nM.

Balslev *et al.* reported the development of a microflow cytometer with on chip integrated dye laser, SU-8 waveguides, and photodiodes, as shown in Fig. 10.5.[12] In this device emitted light from the laser dye is coupled to microchannels through waveguides for exciting liquid samples. Thereafter, the emitted light is collected by another separate set of waveguides coupled to the photodiodes. Absorption of incoming light by the liquid sample reduces the output voltage of the photodiodes indicating the presence of liquid, which can be further used for quantitative measurements such as sample concentration. This device was used for fluorescent measurement of a xylenol orange dye without determining the LOD of the device.

Misiakos *et al.* developed a LOC device which integrates silicon avalanchediode, silicon nitride fibers, and silicon photodiode together as the detection system (see Fig. 10.6).[21] The optical chip was coupled with a PDMS microchannel for transporting liquid samples to the investigation region, formed at

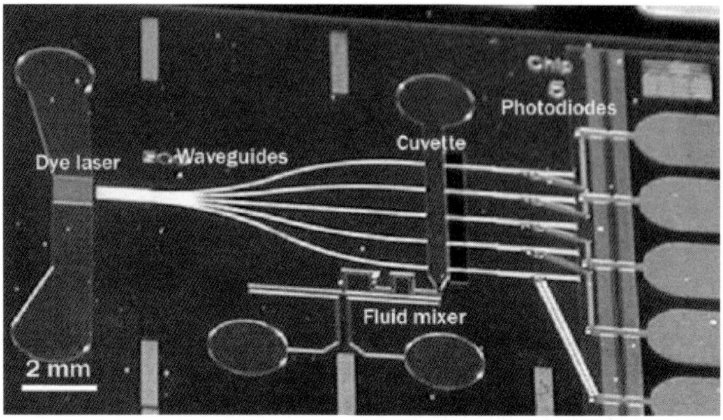

Figure 10.5. Photograph of the lab-on-chip device with integrated microfluidic dye laser, optical waveguides, microfluidic network and photodiodes. The metallic contact pads for the photodiodes are seen on the far right. [S. Balslev, A. M. Jorgensen, B. Bilenberg, K. B. Mogensen, D. Snakenborg, O. Geschke, J. P Kutter and A. Kristensen, Lab-on-a-chip with integrated optical transducers. *Lab Chip* **6**, 213 (2006). Reproduced by permission of The Royal Society of Chemistry.]

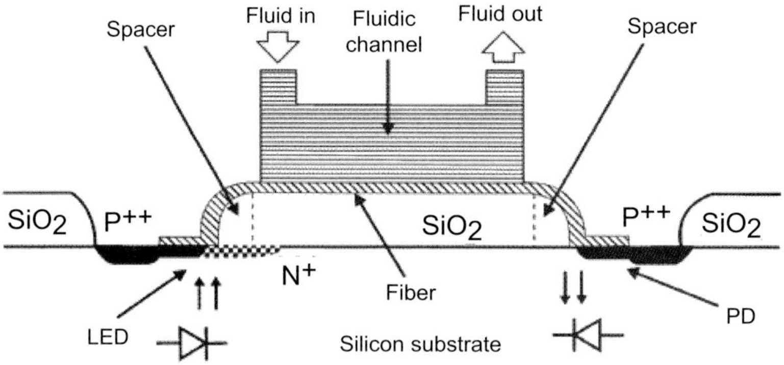

Figure 10.6. Schematic drawing of the monolithic transducer coupled to a microfluidic compartment (not in scale). The fiber bending SiO_2 spacers are emphasized. P++ are the self-aligned LED and photodetector (PD) emitter regions heavily implanted with boron while N+ is the phosphorus preimplanted LED base region. [Reprinted with permission from K. Misiakos, S. E. Kakabakos, P. S. Petrou and H. H. Ruf, Amonolithic silicon opto-electronic transducer as a real-time affinity biosensor. *Anal. Chem.*, **76**, 1366–1373 (2004). Copyright 2004 American Chemical Society.]

the waveguide-channel junction. This device was shown to be capable of fluorescent measurement of protein samples (anti-rabbit IgG solution (1 nM) and a mixture of anti-rabbit IgG (1 nM) and streptavidin (104 pM)) at fluid flow rate of ~5 µl/min, in real-time. Edel *et al.* developed a polymer LED using a simple layer-by-layer deposition procedure on a glass microfluidic device.[23] Utilizing this device, fluorescent dyes (fluorescein and 5-carboxyfluorescein) were detected after on-chip separation at concentrations as low as 1 μm with a mass detection limit of 50×10^{-15} moles.

In another study, Pais *et al.* have developed a PDMS microfluidic chip with integrated green organic LED (OLED) and an organic photodiode (OPD) detector for on-chip fluorescence analysis.[22] In this cytometer, the excitation light was isolated from the fluorescence dye emission by developing two cross-polarization planes oriented orthogonal respect to the light source and optical detector. The LOD of this device was determined to be 100 nM for Rhodamine 6G and 10 μm for fluorescein.

10.3.4 On-chip Integration of Microlenses

Many miniaturized on-chip light sources such as LEDs commonly emit low intensity and non-collimated illumination which requires focusing in order to minimize the LOD of the micro flow cytometer. In order to increase the optical efficiency and detection sensitivity, the fluorescence emission from the sample also needs to be collimated, and the excitation light suitably filtered from the fluorescence emission prior to detection.

In order to achieve this in conventional flow cytometers, several optical filters and lenses are used to adjust optical focus and light alignment. The set up of light source, optical lenses, spectroscopic filters, and detector requires a complex and precise arrangement, which eventually increases the size, complexity, and cost of the overall system. To obviate the need for the optical filters and lenses, and to further minimize the complexity of the cytometers, several research groups fabricated microlenses,[17,24] filters,[16] or cross-polarization planes[22] within the microfluidic chip.

Kuo *et al.* fabricated an array of SU-8 microlenses on a polymethyl methacrylate (PMMA) template.[24] The numerical aperture (NA) of this SU-8 microlens is as high as 0.75, hence offering a high resolution and a high signal-to-noise (S/N) ratio detection scheme. The developed micro-lens array was integrated with a microflow cytometer for counting of human lung cancer cells (15 μm in diameter) labeled with fluorescent dye. The authors noted that with coupling the array of SU-8 microlenses to the microfluidic chip, the SNR of the system can be improved by 400% compared to the system without coupling the microlenses array.

10.4 CONCLUSION AND SUMMARY

The miniaturization and integration of electro-optical components is critically important for the development and realization of a hand-held microflow cytometer. Conventional flow cytometers can take advantage of very sensitive optical detectors, enabling detection of even a single photon, although they are bulky. One the other side, microflow cytometers integrate miniaturized light sources and detectors in the detection system. These components combined commonly offer lower detection sensitivity, and eventually lower LODs, compared to that of

Table 10.1 Summary of performance of microflow cytometers with on-chip integrated optics.

Detection method	Integrated components	Analyte	LOD/ Specifications	Ref.
Fl	Waveguides	Particles, red blood cells	N/A	Fu *et al.*[5]
Fl-Ab	Waveguides	Cy5	320×10^{-18} moles of fluorophore	Ruano *et al.*[11]
Sc	CMOS imager	27–67 μm particle	Velocity ~2 cm/sec	Nieuwenhuis *et al.*[15]
Fl- Ab	CMOS imager	Fluorescein	N/A	Adams *et al.*[16]
Fl- Ab	PIN photodetector microlens	Fluorescein	17 nM	Kamei *et al.*[17]
Sc	CMOS Imager	16 μm Polystyrene Polysphere	N/A	Hartley *et al.*[18]
Sc	CMOS Imager	6–14.6 μm Polystyrene Polysphere	Velocity up to ~0.5 cm/sec	Hosseini *et al.*[19]
Fl	LED, photodiode	Fluorescein	1.96 nM	Novak *et al.*[20]
Fl	Dye laser waveguides photodiodes	Xylenol orange dye	N/A	Balslev *et al.*[12]
Fl	P/N photodetector nitride fibers photodiode	Protein and streptavidin	Rate ~ 5 μl/min	Misiakos *et al.*[21]
Fl	Polymer LED	Fluorescein, 5-carboxy fluorescein	50×10^{-15} moles	Edel *et al.*[23]
Fl	OLED OPED	Rhodamine 6G and fluorescein	100 nM for Rhodamine 6G and 10 μm for fluorescein	Pais *et al.*[22]
Fl	Microlens array	15 μm human lung cancer cells labeled with fluorescent dye	400% improved SNR with coupling microlens array	Kuo *et al.*[24]

Note: Fl=Fluorescence, Ab=Absorbance, Sc=Scattering.

conventional flow cytometers. Hence a need has emerged for the development of miniaturized, highly sensitive detectors, capable of competing with conventional detectors.

Recent advances in this regard, discussed in this chapter, have aimed to move closer to such a device by miniaturizing and integrating some of the more relevant optical components on to a microfluidic chip. To this date, although much progress has been made in this direction, a compact and fully functional cytometer integrating all the essential optical components is still not a reality.

Table 10.1 summarizes the capabilities of the current state of the art microflow cytometers, utilized for analysis of biochemical samples for various biomedical applications.

ACKNOWLEDGMENTS

The research work reported was financially supported by grants from the Natural Sciences and Engineering Research Council (NSERC) of Canada as well as the Canadian Institute for Photonic Innovations (CIPI). Additional support was provided by the Canadian Microelectronics Corporation (CMC) Microsystems and AMIF in the fabrication and assembly of the hybrid cytometric device.

References

[1] A. Y. Fu, C. Spence, A. Scherer, F. H Arnold and S. R. Quake, A microfabricated fluorescence-activated cell sorter. *Nature Biotech.* **19**, 1109–1111 (1999).

[2] S.-Y. Yang, S.-K. Hsiung, Y.-C. Hung, C.-M. Chang, T.-L Liao and G.-B. Lee, A cell counting/sorting system incorporated with a microfabricated flow cytometer chip. *Meas. Sci. Technol.* **17**, 2001–2009 (2006).

[3] Y. Mourzina, A. Kalyagin, and D. Carius, A. R. Offenhäusser, Capillary zone electrophoresis of amino acids on a hybrid poly(dimethylsiloxane)-glass chip. *Electrophoresis*, **26**, 1849–1860 (2005).

[4] S. J. O. Varjo, M. Ludwig, D. Belder and M.-L. Riekkola, Separation of fluorescein isothiocyanate-labeled amines by microchip electrophoresis in uncoated and polyvinyl alcohol-coated glass chips using water and dimethyl sulfoxide as solvents of background electrolyte. *Electrophoresis* **25**, 1901–1906 (2004).

[5] L.-M. Fu, R.-J. Yang, C.-H. Lin, Y.-J Pan and G.-B. Lee, Electrokinetically driven micro flow cytometers with integrated fiber optics for on-line cell/particle detection. *Analytica Chimica Acta* **507**, 163–169 (2004).

[6] J. Han and A. K. Singh, Rapid protein separations in ultra-short microchannels: microchip sodium dodecyl sulfate-polyacrylamide gel electrophoresis and isoelectric focusing. *J. Chromatrography A* **1049**, 205–209 (2004).

[7] C.-X. Zhang and A. Manz High speed free-flow electrophoresis on chip. *Anal. Chem.* **75**, 5759–5766 (2003).

[8] C.-F. Lin, G.-B. Lee, C.-H. Wang, H.-H. Lee, W.-Y. Liao and T.-C. Cho, Microfluidic pH-sensing chips integrated with pneumatic fluid-control devices. *Biosens. and Bioelectron.* **21**(8), 1468–1475 (2006).

[9] M.-H. Wu, J. Wang, T. Taha, Z. Cui, J. P. G. Urban and Z. Cui, Study of on-line monitoring of lactate based on optical fiber sensor and in-channel mixing mechanism. *Biomed. Microdevices* **9**, 167–174 (2007).

[10] N. J. Petersen, K. B. Mogensen and J. P. Kutter, Performance of an in-plane detection cell with integrated waveguides for UV/Vis absorbance measurements on microfluidic separation devices. *Electrophoresis* **23** 3528–3536 (2002).

[11] J. M. Ruano, A. Glidle, A. Cleary, A. Walmsley, J. S. Aitchison and J. M. Cooper, Design and fabrication of a silica on silicon integrated optical biochip as a fluorescence microarray platform. *Biosens. Bioelectron.* **18**, 175–184 (2003).

[12] S. Balslev, A. M. Jorgensen, B. Bilenberg, K. B. Mogensen, D. Snakenborg, O. Geschke, J. P Kutter and A. Kristensen, Lab-on-a-chip with integrated optical transducers. *Lab Chip* **6**, 213–217 (2006).

[13] L. Jiang, and S. Pau, Integrated waveguide with a microfluidic channel in spiral geometry for spectroscopic applications. *Appl. Phys. Lett.* **90**, p. 111108 (2007).

[14] D. A. Chang-Yen, R. K. Eich and B. K. Gale, A monolithic PDMS waveguide system fabricated using soft-lithography techniques. *J. Lightwave Technol.* **23**, 2088–2093 (2005).

[15] J. H. Nieuwenhuis, J. Bastemeijer, A Bossche and M. J. Vellekoop, Near-field optical sensors for particle shape measurements. *IEEE Sensors J.* **3**(5), 646–651 (2003).

[16] M. L. Adams, M. Enzelberger, S. Quake and A. Scherer, Microfluidic integration on detector arrays for absorption and fluorescence microspectrometers. *Sens. and Actuats. A: Physical* **104**, 25–31 (2003).

[17] T. Kamei, B. M. Paegel, J. R. Scherer, A. M. Skelley, R. A Street and R. A. Mathies, Integrated hydrogenated amorphous Si photodiode detector for microfluidic bioanalytical devices. *Anal. Chem.* **75**, 5300–5305 (2003).

[18] L. Hartley, K. V. I. S. Kaler and O. Yadid-Pecht, Hybrid integration of an active pixel sensor and microfluidics for cytometry on a chip. *IEEE Trans. Circuits Systems—I* **54**(1), 99–110 (2007).

[19] Y. Hosseini, L. F. Hartley and K. V. I. S. Kaler, Hybrid integrated CMOS-Microfluidic device for the detection and characterization of particles. In *Proc.1st MNRC 2008*, Ottawa, 49–53.

[20] L. Novak, P. Neuzil, J. Pipper, Y. Zhang and S. Lee An integrated fluorescence detection system for lab-on-a-chip applications. *Lab Chip* **7**, 27–29 (2007).

[21] K. Misiakos, S. E. Kakabakos, P. S. Petrou and H. H. Ruf, A monolithic silicon optoelectronic transducer as a real-time affinity biosensor. *Anal. Chem.*, **76**, 1366–1373 (2004).

[22] A. Pais, A. Banerjee, D. Klotzkin and I. Papautsky, High-sensitivity, disposable lab-on-a-chip with thin-film organic electronics for fluorescence detection. *Lab Chip* **8**, 794–800 (2008).

[23] J. B. Edel, N. P. Beard, O. Hofmann, J. C. deMello, D. D. C Bradley and A. J. deMello, Thin-film polymer light emitting diodes as integrated excitation sources for microscale capillary electrophoresis. *Lab Chip* **4**, 136–140 (2004).

[24] J.-N. Kuo, C.-C. Hsieh, S.-Y. and G.-B. Lee, An SU-8 microlens array fabricated by soft replica molding for cell counting applications. *J. Micromech. Microengg.* **17**, 693–699 (2007).

Chapter Eleven

The Potential of Polymer Photonics for Microflow Cytometry

David Leuenberger* and Marc Ramuz
CSEM Based Mattenstrasse 22, P.O. Box, CH-4016 Basel, Switzerland
**david.leuenberger@csem.ch*

This chapter summarizes the current requirements of a microflow cytometer in terms of illumination source, optical detection and optical system. The state-of-the-art of the available organic photonic components is overviewed. The two parts are converged, and the real potential of organic photonics for microflow cytometry is investigated.

11.1 IMPORTANCE OF POLYMER PHOTONICS TO MICROFLOW CYTOMETRY

Flow cytometry has become a standard technique in cell biology and medicine. Commercially available flow cytometers have grown in complexity and performance, making use of multiple laser sources and an increasing number of detectors.[1]

A new disruptive technology, the bio-system-on-a-chip, holds the promise for new markets such as point-of-care and on-site analysis where portability and price are an issue. Microflow cytometers belong to the new optofluidics category that combines microfluidics and photonics. The main advantages of this approach are size reduction and the possibility of parallelization.

At its simplest, microfluidic flow cytometry chips consist of a microfluidic channel with a flowing liquid core. Detection is accomplished by focusing a laser into the channel and coupling out light (generally via microscope objective) to a photomultiplier tube (PMT), charge coupled device (CCD), or avalanche photo diode (APD). Fluidic control is accomplished via gravity fed systems, syringe pumps, or similar mechanisms.[2] A schematic example of such a system is found in Fig. 11.1. It shows the general architecture of the system including the four building blocks 'light source', 'optical system', 'detection system' and 'microfluidics',

The Microflow Cytometer by J S Kim & F S Ligler

www.panstanford.com
978-981-4267-41-0

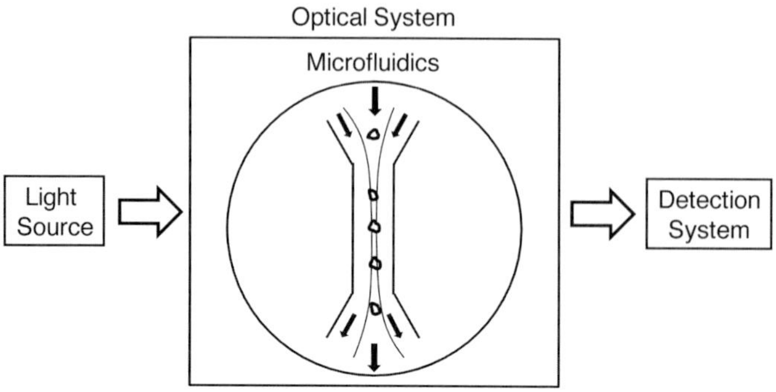

Figure 11.1. Microflow cytometers consist of the following basic building blocks: microfluidic system with a flow focusing unit, excitation source, optical detection system and optical system interfacing the different units. Color reference – pg. 344.

which also corresponds to the breakdown of this chapter. The requirements on the light source, the detector and the optical system will be detailed in the following sections.

Many microflow cytometers still rely on traditional bulky optics when it comes to focusing or extracting light from a small volume and make use of bulky external lasers and PMTs. In a further step, passive optical elements such as optical waveguides and microlenses for excitation and light extraction are added to the chip. As a result the optical alignment of light sources and detectors to the interrogation region in the flow channel is simplified. The issue related to the large footprint of these external sources and detectors still remains. In order to fulfill the promise of a low-cost lab-on-a-chip, new ways have to be found to integrate the light source and/or the detection with the microfluidics in a cost-effective manner.

Whether organic photonics in general, and polymer photonics in particular, will bring both light sources and detectors to where they belong, namely *on* the chip, is the question addressed in this chapter.

11.2 CURRENT STATE OF THE ART OF MICROFLOW CYTOMETRY

11.2.1 Requirements on the Light Source

Flow cytometry heavily relies on fluorescent probes — molecular tags that can be detected with appropriate excitation/emission conditions. Fluorescent probes can be used to detect receptors on cells, to determine the health and physiological state of a cell and even to measure gene expression in individual cells.[3] The main parameters are the excitation wavelength and the optical power of the source.

A huge range of probes suitable for biomedical analysis exists, and new ones are being developed constantly. While the absorption of fluorescent probes is rather broad band, on the order of 50 nm, most instruments rely exclusively on lasers as a source of excitation of these probes. Currently 488 nm is the standard

excitation wavelength in flow cytometry, low cost devices already use alternatives such as 532 and 635 nm laser diodes as well as violet laser diodes, which receive a great deal of attention as the possible next major biomedical laser source.[2] Even the most modern laboratory-scale, multi-laser flow cytometers typically provide no more than six discrete laser wavelengths, and most provide fewer. Coverage of the ultraviolet-to-infrared spectrum is therefore never complete, leaving large gaps in excitation capabilities. Several groups are therefore investigating the use of supercontinuum lasers in this field.[3]

In order to investigate one cell at a time and reduce the background noise, laser beams are tightly focused on the liquid core flow. A typical interrogation volume, *i.e.* the volume defined by the intersection of the core flow and the light cone, is roughly the size of a cell, i.e. in the order of a cube with 10 μm side length.[4] Standard microflow cytometry does not rely on pulsed laser sources in contrast to the related technique of 'scanning microflow cytology'. Concerning the optical power requirement, it is realistic to assume a HeNe laser with a power in the order of 10 mW.[5]

11.2.2 Requirements on the Detection System

Depending on the source and the intensity, a variety of detectors are commonly used in microflow cytometers: PMTs, APDs, CCD cameras, CMOS imaging arrays and PIN photodiodes.[1]

The detector should have external quantum efficiency close to 100% and a high — preferably single photon — sensitivity. The internal gain of a detector is the most important signal amplification step as it increases the signal with the smallest effect on noise.[1] In terms of typical signal power levels, one needs to distinguish between scattered light and fluorescence signal detection. Fluorescence signals can be smaller than 1 nW.[1] It is hard to define a precise figure of merit, but in order to compete with standard APDs, the minimum detectable power should be in the order of 5 pW.[5]

When it comes to speed, state-of-the-art flow cytometers, such as the BD FACSCanto II, can handle up to 10,000 events per second. Therefore the detector should be able to resolve signals of at least 100 kHz.

11.2.3 Requirements on the Optical System Integration

Initially, microfluidic chips were looked at as a mere replacement for the conventional flow cuvette of the cytometer, and all of the optical systems basically remained the same. However utilizing a bulk optical system significantly reduces the miniaturization benefits of the fluidics by requiring time-consuming opto-mechanical alignment steps, sometimes resulting in a "Chip-in-a-lab".

An intermediate solution is to integrate passive optical components to the microfluidic chip. This includes a large variety of optical fibers[6,7] and waveguides as well as on-chip lenses with the goal of transporting the photons from the light

source to the flow channel and from the flow channel to the detector. Ideally, these components are fabricated in the same process step as the microfluidic structures in order to provide self-aligned features suitable for mass-production.

Waveguides offer similar light confinement as optical fibers but in a more robustly integrated way, thereby avoiding the difficulties of alignment, epoxy-fixing or breakage. Other approaches, for instance from Bliss *et al.*[8], involve a combination of fibers and waveguides, interfacing the two by means of fiber-to-waveguide couplers.

The integration of the light collecting system directly on the chip is beneficial, since close-proximity detection can theoretically allow for lower loss and high-numerical aperture (NA) light collection due to the effectively 'immersed' optical system (no on-chip air gaps).[2] Although integration of optical waveguides on microfluidic chips provides several advantages in terms of alignment and total system size, it also comprises a few limitations. The integrated waveguides sources do not provide the very uniform, highly localized illumination compared to external interrogation sources collimated with high-NA microscope objectives. Similarly, light collection by fibers or waveguides does not provide the same localized NA light collection of traditional bench top flow cytometers.[2] In fact, most of the light collected by the fiber originates from locations other than that of the cell.[9]

Some of the drawbacks previously mentioned could be alleviated by adding on-chip lenses to the optical system, resulting in a more focused interrogation beam. This approach also allows an increased NA of the light collection from the cell, leading to higher signal/noise ratios and/or more sensitive detection. Wang *et al.*[10] demonstrated a microchip flow cytometer with integrated optical elements (waveguides, lens and fiber-to-waveguide couplers), all defined in a single layer of SU-8 polymer. In summary, to fully take advantage of the microfluidic approach, *the size of the optical system should scale down to the level of the fluidic system*. There are several technologies available to potentially enable such a reduction in size.

11.3 STATE-OF-THE-ART ORGANIC PHOTONICS

Organic optoelectronic molecules are a fascinating new class of materials. They combine the electrical properties of a semiconductor with the beneficial material properties of plastics.

Organic photonics devices are based on either small semiconducting molecules — mainly deposited by vacuum phase approach — or semiconducting polymers — deposited by liquid phase processes such as printing. Both approaches have specific advantages and disadvantages, which will be explored here. The main focus will be on printable polymer systems, which we believe promise large-volume, low-cost and customizable production.

In the small molecule approach, the basic building blocks (e.g. Alq3) have a low molecular weight and are essentially identical. In contrast, in the polymer approach, the building blocks are long, chain-like molecules with relatively high

molecular weights (a typical conjugated polymer has several hundred repeat units) and significant length and weight variations.

In the remainder of the text, the term organic light emitting diode (OLED) shall comprise both approaches. In order to highlight specific differences, the detailed terms small molecule organic light-emitting diode (SMOLED) and polymer light-emitting diode (PLED) will be employed.

11.3.1 State-of-the-art Organic Light Source

Organic electroluminescence is the electrically driven emission of light from non-crystalline *organic* materials, which was first observed and extensively studied in the 1960s.[11] In 1987, a team at Kodak introduced a double-layer OLED, which combined modern thin-film deposition techniques with suitable materials and structure to give moderately low bias voltages and attractive luminous efficiency.[12] Shortly afterwards, in 1990, a Cambridge research group lead by R. Friend announced a conducting PLED.[13] Since then, there has been increasing interest and research activity in this new field. Enormous progress has been made in color gamut, luminous efficiency and device reliability.

Since an electrical current is needed to stimulate the electroluminescent material, the semiconducting organic material in a light emitting diode has to be contacted with two electrodes. Furthermore, since the electrical conductivity of organic semiconductors is very low, one has to apply a large electric field across the semiconductor in order to pass the required amount of electrical current to stimulate light emission. A very thin ($< 0.2\ \mu$m) film of the organic semiconductor is thus normally sandwiched between two electrodes to form a light-emitting diode. Fig. 11.2 shows the simplified architecture of an OLED.

Glass or a plastic foil such as polyethylene terephthalate (PET) is used as substrate. A thin transparent anode, usually indium tin oxide (ITO), is deposited onto the substrate. The active organic semiconducting material is then applied, followed by the deposition of the cathode. Low work-function, hence reactive metals such as calcium and barium coated by a protection layer of aluminum, are commonly used as cathodes.

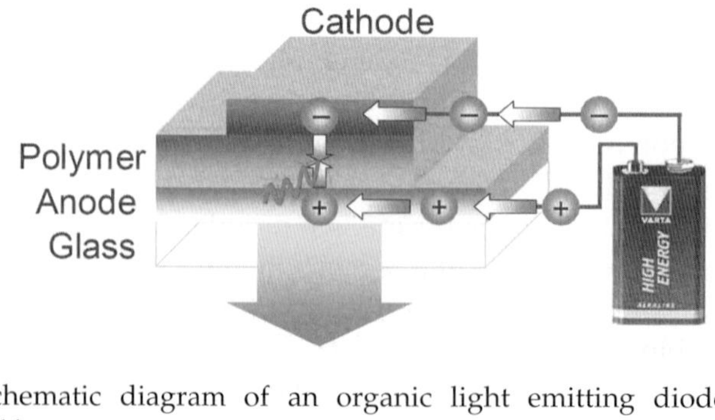

Figure 11.2. Schematic diagram of an organic light emitting diode (OLED). Color reference – pg. 344.

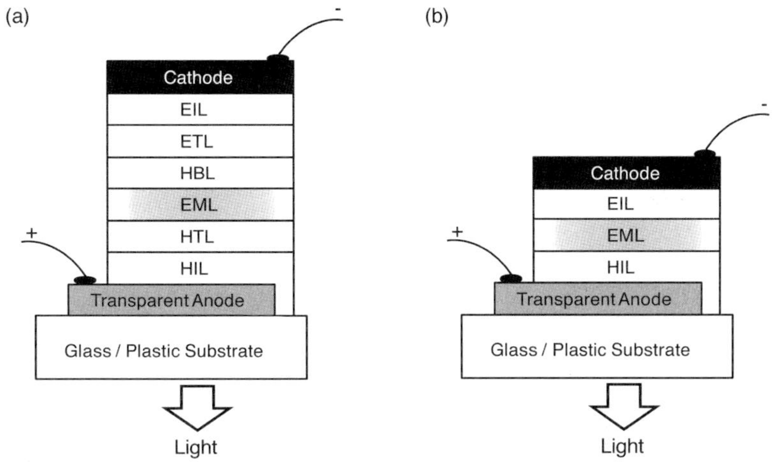

Figure 11.3. Detailed layer structure of (a) small molecule OLED (SMOLED) and (b) polymer light emitting diode (PLED). The stacks consist of a transparent anode (ITO) on a glass/plastic substrate, a hole injection layer (HIL), hole transport layer (HTL), emitting layer (EML), hole blocking layer (HBL), electron transport layer (ETL), electron injection layer (EIL) and a metal cathode.

When a positive voltage is applied to the anode, electrical charges are pushed through the organic thin film and current flows. Negative charge carriers (electrons) are injected into the device by the cathode and travel towards the anode, whereas positive charge carriers (holes) are injected by the anode and are driven towards the cathode. When positive and negative charge carriers meet in the organic thin film, they can "recombine" and emit light. This light escapes the diode usually through the transparent anode. In OLEDs the external quantum efficiency — ratio of the number of photons that escape the diode to the number of electrons that travel through the diode — is 1 to 10 %.

Layer structure, especially in the case of SMOLEDs can be considerably more sophisticated than the one depicted in Fig. 11.2. In order to assure a good charge injection, charge transport and radiative recombination in the emission layer (EML), additional layers such as hole injection layer (HIL), hole transport layer (HTL), hole blocking layer (HBL), electron transport layer (ETL) and electron injection layer (EIL) are added to the stack (11.3.2 Fig. 11.3).

As mentioned, one of the main differences between OLEDS based on small molecules versus polymers lies in processing. The most common way by which thin films of conjugated small molecules are applied is vacuum sublimation. The small-molecule material, which is normally a powder, is heated in vacuum to the point where molecules evaporate at a reasonable rate. A fraction of the molecules in the resulting vapor will fly in the direction of the sample and condense on its surface to form a thin film.

Vacuum deposited thin films of small molecules are normally polycrystalline and show a high degree of structural order. The sublimation process used to

deposit small molecules also ensures that the deposited films are very pure and of high quality. In contrast to small molecules, conjugated polymers are processed from solution. The most widely used deposition techniques are spin-coating, ink-jet printing and gravure printing. These methods produce rather poorly defined films with significant local material variations.

The polymer is first dissolved in a common organic solvent such as xylene. Then this solution is deposited onto the sample surface. As the solvent evaporates a thin amorphous polymer film remains. In the ink-jet method the organic material is dissolved in solvent and flies in a controlled manner out of the ink jet nozzle similar to inkjet printers used at home. The inkjet method applies the organic materials to the areas requiring pixels, precisely allowing for micro-level control of the formation of the film layers.

According to the review paper by Shinar *et al.*[14] the maximum reachable external quantum efficiencies for typical fluorescent SMOLED, fluorescent PLED and phosphorescent SMOLED are 6.5%, 10% and 26%, respectively. The highest efficiency OLEDs now exhibit electrical efficiencies above 20% and power efficiencies exceeding 50 $\mathrm{lmW^{-1}}$,[15] where 1 lm $\equiv$ 1.46 mW at 555 nm (the wavelength to which the human eye is most sensitive) at brightness L $\sim$ 150 $\mathrm{Cdm^{-2}}$, where 1 Cd $\equiv$ 1 lm $\mathrm{sr^{-1}}$.[14] For comparison, the value of 150 $\mathrm{Cdm^{-2}}$ corresponds to the standard brightness of an LCD computer monitor. Kido and coworkers have recently demonstrated 30 $\times$ 30 $\mathrm{cm^2}$ white OLED (WOLED) panels with power efficiencies of 20 $\mathrm{lmW^{-1}}$ at L $>$ 1000 $\mathrm{Cdm^{-2}}$.[16] Schwarz *et al.*[17] have fabricated a white p-i-n OLED with luminous efficiency close to 16 $\mathrm{CdA^{-1}}$ for a brightness of 1000 $\mathrm{Cdm^{-2}}$.

For PLEDs the luminous efficiencies for red, green and blue emitters are 10, 15 and 9.9 $\mathrm{CdA^{-1}}$, respectively.[18]

In terms of stability of OLED devices, it is important to distinguish between the intrinsic lifetime of the electroluminescent polymer in inert atmosphere and the lifetime linked to degradation issues such as polymer and cathode oxidation—hydroxidation due to the oxygen and water in the atmosphere. The latter can be mitigated by proper encapsulation with glass/metal, UV-curable epoxies and the addition of getter materials, i.e. materials with the ability to bind oxygen and other gas traces, inside the sealed cavity.[19] Based on a recent review by Shinar *et al.*[14] red-to-green SMOLEDs and blue SMOLEDs feature continuous operating lifetimes >200,000 hours ($\sim$23 yr) and 100,000 hours ($\sim$11.5 yr), respectively, at 150 $\mathrm{Cdm^{-2}}$. Despite these impressive numbers, there are still serious lifetime issues to be solved, especially at high brightness ($>$ 1000 $\mathrm{Cdm^{-2}}$). For PLEDs the cited lifetimes are somewhat lower; however, they were taken at a higher initial brightness of 1000 $\mathrm{Cdm^{-2}}$. They are 89,000 hours ($\sim$10 yr), 79,000 hours ($\sim$9 yr) and 10,000 hours ($\sim$1 yr) for red, green and blue, respectively.[18] To our knowledge, the lifetimes of the PLED could be at least a factor 2 longer when operated only at 150 $\mathrm{Cdm^{-2}}$, depending on various factors such as duty cycles, peak luminance and pulsed driving.

One of the obstacles for the wide deployment of OLEDs in lab-on-a-chip applications that require narrow excitation lines, are the relatively broad emission spectra of OLEDs, with spectral widths on the order of 75 nm. A possible route to

shrink the linewidth is to fabricate the OLED on top of a distributed Bragg reflector (DBR)[20]. By modifying the thickness of the spacer layer between the DBR and the ITO, the emission wavelength can be tuned in the range of 495–625 nm.

A related topic is organic lasers.[21,22] They offer monochromatic light, are tunable over a wide spectral range and have low pumping thresholds. In many biosensing applications, small linewidth is desired in order to avoid spectral overlap between the excitation source and the marker emission and thus reach a higher signal-to-noise ratio. The quality of organic lasers has improved in such a way that they can be pumped by cost-effective inorganic violet diode lasers. Riedl *et al.*[22] have demonstrated an extremely compact organic thin-film DFB laser that is pumped with a 406 nm laser diode and tunable between 496 and 516 nm. Direct electrical excitation has long been considered the 'holy grail' for all types of semiconductor lasers, and organics are no exception.[23] Until electrically pumped organic lasers become reality, many challenges remain to be mastered.

An alternative to the just described thin-film organic laser might be the optofluidic distributed feedback (DFB) dye laser.[24] It consists of a microfluidic channel with an integrated DFB grating fabricated in polydimethylsiloxane (PDMS). A dye solution which acts as both the core of the optical waveguide and the gain medium can be introduced into the structure through the channel. These lasers are highly compact, widely tunable and robust. They can be pumped either by an external light source or on-chip laser diode.

11.3.2 State-of-the-art Organic Detection System

One of the most interesting structures for an organic photodiode (OPD) is the bulk heterojunction,[25,26] which results when electron donors and acceptors are dissolved in an appropriate solvent and deposited in solution. Bulk heterojunctions have proven to be a very successful concept to overcome the short exciton diffusion length of organic molecules.[27] However, they enhance the disorder and hamper the collection of photogenerated charge carriers. Therefore, creating bicontinuous and interpenetrating networks between the donor and acceptor phase is of key importance (Fig. 11.4). The use of high boiling point solvents and low spin speeds turned out to be a successful approach to increase the molecular order and thus the collection efficiency of charge carriers.[28]

State-of-the-art OPD-blends consist of poly(3-hexylthiophene):phenyl-C61-butyric acid methyl ester, P3HT:PCBM with weight ratio of 1:1. The layer thickness of the OPD is 240 nm and has been optimized for optimal external quantum efficiency, on/off ratio and lifetime.[29,28]

Although OPDs and organic photovoltaics (OPVs) are in many ways similar devices, they nevertheless fulfill a different set of requirements. For OPDs the spectral region of interest is often quite narrow, and the important parameters are the photocurrent and the dark current, with the dark current being only a fraction of the photocurrent. OPVs, on the other hand, must have high conversion efficiency over the entire solar spectrum, which reaches significantly into the near-IR.

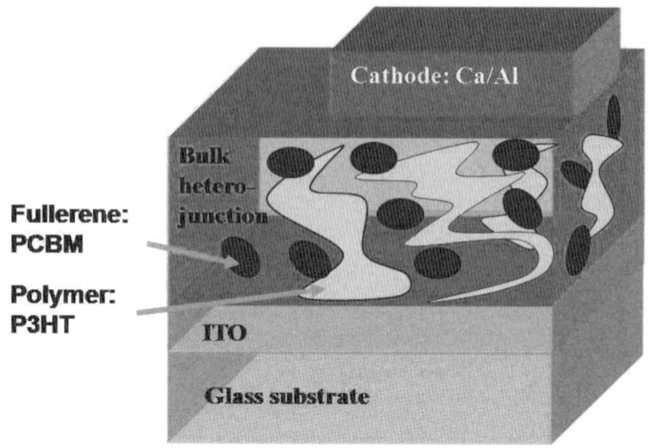

Figure 11.4. Schematic device architecture of a bulk heterojunction organic photodiode (OPD). Color reference – pg. 344.

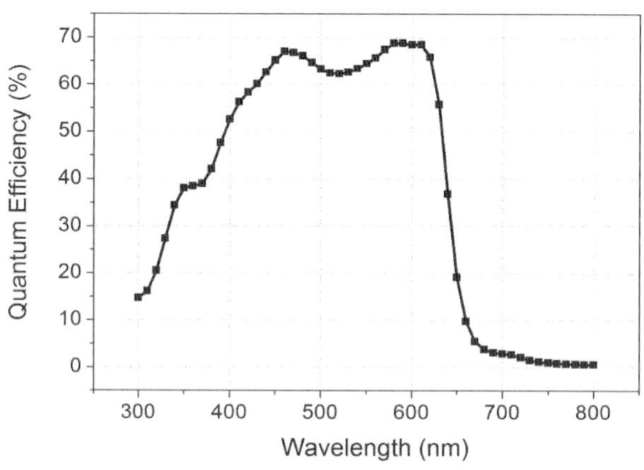

Figure 11.5. EQE at 0 V as a function of wavelength.

The external quantum efficiency (EQE) of such devices is above 60% for the range of 420-630 nm (Fig. 11.5).[29,30,31] Recent publications show a new class of low bandgap polymer with high sensitivity in the red / NIR.[32,33]

Typical frequency response — at 3dB attenuation — of bulk heterojunction OPDs biased at 0 V occurs at a frequency of several 100 kHz.[34,29] Although f_{3dB} increases with increasing reverse bias, the values for the P3HT:PCBM heterojunctions are more than three orders of magnitudes smaller than the ones of optimized small molecule photodiodes.[35] Further device optimization (reducing capacitance) and different driving schemes (e.g. under strong reverse bias) will lead to a somewhat improved response speed of P3HT:PCBM diodes, nevertheless these devices will be limited to frequencies of <1 MHz.

Punke *et al.*[34] have optimized their OPDs for speed by minimizing the pathlengths of the electrodes and by avoiding the parasitic capacitances formed by overlapping electrodes. They measured rise times as small as 1.6 ns and fall times

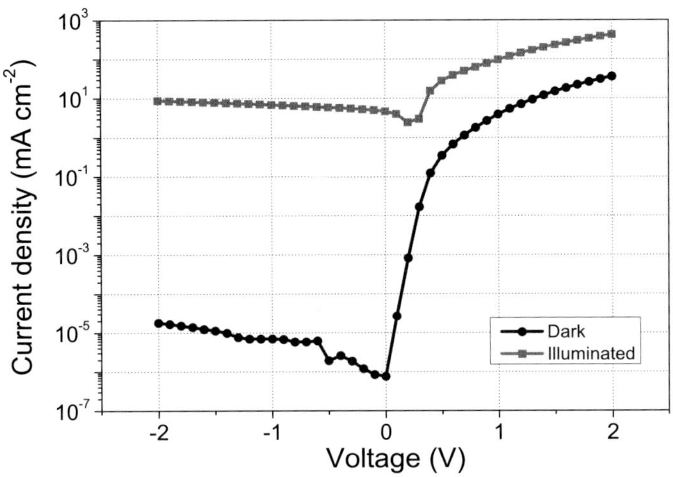

Figure 11.6. I(V) characteristics of a $2 \times 2\,\text{mm}^2$ PD measured in the dark and under ($40\,\text{mWcm}^{-2}$) illumination at 468 nm.

< 40 ns and derived a -3 dB cut-off frequency of ~1 MHz at -5 V bias. State-of-the-art on/off current ratio of 10^6 at -1 V (Fig. 11.6) and dark current densities below $10\,\text{nAcm}^{-2}$ at -1 V have been reported.[29]

Apart from spin-casting, ink-jet printed OPDs achieve very similar performance with EQE > 50% and on/off ratio > 10^3. From a processing and integration point of view, the inkjet technology is very advantageous since the polymers can be easily deposited on the chip with a positional accuracy of much less than 100 micrometers.Another figure of merit, especially when it comes to detecting small fluorescence signals, is detector noise. A low-noise figure is particularly important for the detection of weak side scattered light and fluorescence emission. It allows reducing detection levels and increasing dynamic range. For instance in OPDs fabricated in our lab, we obtained a specific detectivity of $7 \times 10^{12}\,\text{cm} \times \text{Hz}^{-1/2}\text{W}^{-1}$.[29]

Operating lifetime has also improved over the last few years.[36,29] As seen on Fig. 11.7, the photocurrent is stable for more than 1500 hours. After approximately 1800 hours the photocurrent starts to increase gradually. The on/off ratio decreases monotonically but is larger than 10^3 over the entire measuring period of 3000 hours (four months).

11.3.3 State-of-the-art Optical System Integration Using Organic Photonics

While there have been publications on organic photonics used in lab-on-a-chip applications,[2] to our knowledge there has not been a single peer-reviewed publication on a flow cytometer completely based on organic photonics. Knowing that the real potential of organics and flow cytometry lies in the miniaturization and integration, we would like to give the reader insight how far this field has advanced

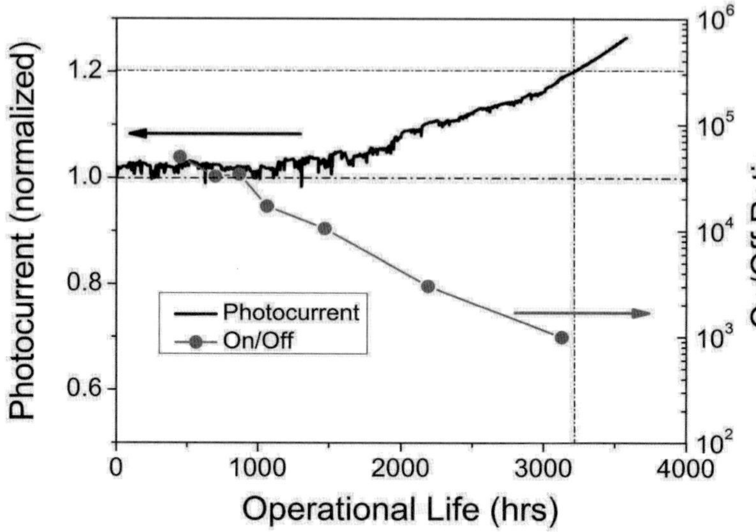

Figure 11.7. Stability of the PDs without PEDOT:PSS layer. The normalized photocurrent, measured at –1 V and wavelength illumination of 468 nm, is plotted as a function of operating time at room temperature in N_2 atmosphere.

in related application areas. Some of these areas might also be relevant for an integrated flow cytometer.

When it comes to integration of organic photonics on lab-on-a-chip two possible schemes are: (I) sandwich design and (II) waveguide design. The first scheme defines functional layers such as excitation, microfluidics and optical filtering layer and stacks them on top of each other. It has certain advantages, with respect to wafer-scale fabrication and parallelization of the detection. The second approach, which is the one our group pursues, builds everything around a waveguide structure. It enables better SNR and is considerably more flexible at the expense of lateral size, e.g. it is possible to increase the optical interaction length with the analyte and to integrate complex optical functions such as waveguides, dispersive elements and plasmonics.

In the following paragraphs different examples of both categories schemes are reviewed:

A simple yet powerful scheme of the stacked type for fluorescence-based assays has been proposed by Pais *et al.*[37,38] (Fig. 11.8). It is based on a high-sensitivity, cost-effective, cross-polarization scheme to filter out excitation light from a fluorescent dye emission spectrum. The cross-polarizers suppress the polarized excitation signal by 22 dB with respect to the randomly polarized fluorescent signal.

In order to detect oxygen and biological agents, a similar back detection geometry scheme was chosen by Shinar *et al.*[39,14] In this approach interdigitated OLEDs and PDs are placed on the same plane, separated from the sensing layer by an optical filter. The multi-analyte concentration is determined using a robust fluorescence decay technique with a pulsed OLED.

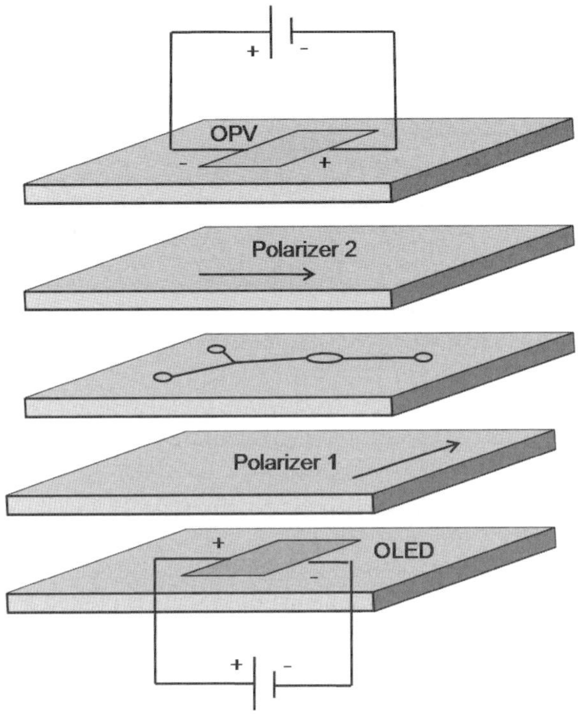

Figure 11.8. Schematics of integrated excitation/detection system. [Figure adapted from Banerjee *et al.*[38]] Color reference – pg. 345.

With a hybrid device using a silicon-based photodiode and an OLED, Shin *et al.*[40] have demonstrated an integrated fluorescence detector that achieved a limit of detection of 1 μM.

Hofman *et al.*[41] have used CuPc-C60-based thin film small molecule OPDs to successfully monitor chemiluminescence reactions. Their OPDs had an external quantum efficiency of ~30% in the 600–700 nm wavelength range and an active area of $2 \times 8\,\text{mm}^2$.

Another example uses an integrated PPV-based PLED excitation source for micro-scale fluorescence detection.[42] PLEDs also have been used as excitation source in micro-scale capillary electrophoresis.[43]

A company active in lab-on-a-chip is BioIdent.[44] They have developed a PhotonicLab™ Platform consisting of the combination of printed opto-electronic components with microfluidic systems. The novel concept allows the integration of illumination and detection capabilities onto microfluidic-based devices using printing technologies (Fig. 11.9).

The ultrathin photodiodes with an overall thickness of only 300 to 500 nm show quantum efficiencies better than 0.5 and linear light-response over 6 orders of magnitude. The pixel size can range from 50 to over 1000 μm and inkjet fabrication allows tailoring the sensor layout to the needs of the specific application.[45] An

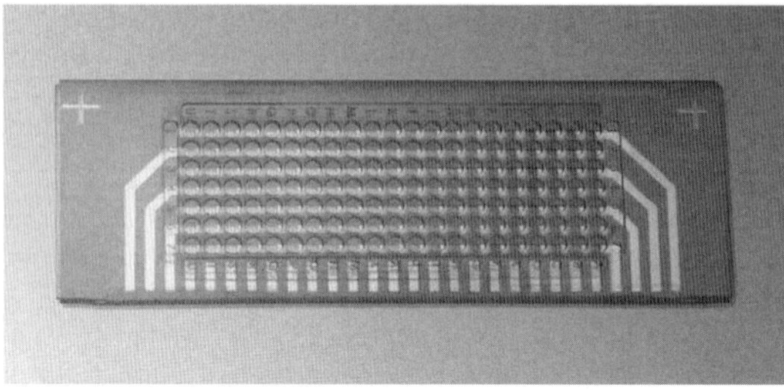

Figure 11.9. Disposable nanotiterplate with fully integrated optical readout system (© BIOIDENT Technologies, Inc.). Color reference – pg. 345.

equivalent OLED array can be generated using the same fabrication-procedure but with different organic materials. As a consequence, in principle any combination of light emitting and light detecting diodes can be printed on a variety of substrates for sample illumination and signal detection.[44]

A good showcase for the integration of polymer photonics in lab-on-a-chip applications is the European project SEMOFS (Surface Enhanced Micro Optical Fluidic System) which aims at the development of a fully integrated disposable biosensor. The detection is based on a surface plasmon (SP) scheme (Fig. 11.10): Light from an OLED is coupled into a single-mode waveguide and propagates towards the microfluidic flow cell containing the SP stack.

For a certain narrow wavelength range, the waveguide mode couples evanescently to a surface plasmon and is subsequently lost. This phase matching

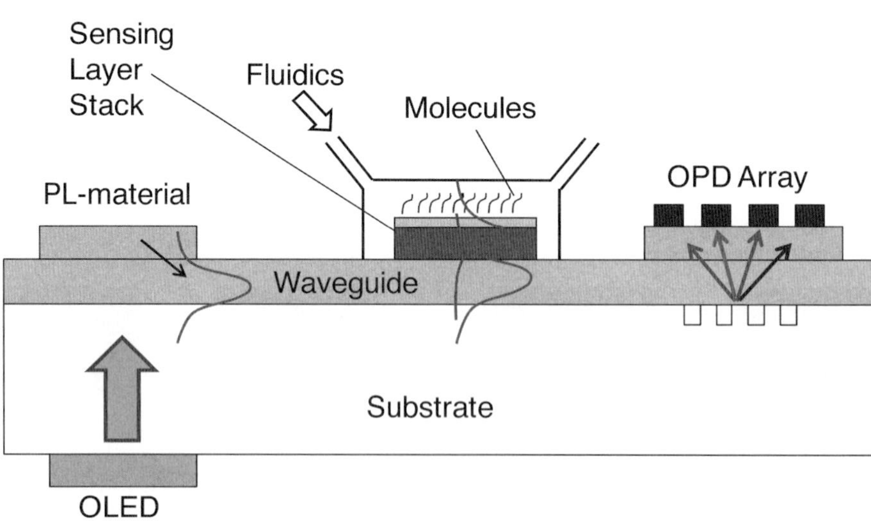

Figure 11.10. Schematics of microfluidics with integrated organic light source and organic mini-spectrometer. Color reference – pg. 346.

condition relates to a dip in the transmission spectrum which is detected by the organic mini-spectrometer. Since the dispersion of the SP mode is very sensitive to refractive index changes close to the metal surface, the adsorption of biomolecules on the functionalized surface of the plasmon stack translates into a shift of the phase matching condition and thus a shift in the transmission dip.

One of the key challenges for coupling light from an OLED into a waveguide is the fact that the waveguide has to be single-mode in order to have an unambiguous transmission dip. The OLED is a Lambertian emitter, i.e. it appears to have equal brightness independent of the observation angle. In contrast a single-mode waveguide allows propagation only from a small solid angle. The difficulty arises from the geometrically invariant radiance (power/area/steradian), *i.e.* independently of how tightly focused the photons are, the radiance basically remains the same. In order to get an intense focal spot, it is best to use a light source that has a high luminous intensity. Subsequently, two different schemes to couple the light from the OLED into a waveguide have been tested within this project. The first approach consists of depositing the OLED directly on top of the waveguide (Fig. 11.11(a)).

The approach is based on evanescent coupling of the waveguide mode and the dipoles inside the OLED. In this scheme our group has demonstrated coupling efficiencies as high as 3.2%.[46] There are two drawbacks in this geometry[47]: (a) The TM mode is suppressed due to the absorption of the metal cathode of the OLED itself, which is not acceptable for a surface plasmon resonance detection scheme; (b) the OLED contains optical loss layers such as the transparent anode (ITO based) and the metallic cathode. Thus, only light coupled into the waveguide within a

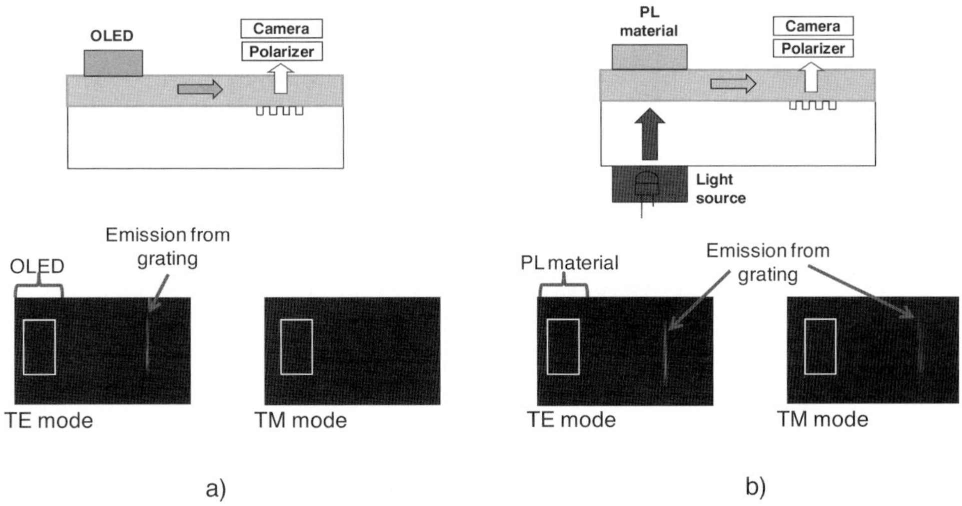

Figure 11.11. Two distinct schemes for coupling light from an OLED into single-mode waveguide. (a) Direct evanescent field coupling (TE mode only). (b) Indirect coupling by optically pumping a PL layer on top of waveguide. Color reference – pg. 346.

distance of 20 μm from the OLED edge propagates, the rest is absorbed by the electrodes.

In a second route, we use an indirect OLED-to-waveguide coupling in the way that the OLED emission was used to pump a PL-material layer located on top of the waveguide,[47,48] as depicted in Fig. 11.11b. In this configuration, loss layers (electrodes) of the OLED are kept at a safe distance from the waveguide. Thus, this architecture allows coupling of both TE and TM mode into the waveguide. Fig. 11.11b shows the extracted light at the out-coupling grating under TE or TM polarization. Another advantage of this second architecture is that the power coupled into the waveguide scales with the area of the illuminated PL-layer.A novelty in the device sketched in Fig. 11.10 is the organic mini-spectrometer that could be monolithically integrated on the biochip. It consists of a grating, etched in the glass substrate before depositing the waveguide, and a dense array of OPDs (Fig. 11.12).

The spectral resolution, $\Delta\lambda = \frac{g}{d}\Lambda$, is defined by the periodicity Λ, the length $g$of the grating, and as well by the distance $d$between the grating and the OPDs. Of course, the spectral resolution is directly related to the width of each individual OPD. Various masks were designed for spectral resolution ranging from 50 nm to 5 nm. Highest spectral resolution mask, as seen in Fig. 11.12(b), consists of 10 μm $\times$ 3.4 mm (width $\times$ length) OPD pixel with 5 μm spacing between two adjacent OPD pixels.

The guided light is diffracted by the grating and was detected by the organic spectrometer. In a first experiment, the PL-material has been pumped by a green OLED based on iridium complex emitter (Fig. 11.13).

The obtained spectral resolution was rather low ($\sim$10 nm) however proof of concept has been achieved. Replacing the OLED with an inorganic blue LED improved the spectral resolution of the organic spectrometer to below 10 nm FWHM (Fig. 11.13). Scaling up the PL-layer length from 2 mm to 6 mm will enhance the optical power in the waveguide by a factor of 5. Thus, a spectral resolution below 5 nm should be in principle achievable. Higher resolution organic spectrometers are currently under preparation.

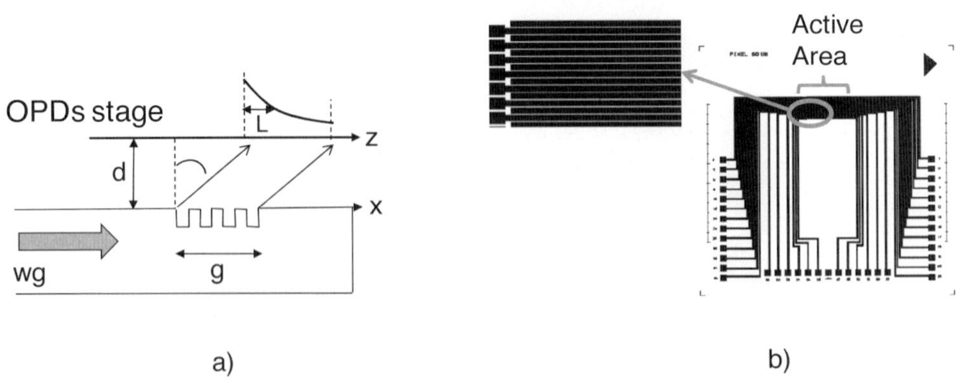

Figure 11.12. (a) Sketch of a simple model to calculate the spectral resolution. (b) ITO mask for organic spectrometer.

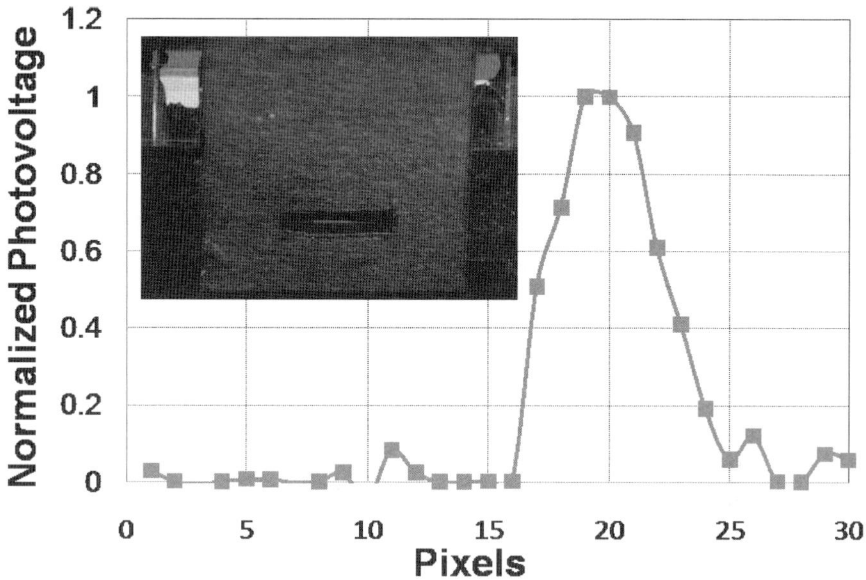

Figure 11.13. Organic spectrometer response when PL-material MEH-PPV pumped by green OLED. Inset: Photograph showing green OLED emission and red PL at output grating. Color reference – pg. 347.

The biochip developed within the SEMOFS project highlights the relatively simple deposition processes of organic material for integrating light source and light detection in a lab-on-a-chip device. However, it will be a complex task to converge all the different building blocks including the microfluidics and the surface chemistry to a fully integrated device. Figure 11.4 illustrates the end goal of this activity: an integrated organic opto-fluidic device.

Figure 11.14. Top view of the SEMOFS chip with microfluidic components. Color reference – pg. 347.

11.4 OPPORTUNITIES AND CHALLENGES FOR THE APPLICATION OF ORGANIC PHOTONICS IN MICROFLOW CYTOMETRY

As mentioned earlier there are actually no publications that propose a flow cytometer entirely based on organic photonic components. A valid question is whether time is not ready yet or if there is some fundamental issue. A rough calculation shows that the OLED illumination is going to be the most challenged part. An OLED is by nature a large area emitter which emits in a half sphere. As such it is extremely difficult to focus it tightly without using very bulky optics. Assuming a conventional flow cytometer with a 10 mW laser focused on a flow channel on an area of $10 \times 10\ \mu m^2$ yields an intensity of 10^7 mWcm^{-2}. An OLED with a brightness of 1000 cdm^{-2} on the other hand emits roughly 1 mWcm^{-2} into a half space (1 lm = 1.46mW at 555 nm). There is a mismatch of 7 orders of magnitude in intensity, which is extremely challenging to compensate with external optics. A possible workaround is to make use of near-field light concentration methods such as high-quality factor optical resonators[49] or light-harvesting plasmonic structures.[50]

The detection side looks promising. As shown in the section on state-of-the-art of organic detection system solution bulk heterojunction OPDs provide external quantum efficiencies between 60% and 70% over the whole visible range. In terms of detector speed, state-of-the-art OPDs should be able to detect signals in the 100 kHz allowing for high-speed screening.[29] Speed-optimized OPDs have a frequency response up to 1 MHz.[34] In terms of noise figure, they reach a specific detectivity of 7×10^{12} cm $\times$ Hz$^{-1/2}$W^{-1}, which is comparable to common inorganic silicon photodiodes like for example a Hamamatsu S2551 with a specific detectivity of 1.5×10^{13} cm $\times$ Hz$^{-1/2}$W^{-1}. A classical method to externally amplify the detector without amplifying the inherent noise is the lock-in technique. Tung *et al.*[9] have applied it to standard PIN photodiodes but it should, in principle, also work for OPDs.

While attractive, the approach of solely replacing the external APD, PIN diode by an OPD would also mean not to realize the full potential of organic photonics. A big potential lies in the integration possibilities, i.e. the possibility to place a single OPD or an OPD array exactly where it is needed. This is especially true for liquid processed polymer organic photonics. The tool of choice in that context is inkjet printing which allows precise layer deposition with a spatial resolution better than 100 μm. We think it will prove very fruitful to investigate additional functionalities.

The ability to incorporate OPD in close proximity to the microfluidic channels would eliminate the need for complicated collection optics including bulk lenses, microscope objectives and fiber optics. The ability to easily fabricate OPD *arrays* opens up unprecedented possibilities in increasing the signal quality by performing multiplexed data collection and employing time-delay cross-correlation for signal enhancement.[51] Reducing the size of the optical system will also reduce the optical path length and therefore absorption losses, an important consideration in

fluorescence measurements.[2] OPD arrays can also perform imaging functions and thus replace bulky CMOS and CCD chips.

In the section on state-of-the-art optical system integration using organic photonics, we have seen a few very promising examples of system integration of OLEDs and OPDs for sensing applications. A highlight and potentially useful are polymer OPD arrays in conjunction with microptical elements such as diffraction gratings as in the case of the first organic mini-spectrometer.[47] Micro-optical elements such as gratings and microlens arrays may be fabricated by high-throughput low cost technologies, preferably in the same step as the patterning of the microfluidics channels.

In summary, we have seen that organic photonics are low-cost, environmentally friendly and thus potentially disposable. Because of the flexibility in processing and the ease of integration with various types of substrates including plastics, it could potentially allow the fabrication of very compact, lightweight and therefore portable devices. These features make organic photonics potentially attractive for application areas such as environmental sensors, personalized diagnostic tools deployed on large scales and screening tools for professional biologists. For certain applications, disposability might actually not just be a cost argument, but rather a requirement, e.g. in biological testing when working with pathogens.

Classical flow cytometry is a very mature field and hence the demands on the light source, the detector and the optical system are very high. Therefore it is reasonable to assume that microflow cytometry might not be the entry point of the emerging organic photonics technology. However we believe in the potential of organic photonics in related fields of diagnostics and screening, and that with increasing maturity, the technology will also penetrate microflow cytometry.

ACKNOWLEDGEMENTS

The work about the integrated plasmonic bio-sensor has been partially supported by the European project SEMOFS (IST-FP6-016768). The authors also acknowledge the support by Zeptosens and Prof. Keppner.

References

[1] D. A. Ateya *et al.*, The good, the bad, and the tiny: A review of microflow cytometry. *Anal. Bioanal. Chem.* **391**, 1485–1498 (2008).

[2] J. Godin *et al.*, Microfluidics and photonics for bio-system-on-a-chip: A review of advancements in technology towards a microfluidic flow cytometry chip. *J. Biophoton.* **1**, 355–376 (2008).

[3] W. G. Telford, More flexibility for flow cytometry. *BioPhotonics* 34698 (2008).

[4] R. Scott, P. Sethu, and C. K. Harnett, Three-dimensional hydrodynamic focusing in a microfluidic coulter counter. *Rev. Sci. Instrum.* **79**, 46104 (2008).

[5] L. Fu *et al.* Electrokinetically driven micro flow cytometers with integrated fiber optics for on-line cell/particle detection. *Anal. Chim. Acta* **507**, 163–169 (2004).

[6] M. L. Chabinyc *et al.*, An integrated fluorescence detection system in poly(dimethylsiloxane) for microfluidic applications. *Anal. Chem.* **73**, 4491–4498 (2001).

[7] N. Pamme, R. Koyama and A. Manz, Counting and sizing of particles and particle agglomerates in a microfluidic device using laser light scattering: application to a particle-enhanced immunoassay. *Lab Chip* **3**, 187–192 (2003).

[8] C. L. Bliss, J. N. McMullin and C. J. Backhouse, Rapid fabrication of a microfluidic device with integrated optical waveguides for DNA fragment analysis. *Lab Chip* **7**, 1280–1287 (2007).

[9] Y. Tung *et al.*, PDMS-based opto-fluidic micro flow cytometer with two-color, multi-angle fluorescence detection capability using PIN photodiodes. *Sens. Actuators, B* **98**, 356–367 (2004).

[10] Z. Wang *et al.*, Measurements of scattered light on a microchip flow cytometer with integrated polymer based optical elements. *Lab Chip* **4**, 372–377 (2004).

[11] M. Pope, H. Kallmann and P. Magnante, Electroluminescence in organic crystals. *J. Chem. Phys.* **38**, 2042 (1963).

[12] C. W. Tang and S. A. VanSlyke, Organic electroluminescent diode. *Appl. Phys. Lett.* **51**, 913–915 (1987).

[13] J. H. Burroughes *et al.*, Light-emitting diodes based on conjugated polymers. *Nature* **347**, 539–541 (1990).

[14] J. Shinar and R. Shinar, Organic light-emitting devices (OLEDs) and OLED-based chemical and biological sensors: an overview. *J. Phys. D: Appl. Phys.* **41**, 133001 (2008).

[15] C. Adachi *et al.*, Nearly 100% internal phosphorescence efficiency in an organic light emitting device. *J. Appl. Phys.* **90**, 5048–5051 (2001).

[16] R. F. Service, Organic LEDs look forward to a bright, white future. *Science* **310**, 1762–1763 (2005).

[17] G. Schwarz *et al.*, Highly efficient white organic light emitting diodes comprising an interlayer to separate fluorescent and phosphorescent regions. *Appl. Phys. Lett.* **89**, 083509 (2006).

[18] T. Yamada *et al.*, Invited paper: Recent progress in light-emitting polymers for full color OLEDs. *SID 08 DIGEST* 404 (2008).

[19] S. A. Choulis *et al.*, The effect of interfacial layer on the performance of organic light-emitting diodes. *Appl. Phys. Lett.* **87**, 113503 (2005).

[20] A. Dodabalapur *et al.*, Physics and applications of organic microcavity light emitting diodes. *J. Appl. Phys.* **80**, 6954 (1996).

[21] C. Karnutsch *et al.*, Improved organic semiconductor lasers based on a mixed-order distributed feedback resonator design. *Appl. Phys. Lett.* **90**, 131104 (2007).

[22] T. Riedl *et al.*, Tunable organic thin-film laser pumped by an inorganic violet diode laser. *Appl. Phys. Lett.* **88**, 241116 (2006).

[23] J. Hecht, The pump is the challenge. *Laser Focus World* **44**, 72–75 (2008).

[24] D. Saltis, S. R. Quake and C. Yang, Developing optofluidic technology through the fusion of microfluidics and optics. *Nature* **442**, 381–386 (2006).

[25] G. Yu *et al.*, Polymer photovoltaic cells: Enhanced efficiencies via a network of internal donor-acceptor heterojunctions. *Science* **270**, 1789–1791 (1995).

[26] J. J. M. Halls *et al.*, Efficient photodiodes from interpenetrating polymer networks. *Nature* **376**, 498

[27] P. Peumans, A. Yakimov and S. R. Forrest, Small molecular weight organic thin-film photodetectors and solar cells. *J. Appl. Phys.* **93**, 3693 (2003).

[28] J. Huang, G. Li and Y. Yang, Influence of composition and heat traitment on the charge transport properties of poly(3-Hexylthiophene) and [6,6]-phenyl C61-butiric acid methyl ester blends. *Appl. Phys. Lett.* **87**,112105 (2005).

[29] M. Ramuz *et al.*, High sensitivity organic photodiodes with low dark currents and increased lifetimes. *Org. Electron.* **9**, 369–376 (2008).

[30] P. Schilinsky, C. Waldauf and C. Brabec, Performance analysis of printed bulk heterojunction solar cells. *Adv. Funct. Mater.* **16**, 1669 (2006).

[31] F. Padinger, R. S. Rittberber and N. S. Sariciftci, Effects of postproduction treatment on plastic solar cells. *Adv. Funct. Mater.* **13**, 85–88 (2003).

[32] M. M. Wienk *et al.*, Narrow-bandgap diketo-pyrrolo-pyrrole polymer solar cells: The effect of processing on the performance. *Adv. Mater.* **20**, 2556–2560 (2008).

[33] J. Peet *et al.*, Efficiency enhancement in low-bandgap polymer solar cells by processing with alkane dithiols. *Nat. Mater.* **6**, 497–500 (2007).

[34] M. Punke, Dynamic characterization of organic bulk heterojunction photodetectors. *Appl. Phys. Lett.* **91**, 71118 (2007).

[35] P. Peumans, V. Bulovic and S. Forrest, Efficient, high-bandwidth organic multilayer photodetectors. *Appl. Phys. Lett.* **76**, 3855 (2000).

[36] C. Schilinsky *et al.*, Polymer photovoltaic detectors: progress and recent developments. *Thin Solid Films* **451**, 105–18 (2004).

[37] A. Pais *et al.*, High-sensitivity, disposable lab-on-a-chip with thin-film organic electronics for fluorescence detection. *Lab Chip* **8**, 794–800 (2008).

[38] A. Banerjee *et al.*, A polarization isolation method for high-sensitivity, low-cost on-chip fluorescence detection for microfluidic lab-on-a-chip. *IEEE Sensors J.* **8**, 621–627 (2008).

[39] R. Shinar *et al.*, Luminescence-based oxygen sensor structurally integrated with an organic light-emitting device excitation source and an amorphous Si-based photodetector. *J. Non-Cryst. Solids* **352**, 1995–1998 (2006).

[40] K. Shin *et al.*, Characterization of an integrated fluorescence-detection hybrid device with photodiode and organic light-emitting diode. *IEEE Electron Device Lett.* **27**, 746-748 (2006).

[41] O. Hofmann *et al.*, Thin-film organic photodiodes as integrated detectors for microscale chemiluminescence assays. *Sens. Actuators, B* **106**, 878–884 (2005).

[42] O. Hofmann *et al.*, Towards microalbuminuria determination on a disposable diagnostic microchip with integrated fluorescence detection based on thin-film organic light emitting diodes. *Lab Chip* **5**, 863.

[43] J. B. Edel, *et al.*, Thin-film polymer light emitting diodes as integrated excitation sources for microscale capillary electrophoresis. *Lab Chip* **4**, 136 (2004).

[44] Bioident Technologies, Inc. at <http://www.bioident.com/>

[45] R. Pieler, E. Füreder and M. Sonnleitner, Printed photonics for lab-on-chip applications. *Proc. SPIE* **6739**, 673919 (2007).

[46] M. Ramuz *et al.*, Coupling light from an OLED into a single-mode waveguide: Towards monolithically integrated optical sensors. *J. Appl. Phys. (submitted).*

[47] M. Ramuz *et al.*, OLED and OPD-based mini-spectrometer integrated on a single-mode planar waveguide chip. *EPJ Applied Physics (to be published).*

[48] M. Ramuz *et al.*, Method for efficient coupling of a light source into an optical waveguide and its applications for optical sensing.

[49] N. A. Mortensen, S. Xiao and J. Pedersen, Liquid-infiltrated photonic crystals: Enhanced light-matter interactions for lab-on-a-chip applications. *Microfluid. Nanofluid.* **4**, 117–127 (2008).

[50] P. J. Schuck *et al.*, Improving the mismatch between light and nanoscale objects with gold bowtie nanoantennas. *Phys. Rev. Lett.* **94**, 17402 (2005).

[51] V. Lien *et al.*, High-sensitivity cytometric detection using fluidic-photonic integrated circuits with array waveguides. *IEEE J. Sel. Top. Quantum Electron.* **11**, 827 (2005).

Chapter Twelve

Electrical Detection in Microfluidic Flow Cytometers

Marco Di Berardino
Leister Process Technologies, Axetris Division, Schwarzenbergstrasse 10-12, CH-6056 Kaegiswil, Switzerland
marco.diberardino@leister.com

Electrical detection in microfluidic devices provides a means of enumerating, sizing, and characterizing particles or cells without fluorochromes or other dyes. A multi-parametric measurement is supplied by using alternate current over a broad radio frequency range and simultaneous impedance measurement at various frequencies. The sensitivity of the device is dependent on the microchannel dimensions and position of the microelectrodes. Results obtained on different cell models show that the system can be a valuable alternative to traditional fluorescence-based cytometers for many applications, where costs, speed and ease of use are more important than specificity and multiparametry.

12.1 INTRODUCTION

One of the first parameters measured using a flow cytometric approach was the electronic cell volume or Coulter volume.[1] Early devices were able to count and size small particle or cells. The measurement based initially on a direct current (DC) electrical resistance change caused by particles or cells passing through a small aperture placed between two electrodes. From the notion that all living cells possess electrical properties which are characteristic of their physiological status, it became evident that such properties, typically expressed as membrane capacitance (C_{mem}, μF/cm^2), membrane resistance (R_{mem}, Ω cm^2), and cytoplasmic resistivity (R_i, Ω cm), could also be addressed by electrical sensing. These parameters were interrogated with improved instruments, that measured with alternate current (AC) the impedance of cells in the radio frequency range, confirming the high potential of this technology.[2,3] The small progress obtained since then, however, does also emphasize the challenge and technical limitations to obtain "high-resolution"

The Microflow Cytometer by J S Kim & F S Ligler

www.panstanford.com
978-981-4267-41-0

data from this technology. Moreover, as an alternative to the electrical methods, fluorescent-based cell analysis and sorting devices (so-called FACS, fluorescence activated cell sorters) emerged and constantly improved in the past 30 years. With a plethora of specific fluorescent markers these instruments provided accurate high-throughput screening and sorting, and thereby degraded impedance measurements to a rather complementary technology.

Nevertheless, conventional flow cytometry does also bear several disadvantages: on the one hand the equipment is, despite the technological advances in the last decade, still rather expensive and complex to operate. On the other hand, it demands the use of fluorescent cell markers or antibodies, which complicates the whole analysis even more, because (i) the labeling procedure takes time and increases the gap between sampling and analysis, (ii) cells are modified through the labeling procedure and might not be used for downstream processes, and (iii) cell markers are either quite expensive or even not yet available for certain applications.

With the advent of micro-fabrication technologies in the last decade, a new attempt was started in the development of impedance-based single-cell analysis. This novel technology addressed some issues of conventional flow cytometry, such as portability, low cost operation, and compatibility with high-throughput and low-volume analysis, and promised a boost in sensitivity compared to that of macro-scale impedance devices. The first micro-Coulter devices were presented by Larsen *et al.*[4] and Koch *et al.*,[5] but were still limited to DC or low AC frequency measurements. An impedance spectrum of single cells in micro-machined channels was measured first by Ayliffe *et al.*,[6] but only later studies manifested the high potential behind these micro-fabricated devices.[7–11]

Of course, other cell analysis technologies using impedance measurements have long ago attained commercial status. Besides the already mentioned Coulter counters, however, only semi- and fully automated patch-clamp devices are operating at the single cell level. These instruments are mainly used for electrophysiological measurements on ion channels, which pharmaceutical companies consider with other membrane proteins as the most promising drug targets. Despite the low throughput of the patch-clamp technology, it is still regarded as the gold standard and as an indispensable tool in the drug discovery process.

For about 20 years, impedance measurements provided reliable results for online biomass monitoring in cell suspensions, for example in fermenters. The measurements are done with AC in the radio-frequency range and take advantage of the fact that only living cells are contributing to the signal.[12] A similar approach, but directed towards adhesive cell populations growing in microplates with integrated sensor arrays patterned onto the bottom of the wells, monitors confluency and can be indirectly used, for instance, for cytotoxicity studies.[13,14] Both methods, however, deliver only averaged values of the whole cell population and are not recording data at the single-cell level.

A detailed description of these technologies would definitely go beyond the scope of this book, even more, since there is relatively little coherence to classical flow cytometry. This chapter, therefore, will focus on electrical impedance-based single-cell analysis performed on micro-fabricated devices in a flow-through manner. It will, on the one hand, emphasize advantages, issues and challenges of miniaturizing microfluidic channels and electrodes. On the other hand, it will illustrate some aspects of moving from an academical feasibility study to a commercial product, and thus bestow to single-cell microflow impedance analysis more than just a proof-of-principle status.

12.2 IMPEDANCE MICROFLOW CYTOMETRY

The pretensions of impedance microflow cytometry as a single-cell analysis tool are easily benchmarked to the present achievements and nourished by the prospect that microtechnologies might provide smaller (portable), faster, simpler, and cheaper solutions. It is mandatory, therefore, to supply a realistic view of the potential of this technology, be it as an inheritance from Coulter counter technology or as alternative to fluorescence-based flow cytometry. Ideally, from an academic perspective, a novel device would combine all possible features that are presently addressed by those technologies: counting and sizing cells, determining cell concentrations, and characterizing and sorting cells, and of course performing all these operations in a fast, high-throughput and label-free manner. Whether all these expectations can be met is unclear. This section shall give a basic overview about the technology.

12.2.1 Principles of Measurement

Basically, the impedance measurement consists of determining the impedance change ΔZ caused by a particle or cell passing through a sensing region in a microfluidic channel filled with a conductive liquid. The sensing region is defined by the microelectrodes placed typically on the channel walls. The variation of impedance is best determined by a differential impedance measurement, which compares the cell-derived impedance with the reference impedance of the surrounding medium (Fig. 12.1). Preferably, the reference measurement is performed close to the cell measurement in the same microchannel and not in a separate reference channel, such that cell and reference impedances are measured under the very same conditions. By switching reference and measurement electrodes together, it is also possible to detect and correct drifts of the electrode properties, since the signal amplitude of both electrode pairs should be the same. Moreover, from the signal amplitudes, it is also possible to determine the flow speed of the particles and information about the flow profile of the measured particles. This information provides also a means for controlling the active flow element (pump) of the device.

In order to understand the electrical response of a cell in suspension, theoretical models have been previously drawn and analyzed. For the determination of the electrical impedance of a cell suspension, Pauly and Schwan modeled cells as spherical cytoplasm encircled by a thin shell.[15] This single-shell model

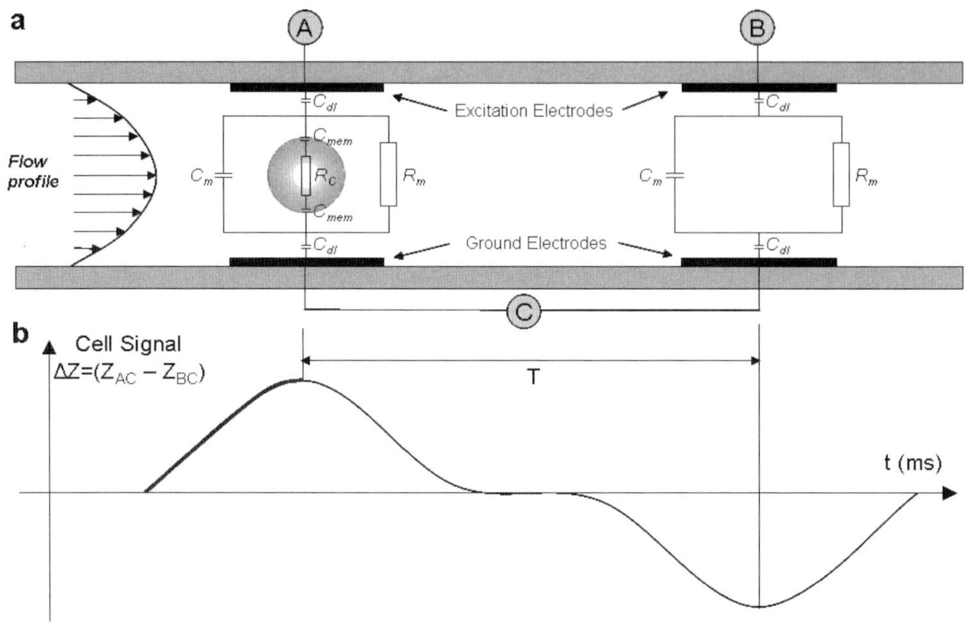

Figure 12.1. (a) Simplified electrical model for a single cell suspended in a microfluidic channel supplied with two pairs of parallel facing microelectrodes, the measurement (AC) and reference electrodes (BC). (b) Differentially measured impedance signal $\Delta Z = Z_{AC} - Z_{BC}$. The particle speed can be calculated through the transit time T and the known electrode distance. Color reference – pg. 348.

characterized electrical properties of cells like mammalian erythrocytes, which contain neither nuclei nor other cellular organelles. Asami *et al.* demonstrated that the dielectric behavior of lymphocytes (having a nucleus) differed from that model and proposed a double-shell model that reflected this fact.[16] Fig. 12.1 illustrates an equivalent circuit model, as proposed by Gawad *et al.*[8] or Sun *et al.*,[11] in which the low conductivity membrane of a viable cell is modeled as a capacitor (C_{mem}) and its cytoplasm as resistor (R_c), while the impedance of the medium is represented by a resistor (R_m) and capacitor (C_m) in parallel. C_m and the capacitance of the cytoplasm (not drawn) can be neglected under the measuring conditions (cytoplasm and medium are normally conductive). Additionally, the model considers also the capacitive electrode-medium interface effect, known as double layer capacitance (C_{dl}).

If, in this system, an AC electric field is applied, biological cells become polarized as a result of charge accumulation at the limits between the insulating plasma membrane and the aqueous medium. This interfacial polarization goes along with the formation of dipoles, which are characterized by a specific time constant and are therefore dependent on the frequency of the applied AC. Thus, at low AC frequencies, maximum polarization can easily be achieved in a short time, while at higher frequencies, the rate is too high for the dipoles to attain full polarization. This phenomenon has an important impact when measuring the impedance of a cell suspension or of a single cell. Since the insulating properties of cell membrane represent a substantial obstacle to current flow at low frequencies (below 500 kHz), the cell is principally non-conducting, reflecting essentially the Coulter volume measurement.[2] However, at intermediate frequencies (around 500 kHz to 6 MHz) the plasma membrane polarization decreases leading to a decrease of the capacitance of the suspension—an effect known as β-dispersion or dielectric relaxation. Measurements in this range normally provide information about the electrical properties of the plasma membrane. At high frequencies (6–20 MHz), a polarization of the plasma membrane is almost nonexistent. Under these conditions, the cell membrane does not represent a barrier to the current, and the measurements rather provide information about the cytoplasmic conductivity.

Figure. 12.2 shows scatter plots of an impedance measurement of polystyrene beads and red blood cells. The complex value of the impedance signal is usually

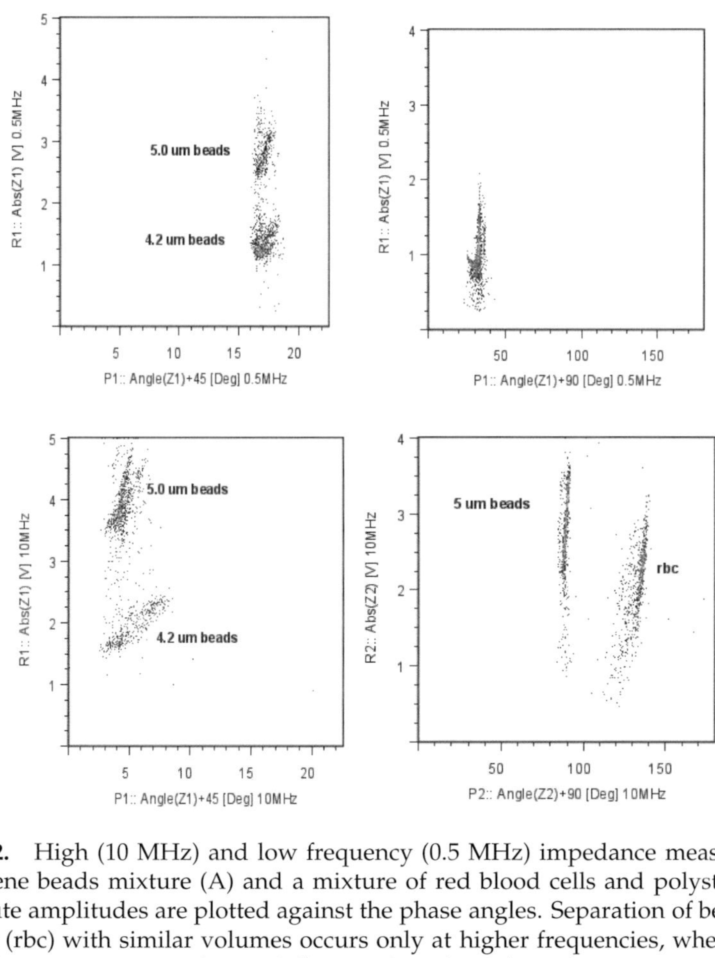

Figure 12.2. High (10 MHz) and low frequency (0.5 MHz) impedance measurements of a polystyrene beads mixture (A) and a mixture of red blood cells and polystyrene beads (B). Absolute amplitudes are plotted against the phase angles. Separation of beads and red blood cells (rbc) with similar volumes occurs only at higher frequencies, where the phase angles of the respective impedances differ significantly. Color reference – pg. 348.

plotted either as real part versus imaginary part or as absolute value versus phase angle. In contrast to fluorescence-based flow cytometry, which relies on light intensities and provides values at the logarithmic scale, impedance measurements produce linear voltage values. Another valuable option to exhibit impedance data is the use of the so-called opacity, which represents the ratio of the impedance magnitude at a high frequency to a low frequency ($|Z_{high}|/|Z_{low}|$) and determines the difference in particle resistivity at high and low frequencies.[3] Opacity has been shown to be essentially independent of the cell size and cell position between the measurement electrodes in the microfluidic channel.[8] As shown in Fig. 12.2, polystyrene beads behave as insulating particles (capacitors) over the whole frequency spectrum and are easily discriminated in the impedance microflow device described in Fig. 12.1. Cells also behave as capacitors at the low frequency range and can be discriminated from beads with a similar size (volume) only at higher frequencies.

The microfluidic devices described so far for flow-through impedance measurements were equipped with external function generators (to drive the excitation electrodes) and separate lock-in amplifiers (for the demodulation and amplification of each frequency), which allowed for manual AC single-frequency measurements through sequential sweeping over a broad frequency range (usually from 100 kHz to 15 MHz). Measuring the impedance of a cell at two different frequencies simultaneously meant using at least two lock-in amplifiers and some function generators (also for DEP, see section 12.2.3), making the system large, expensive and not really suitable for commercialization. An alternative to this set-up was presented by Sun *et al.*, who demonstrated the use of the maximum length sequences (MLS) technique to discriminate polystyrene beads of different size without the need of expensive hardware.[18] Unfortunately, the maximum measurement frequency was limited to 500 kHz and performed on beads only. Results with measurements of biological cells at higher frequencies have not yet been published.

12.2.2 Chip Design

For obtaining the best results for electrical impedance measurements, the most crucial parameter is undoubtedly chip design. Critical issues are sensitivity and reproducibility of the sensor as well as interfacing the microfluidic device with peripheral modules (electronics, fluidic system, etc.).

12.2.2.1 Chip Microfabrication

Principally, the chip manufacturing process used for many microflow impedance devices consists of patterning the microelectrodes onto glass wafers, then structuring the channel walls into an overlaying polymer layer. The metal of choice of the microelectrodes is platinum, which is much more resistant than gold to electrochemical degradation.[19] Depending on the desired electrode placement (see section 12.2.2.2), the micro-channels are either sealed with a PDMS cover (coplanar design) or by thermal polymer bonding of two axially symmetric chips after aligning the facing structures (parallel facing electrodes, for details see Cheung *et al.*[9]). Adhesive polymers used for structuring the channels are usually polyimide, SU-8 or BCB (benzo-cyclo-butene), which have proven to be chemically resistant and to offer optical transparency and excellent dielectric properties. Compared to the PDMS cover solution, polymer bonding is a time-consuming and expensive process. To access the channels requires drilling holes in the glass substrate, which is an additional step reducing the yield per wafer. On the other hand, bonded chips can withstand higher fluid pressures, which would benefit high-throughput applications that demand higher flow rates.

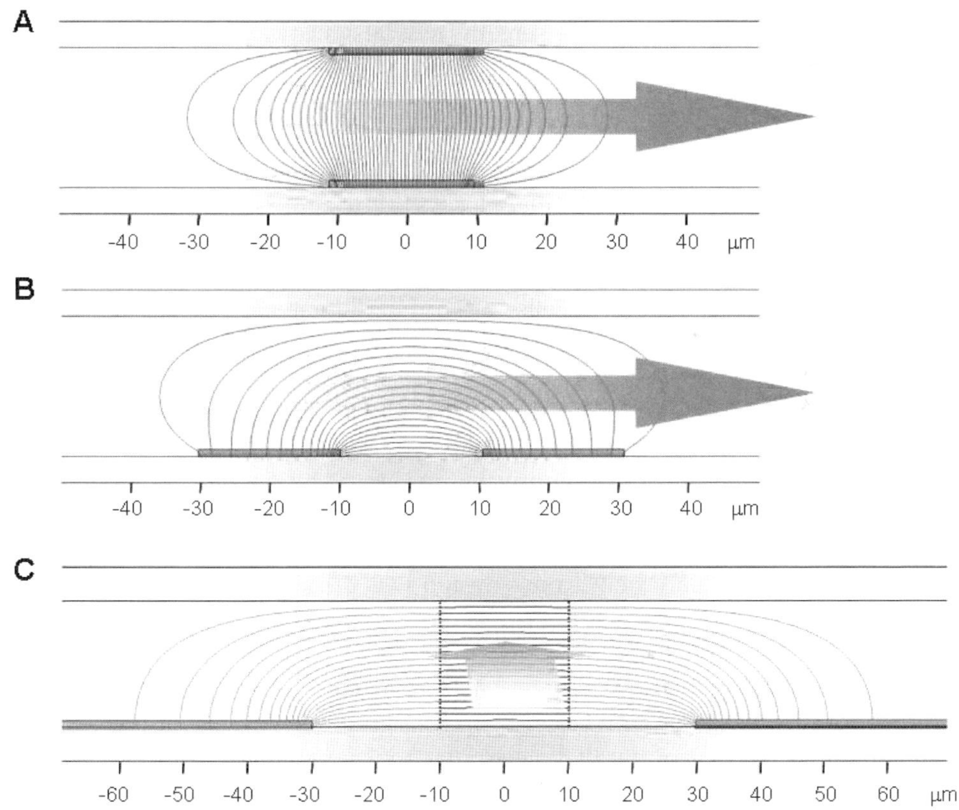

Figure 12.3. Schematic electric field density simulations performed with Comsol Multiphysics v3.5 on typical electrode designs for impedance measurements. (A) and (B) are side views into the measuring channel with facing parallel electrodes and coplanar electrodes (width = 20 µm), respectively. (C) Front view into polymer-structured microchannel (yellow side walls) of a liquid electrode design, in which the microelectrodes are not placed in the measurement channel. [Y.-C. Tung, Y.-S. Torisawa, N. Futai and S. Takayama, *Lab Chip* **7**, 1497 (2007). Reproduced by permission of The Royal Society of Chemistry.] Color reference – pg. 349.

12.2.2.2 Microelectrode Design

The electrode geometries within a microfluidic channel define the electric field distribution and consequently influence many aspects of the impedance measurement. Moreover, the position of the microelectrodes can have an important impact on the fabrication process of the chips. Fig. 12.3 illustrates some typical electrode arrangements and the respective field distribution. Typically, these microelectrodes are situated within the microchannel and are in direct contact to the medium and potentially to the cells as well. Sun *et al.* have compared the sensitivity of parallel facing electrodes versus coplanar electrodes using the Schwartz-Christoffel Mapping (SCM) method.[11] It was demonstrated that the sensitivity of the sensor (impedance change caused by a cell) increases with a smaller detection volume, defined by the electrode width and channel height, providing evidence that the parallel electrode design was more sensitive. Reducing the electrode width for improving the sensitivity is also critical, because small electrodes have also a lower interfacial capacitance (C_{dl}), increasing thus the system impedance and limiting the sensitivity in the low frequency range. Moreover, smaller electrodes are subject to a greater degree of electrochemical degradation than larger ones.

Another advantage of the parallel electrodes design as compared to the coplanar approach is that it provides a higher field homogeneity over the channel height. This field distribution is more tolerant to the position of the cell in the detection volume. Thus, for most cells (> 5 µm in diameter), it was demonstrated in our laboratory (data not shown) that focusing the particles in the middle of the channel had a narrowing effect on the transit time profile, but did not affect the signal magnitude, as observed for the coplanar electrodes design.[7] These observations are supported by particle flow velocity studies, which show that particle speed is constant over almost the entire channel cross section, declining only close to the cannels walls.[20]

A different approach (Fig. 12.3(c)) for obtaining a more homogenous field distribution with a coplanar electrodes design was presented by Demierre *et al.*[21] These so-called liquid electrodes provide a more regular field in the detection volume. Furthermore, the large metal electrodes have a longer life-time, no direct contact with the cells, and a larger interfacial capacitance, extending thus the measurement range to lower frequencies. Even though the sensitivity of this solution is still smaller than that of parallel electrodes, the signal amplitudes are also less susceptible to the position of the cell in the detection volume. Küttel *et al.* have already shown that this design approach promises sensitivities for many specific applications.[22]

12.2.2.3 Cell Focusing

An accurate positioning of the cell in the detection volume of the microfluidic channel is mandatory for reproducible impedance measurements (depending on the electrode design, see section 12.2.2.1). Several techniques for focusing cells in a microfluidic channel have been proposed. The most frequently adopted method is hydrodynamic focusing using a sheath fluid. This method is rather complex in terms of fabrication and two-dimensional fluid control if applied for a microfluidic device. For a coplanar electrode design, however, it is the only way for providing precise cell alignment. Rodriquez-Trujillo *et al.* have presented an elegant solution for two-dimensional, adjustable hydrodynamic focusing.[23] Importantly, using (non-conducting) deionized water as sheath fluid can also be exploited to reduce the (conducting) impedance detection volume, increasing thereby the sensitivity of the chip.

An alternative electrokinetic approach, called dielectrophoresis (DEP), has also been shown to effectively focus particles in microfluidic devices.[24,25] One advantage of this solution is that there is no need of sheath fluid, simplifying considerably the peripheral fluidic system. The dielectrophoretic force F_{DEP}, moreover, directly acts on the particles rather than on the fluid. Jones *et al.* have shown that this force is proportional to the particle volume and to the gradient of the electric field intensity squared.[26]

Depending on the polarizability of the cells, which is related to the frequency of the applied AC field, DEP can attract (positive DEP) or repel (negative DEP) particles. For focusing purposes, negative DEP is required to push cells in the center of the channel. This can be only achieved using electrodes that are placed one opposite to the other. Figure 12.4 illustrates this focusing solution for the parallel electrodes design. Since different cells can experience a different polarizability in the same electric field, DEP can also be exploited for sorting purposes as separation device, either without[27] or more selectively, after impedance analysis (Fig. 12.4).

The smaller the particle to be analyzed is, the more important it becomes to center the particle in the channel. Unfortunately, the electric field gradients required for DEP focusing are relatively high, even higher for smaller particles. High voltages can heat up the medium and lead to the formation of gas bubbles which perturb the subsequent impedance measurement. Furthermore, a temperature rise in the medium can possibly damage the cells and affect the results of the impedance measurement. The temperature rise is proportional to the conductivity of the medium (σ_m),[28] which is usually a phosphate buffer (such as PBS) with $\sigma_m \approx 1.5\,\mathrm{Sm}^{-1}$. If higher voltages are applied,, this could easily have negative effects on the viability of the cells. Alternatively, impedance measurements demand a medium with a relative high σ_m for optimal sensitivity. Therefore, using DEP for focusing the cells upstream of an impedance measurement is always a trade-off among the parameters electric field strength, buffer conductivity and electrode design.

Another rather recent technology for manipulating cells is by means of ultrasonic standing waves (USW), which can be used for focusing, separating and concentrating cells.[29,30] USW have already been combined with DEP but not with impedance measurements, and therefore, it is not known yet whether USW could affect the highly sensitive impedance measurements.

12.3 FUTURE DEVELOPMENTS IN IMPEDANCE-BASED MICROFLOW CYTOMETRY

Presently, the technology related to chip-based impedance flow cytometry has reached a stage that goes beyond basic research. Despite all the efforts spent in this research area, there are no single-cell analysis devices that have attained commercial status. A first step in this direction has been done by Leister Process Technologies, which has developed an advanced prototype (Fig. 12.6) that is close to commercialization. In the following section, developments that were necessary for reaching that stage will be presented. In a first phase it is essential that the technology prove sensitive enough for cell characterization applications. Among the described potential chip designs available, Leister has chosen the parallel facing electrode approach (Fig. 12.1 and 12.3(A)), because this is currently the most sensitive solution. All data presented in this article were obtained with this chip design and 20 × 20 μm channel dimensions. In a second phase, the prototype has to be optimized such that it can persist in a quite competitive environment, addressing consequently robustness, reliability and ease of use of the system. Future improvements can rely also on concurrent academic activities that will boost throughput and sensitivity of impedance-based microflow cytometry towards new dimensions.

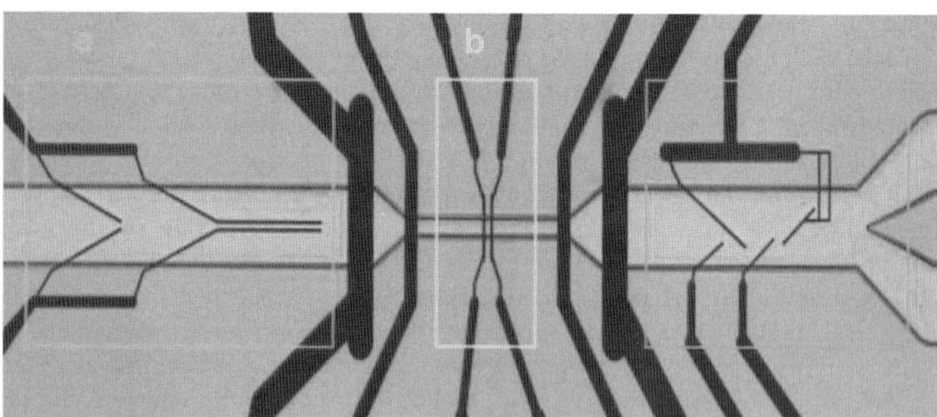

Figure 12.4. Schematic top view of a typical impedance chip used at Leister. The parallel electrodes design is based on studies made by Cheung *et al.*[9] The chip is divided in three functional zones: (a) Focusing region, where particles are centered by negative DEP in the microfluidic channel (200 × 20 μm, w × h), (b) measurement region (channel dimensions are 20 × 20 μm), where the differential impedance measurement occurs, and (c) a sorting region, in which negative DEP can be used for a potential sorting of particles by switching the respective electrodes. The large electrodes in between these three regions are shield electrodes required for avoiding disturbances of the impedance signal caused by the focusing or sorting functionality. Color reference – pg. 349.

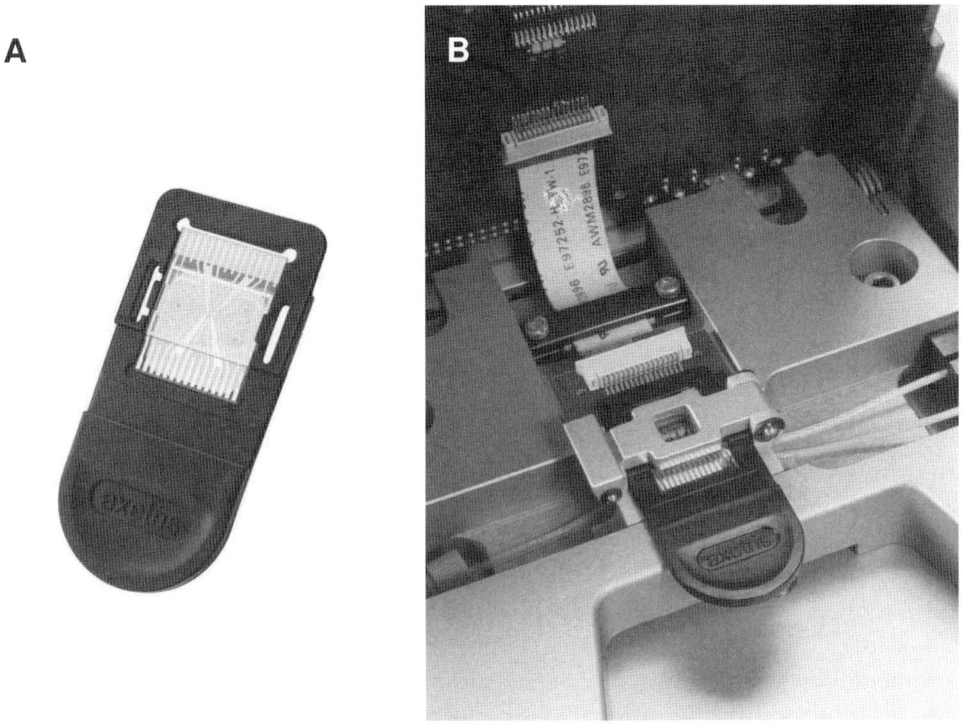

Figure 12.5. Interfacing microfluidics with the macro-world. (A) Impedance chip with parallel electrodes design in chip holder for simplified handling. (B) Interconnection of chip with fluidic system and electronics.

12.3.1 Interfacing Microfluidics

12.3.1.1 Chip Handling

As mentioned in the previous section, the sensitivity of the impedance sensor is mainly dependent on the detection volume. Consequently, in order to obtain the optimal sensitivity for a specific application, the channel dimensions should be adapted to the size of the cells under investigation. It is almost impossible, for example, to distinguish a small bacterial cell (1 μm in diameter) from the noise level using a chip with a channel aperture of $20 \times 20\,\mu m$. A user-friendly and reliable interchangeability of the chips with various channel dimensions is, for such cases, a must. Fig. 12.5 shows an impedance chip (dimensions $20 \times 15\,mm$) framed in a versatile chip holder, which can easily be connected to the systems fluidics and electronics. The interconnection to the fluidic system has to be tight and withstand pressures of several bars. Unfortunately, there are as many innumerable fluidic connector solutions to microfluidic devices as groups working in the field, because adaptations to the specific requirements are inevitable.

12.3.1.2 Fluidic System

For reproducible impedance measurements in a microfluidic environment, a slow and regular flow (pulsation-free) is essential. In our standard measurement channel ($20 \times 20\,\mu m$), flow rates of about 0.5–2.0 μl/min are used, corresponding to a particle speed around 60 mm/sec, for which the flow is still laminar. At these flow rates, it can be assumed that the shear stress is not affecting the cells, as was shown for erythrocytes traveling with 2.5 m/sec in a 100 μm channel.[31] The channel dimensions make high demands to the pumping system. Even though channel blockage can be prevented by using mesh filters or filtering the sample before loading, it is still possible that cell aggregations temporarily clog the chip. In such cases the pump must supply overpressures of several bars in order to get rid of transient channel obstructions. Overpressures are also demanded for flushing/cleaning the system in a reasonable time frame. The pressure-driven flow used by FACS instruments might not be the desired solution for a system that should become portable and for which sheath fluid focusing is not required. Micro annular gear pumps or syringe pumps are alternatives that could meet the requirements of flow-rate control.

A critical aspect of fluid transport in microfluidic devices is that, at these rather low flow rates, particles and cells tend to sediment according to their size. Sedimentation occurs even before the sample has reached the chip, which can significantly modify the original cell concentration of the sample. Since a reliable determination of cell concentration is almost a must for all flow cytometers, solutions are demanded which avoid long tubing connections to the chip and other intermediate fluidic elements (*i.e.* valves) that increase the dead volume. Lee *et al.* presented a microfluidic solution independent of flow rate for the determination of cell concentration by impedance measurement.[32] However, this solution works only if the concentration at the inlet is equal to that of the sample. The possibility to record the transit time of the particles with our device provides a means of indirectly determining the flow speed and, thus, the cell concentration of the sample without the need of a flow sensor.

Considering the low flow rates used for impedance measurements, the measurable sample volume comes into question. A device that would necessitate absolute sample volumes larger than 200 μl for analysis would be difficult to implement.

12.3.2 Data Acquisition and Analysis

Even though it is typically a specific application that defines the required performance of a flow cytometer, instruments manufacturer tend to praise their cytometers (besides the number of lasers and colors) according to the maximum cell analysis or cell sorting rate. Popular values for cell analysis are currently around 100,000 events per second. So far, the microflow impedance cytometers reach values around 1,000 events per second for counting and even less for characterizing cells.[9] Normally, this is not an intrinsic limitation of electrical sensing,

Figure 12.6. Impedance microflow cytometer prototype. Impedance measurement is performed at four frequencies simultaneously in the range from 300 kHz to 20 MHz. Color reference – pg. 350.

but rather a result of using commercially available electronics and data acquisition boards based on analog technology.

One of the most important milestones for developing an impedance-based microflow cytometer was the replacement of the expensive lock-in amplifiers and function generators with dedicated electronic boards and thereby switching to digital data processing. This was achieved through direct digital synthesis for frequency generation and RF mixers for the demodulation by using Field Programmable Gate Array (FPGA) technology. Data acquisition via analog acquisition board was replaced with a USB interface. Beside the reduction of the noise level, which improved the sensitivity of the system significantly, the prototype provided simultaneous measurements at four different frequencies. Hence, size (volume), membrane capacitance and cytoplasmatic conductivity can be interrogated simultaneously for every single cell. Similar to fluorescence-based cytometry, in which the use of different dyes and markers provide a multiparametric measurement, such multifrequency analysis supplies extensive information about the electrical properties of a cell. Now, the limits of performance are defined by the sampling rate and the decimation filters set in the FPGA, and by the electrode design used. Thus, currently it is possible to measure with a sampling rate of about 200 kHz, which allows for a theoretical analysis rate of about 40,000 events/sec. The differential impedance measurement obtained with two electrode pairs is beneficial for ground level stability and signal amplification. The performance is limited by the fact that only one cell at the time should reside within the microchannel volume, defined by the distance of the electrode pairs, for a correct impedance measurement. With the implementation of an USB interface and the resulting elimination of AC filters, it is now possible to obtain high quality signals without a differential measurement enabling the use of only one electrode pair and consequently an increase of the maximal cell density. Moreover, the dynamic range was amplified by an order of magnitude. The whole device is reduced to a compact desktop instrument, which means a first step towards portability. Finally, the acquired data can be converted into the standard fcs3-format – the latest version of the standard data file format, providing the specifications required to completely describe data obtained from flow cytometric measurements[33] and thus enabling the use of commercially available software tools.

12.3.3 Applications

Until recently, impedance measurements on single cells were only used for cell counting and sizing applications. Miniaturization would not justify all the efforts put into this novel technology if it was not possible to obtain additional information out of these measurements. Even though counting and sizing is still possible with microflow impedance cytometry, this section will focus on the new possibilities that it promises.

The most important advantage of impedance-based microflow cytometry is the significant rise in sensitivity using the microfluidic approach. This, combined with the multi-frequency measurement, extends the application field beyond simple counting and sizing. Many cell characterization applications can now be addressed with the present impedance flow cytometer prototype. Another benefit of electrical label-free measurements is that cells do not need an extensive and time-consuming labeling procedure before measurement. Thus, the impedance measurement can be executed right after sampling and, not rarely, directly in the cultivation medium. Moreover, fluorescent dyes are still quite expensive and significantly affect assay costs. After impedance analysis, the cells can easily be recovered and used for downstream processes, since measurement and focusing do not alter the cells. Label-free also means no specificity. The technology is consequently best suited for near-inline routine and quality control applications in which the analyzed system is already known. Some examples in the fields of cell differentiation, viability, microbiology and parasitology (see also Figs. 12.7 and 12.8) indicate in which application fields impedance microflow cytometry could first gain ground.[34]

Another potential application field that could be addressed consists of all those applications, in which conventional flow cytometry fails to provide a solution because there are no or only unspecific fluorescent dyes. For some bacterial species, for example, it was shown that standard propidium iodide (PI) staining failed as dead cell indicator.[35] Alterations of mechanical membrane properties are known to play a role in the pathophysiology of hematologic diseases such as sickle cell ane-

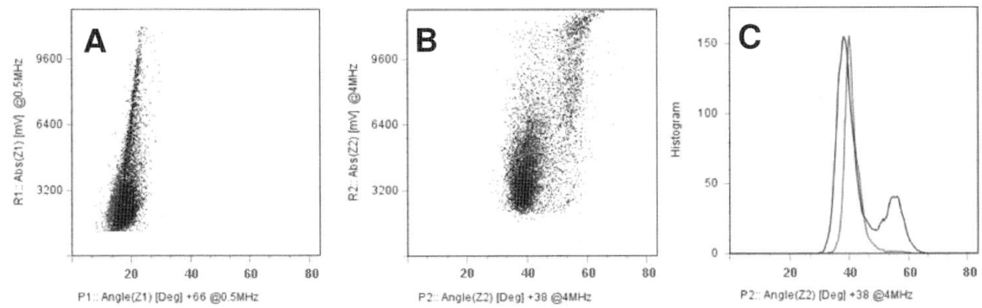

Figure 12.7. Cell differentiation study with leukemic cell line U937. The monoblast-like cells were treated with TPA (Tetradecanoylphorbol-acetate) for 48 h and the impedance signal determined at 0.5 (A) and 4.0 MHz (B). Red dots represent untreated, blue dots TPA-treated cells. At higher frequencies, the treated cell culture splits into two populations, one of which represents the differentiated macrophages with a fraction of about 24%. (C) Histogram from measurement at 4 MHz. Color reference – pg. 350.

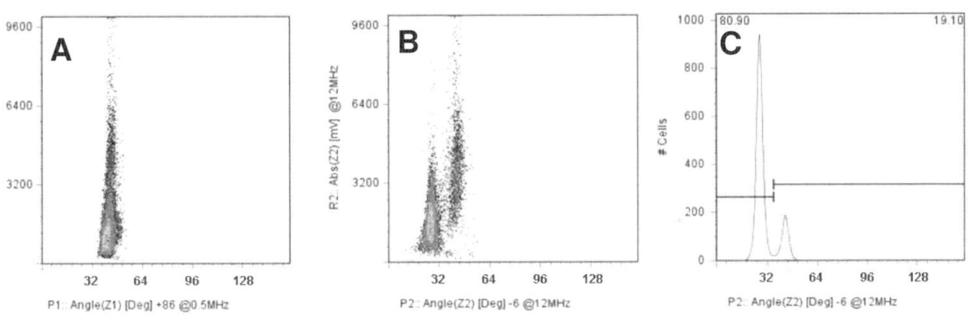

Figure 12.8. Viability study with bakers yeast *Saccharomyces cerevisiae carlsbergensis*. An aliquot of an exponentially growing yeast culture was heat inactivated for 10 minutes at 70°C and then mixed with an aliquot of living cells. The mixture was then measured on Axetris' microflow cytometer at 0.5 MHz (A) and 12 MHz (B). Dead cells show a clear phase shift at 12 MHz and can thus be separated from vital cells. (C) Histogram from measurement at 12 MHz indicating the ratio of dead versus vital cells. Color reference – pg. 350.

mia, sepsis, and diabetic retinopathy. These diseases were analyzed by biophysical flow cytometry because they cannot be measured with FACS instruments.[36] Impedance microflow cytometry could prove to be a more reliable diagnostic tool for these applications as well as for those cases in which the permeability of the cells is increased by parasite infection.[37,38] Data obtained by Küttel *et al.* on *Babesia*-infected bovine red blood cells indicate the potential of our system for future diagnostic applications.[22]

Some applications are not accessible for impedance-based cytometry, because of the lack of specificity that results from the label-free measurement. Thus, rare event analyses are currently not believed to enrich the application portfolio of an impedance-based cytometer. Another limitation of the micro-devices is the low flow rate, which precludes the analysis of large sample volumes.

12.4 CRITICAL ISSUES

The results obtained so far with our impedance microflow cytometer are promising and indicate that this technology is suitable for single-cell characterization. Nevertheless, there are still some technological issues that need to be solved. The sensitivity of the impedance sensor represents a crucial factor. If small cells like bacteria need to be analyzed, reduction of the detection volume is necessary. Narrowing the measurement channel to dimensions smaller than 20 μm × 20 μm does also significantly affect the microfluidic flow properties (pressure, particle velocity, etc.) and might require adaptations to pump, cell focusing concept and electrode design. Moreover, the fabrication of the most sensitive impedance sensor (parallel facing electrodes) is still a highly complex process. Another more general problem of impedance microflow cytometry is the fluidic and electric interconnection of the chip to the periphery of the instrument. The analysis of different types of cells (bacteria, yeast or mammalian cells) with the same instrument will demand the use of sensors with optimal sensitivity and as a consequence a robust and reliable exchange of sensors, the implementation of which could prove to become a challenge.

12.5 CONCLUSIONS AND OUTLOOK

This chapter provides evidence that impedance microflow cytometry has moved beyond the purely academic environment. Besides counting and sizing functionalities, this non-invasive and label-free technology characterizes cells through their electrical properties using a multi-frequency measurement over a broad radio frequency range. Many applications that rely on the capabilities of conventional flow cytometry are now accessible with a simpler device. Nevertheless, impedance-based flow cytometry will not compete, in the short- and mid-term, with this established technology, mainly because it does not offer specificity. Yet, many single cell analyses do not demand this level of specificity and fast and cost-effective impedance measurements could provide a valuable alternative to fluorescence-based flow assays. It is also conceivable to combine the impedance and fluorescence measurement on the same device, for which the presented impedance sensors would likely work. This could enable the operator to choose between the most effective measurements or even to use the impedance analysis as a prediagnostic measurement before starting a time-consuming labeling procedure.

The potential for improvement in the field of Microtechnology is anything but exhausted. The sensitivity of the sensors is likely to be increased further as a result of enhanced chip fabrication techniques or through the development of more powerful electronics components. This could also boost the development of a sorting capability on the impedance chip, which would be an important advantage compared to FACS. The integration of other components on the chips, such as micropumps and microvalves, is no longer an unrealistic intellectual game. These microdevices will shrink the current peripheral modules interfacing with the microfluidic sensors and propose solutions aimed at generating portable devices. In the near future, conventional flow cytometers will become less expensive and much easier to use and thus find their place in almost every cell analysis laboratory, while cell sorters will be rather confined to specialized FACS facilities. Impedance microflow cytometers could play an important role in this trend, however, not the technology but the application and the need to solve a specific problem will decide about its commercial success.

References

[1] W. H. Coulter, High speed automatic blood cell counter and cell size analyzer. *Proc. Natl. Electron Conf.* **12**, 1034–1040 (1956).

[2] R. A Hoffman and W. B. Britt, Flow-system measurement of cell impedance properties. *J. Histochem. Cytochem.* **27**, 234–240 (1979).

[3] R. A. Hoffman, T. S. Johnson and W. B. Britt, Flow cytometric electronic direct current volume and radiofrequency impedance measurements of single cells and particles. *Cytometry A* **1**, 377–384 (1981)

[4] U. D. Larsen, G. Blankenstein and S. Ostergaard, Microchip Coulter particle counter. *Proceedings in Transducers* 1319 (1997).

[5] M. Koch, A. G. R. Evans and A. Brunnschweiler, Design and fabrication of a micromachined Coulter counter. *Proceedings in Micromechanics Europe* 155–158 (1998).

[6] H. E. Ayliffe, A. B. Frazier and R. D. Rabbitt, Electric impedance spectroscopy using microchannels with integrated metal electrodes. *IEEE J. MEMS*, **8**, 50–57 (1999).

[7] S. Gawad, L. Schild and P. Renaud, Micromachined impedance spectroscopy flow cytometer for cell analysis and particle sizing. *Lab Chip* **1**, 76–82 (2001).

[8] S. Gawad, K. Cheung, U. Seger, A. Bertsch and P. Renaud, Dielectric spectroscopy in a micromachined flow cytometer: Theoretical and practical considerations. *Lab Chip* **4**, 241–251 (2004).

[9] K. Cheung, S. Gawad and P. Renaud, Impedance spectroscopy flow cytometer: On-chip and label-free cell differentiation. *Cytometry A* **65**, 124–132 (2005).

[10] H. Morgan, D. Holmes and N. G. Green, High speed simultaneous single particle impedance and fluorescence analysis on a chip. *Curr. Appl. Phys.* **6**, 367–370 (2006).

[11] T. Sun, N. G. Green S. Gawad and H. Morgan, Analytical electric field and sensitivity analysis for two microfluidic impedance cytometer designs. *IET Nanobiotechnol.* **1**, 69–79 (2007).

[12] J. P. Carvell and J. E. Dowd, On-line measurements and control of viable cell density in cell culture manufacturing processes using radio-frequency impedance. *Cytotechnology* **50**, 35–48 (2006)

[13] K. Solly, X. Wang, X. Xu, B. Strulovici, and W. Zheng, Application of real-time cell electronic sensing (RT-CES) technology to cell-ased assays. *Assay Drug Dev. Technol.* **2**, 363–372 (2004).

[14] J. M. Atienza, J. Zhu, X. Wang X. Xu, and Y. Abassi, Dynamic monitoring of cell adhesion and spreading on microelectronic sensor arrays. *J. Biomol. Screen.* **10**, 795–805 (2005).

[15] H. Pauly and H. P. Schwan, Über die Impedanz einer Suspension von kugelförmigen Teilchen mit einer Schale. *Z. Naturforsch.* **14B**, 125–131 (1959).

[16] K. Asami, Y. Takahashi and S. Takashima Dielectric properties of mouse lymphocytes and erythrocytes. *BBA* **1010**, 49–55 (1989).

[17] T. Sun, N. G. Green and H. Morgan, Analytical and numerical modelling methods for impedance analysis of single cells on-chip. *NANO: Brief Reports and Reviews* **3**, 55–63 (2008).

[18] T. Sun, D. Holmes, S. Gawad, N. G Green and H. Morgan, High speed multi-frequency impedance analysis for single particles in a microfluidic cytometer using maximum length sequences. *Lab Chip* **7**, 1034–1040 (2007).

[19] E. Holland-Moritz, J. Gordon, K. Kanazawa and R. Sonnenfeld, Reversible oxidation rougheing of Au(111) in aqueous salt solutions. *Langmuir* **7** 1981–1987 (1991).

[20] T. Kim and Y. A. Cho, particle flow velocity profiler using in-channel electrodes with unevenly divided interalectrode gaps. *J. MEMS* **17**, 582–589 (2008).

[21] N. Demierre, T. Braschler, P. Linderholm, U. Seger and P. Renaud, Characterization and optimization of liquid electrodes for lateral dielectrophoresis. *Lab Chip* **7**, 355–365 (2007).

[22] C. Küttel, E. Nascimento, N. Demierre, T. Silva, T. Braschler, P. Renaud and A. G. Oliva, Label-free detection of *Babesia bovis* infected red blood cells using impedance spectroscopy on a microfabricated flow cytometer. *Acta Tropica* **102**, 63–68 (2007).

[23] R. Rodriquez-Trujillo, O. Castillo-Fernandez, M. Garrido, M. Arundell, A. Valencia and G. Gomila, High-speed particle detection in a micro-Coulter counter with two-dimensional adjustable aperture. *Biosens. Bioelectron.* **24**, 290. (2008).

[24] S. Fiedler, S. G. Shirley, T. Schnelle and G. Fuhr, Dielectrophoretic sorting of particles and cells in a microsytsem. *Anal. Chem.* **70**, 1909–1915 (1998).

[25] T. Müller, G. Gradl, S. Howitz, S. Shirley, T. Schnelle and G. A Fuhr, 3-D microelectrode system for handling and caging single cells and particles. *Biosens. Bioelectron.* **14**, 247–256 (1999).

[26] T. B. Jones, *Electromechanics of Particles*, Cambridge University Press, Cambridge (1995).

[27] E. M. Nascimento, N. Nogueira, T. Silva, T. Braschler, N. Demierre, P. Renaud and A. G. Oliva, Dielectrophoretic sorting on a microfabricated flow cytometer: Label free separation of *Babesia bovis* infected erythrocytes. *Bioelectrochemistry* **73**, 123–128 (2008).

[28] A. Ramos, H. Morgan, N. G. Green and A. Castellanos, AC Electrokinetics: A review of forces in microelectrode structures. *J. Phys. D: Appl. Phys.* **31**, 2338–2353 (1998).

[29] M. Wiklund, C. Günther, R. Lemor, M. Jäger, G. Fuhr and H. M. Hertz, Ultrasonic standing wave manipulation technology integrated into a dielectrophoretic chip. *Lab Chip* **6,** 1537–1544 (2006).

[30] O. Manneberg, B. Vanherberrghen, J. Svennebring, H. M Hertz, B. Önfelt and M. A. Wiklund, three-dimensional ultrasonic cage for characterization of individual cells. *App. Phys. Lett.* **93**, 063901 (2008).

[31] V. Kachel. *Flow Cytometry and Sorting*, Melamed, M. R., Lindmo, T. and Mendelsohn, M. L., Eds. Wiley-Liss: New York, 45–80 (1990).

[32] D. W. Lee, S. Yi and Y. Cho, A flow rate independent cell concentration measurement chip using electrical cell counters across a fixed control volume. *J. MEMS* **17**, 139–146 (2008).

[33] Data File Standard for Flow Cytometry, Version FCS3.0, http://www.isac-net.org/content/view/101/150/ (2.01.2009).

[34] G. Schade-Kampmann A. Huwiler M. Hebeisen T. Hessler and M. Di Berardino, On-chip non-invasive and label-free cell discrimination by impedance spectroscopy. *Cell Prolif.* **41**, 830–840 (2008).

[35] L. Shi, S. Günther, T. Hübschmann, L. Y. Wick and H. Harms, S. Müller, Limits of propidium iodide as a cell viability indicator for environmental bacteria. *Cytometry A* **71A**, 592–598 (2007).

[36] M. J. Rosenbluth, W. A Lam and D. A. Fletscher, Analyzing cell mechanics in hematologic diseases with microfluidic biophysical flow cytometry. *Lab Chip* **8**, 10621070 (2008).

[37] A. Abdulnaser, D. A Hill and S. A. Desai, Babesia and plasmodia increase host erythrocyte permeability through distinct mechanism. *Cell. Microbiol.* **9**, 851–860 m (2007).

[38] P. Gascoyne, C. Mahidol, M. Ruchirawat, P. Satayavivad, P. Watcharasit and F. F. Becker, Microsample preparation by dielectrophoresis: isolation of malaria. *Lab Chip* **2**, 70–75 (2002).

Chapter Thirteen

Microflow Cytometer Electronics

Jeffrey S. Erickson,[1,*] Dustin J. Kreft,[2] and Matthew D. Kniller[3]

[1]*Center for Bio/Molecular Science and Engineering, Naval Research Laboratory, 4555 Overlook Avenue SW, Washington, DC 20375, USA*
[*]*jeffrey.erickson@nrl.navy.mil*
[2]*Department of Electrical and Computer Engineering 2439 Engineering Hall, 1415 Engineering Drive, Madison, WI 53706, USA*
[3]*Nova Research, Inc., 1900 Elkin St., Ste. 230, Alexandria, VA 22309, USA*

The development of portable and autonomous microflow cytometers is an important area in microfluidics research. One key component in these new instruments is a small, low power electronics package capable of performing data collection, analysis, and other operations necessary to run the instrument. The goal of this chapter is to present a general discussion of the parts required to build flow cytometer electronics that run without an attached personal computer.

13.1 IMPORTANCE OF ELECTRONICS IN FLOW CYTOMETRY

Electronics and data analysis software play an important role in any flow cytometer by translating detector signals into meaningful information. In the original hemocytometers built before 1950, data for blood was collected by manually counting cells under a microscope in a known volume of fluid. Test results were derived from a few simple calculations. Unfortunately, this manual procedure was both tedious and time consuming, and due to the sheer number of counts that were required to see rare events and get good statistics it was not a particularly accurate one. With the invention of the flow cytometer and the Coulter counter, light or electrical pulses were captured by an automated detector rather than the human eye. System throughput greatly increased at the cost of electronic complexity. In the earliest cytometers, only binary signals were important: either a cell passed through the instrument or it did not. One could simply set a threshold value and count the number of voltage spikes that exceeded it. With the later introduction of non-binary measurements such as light scattering, the magnitude of the signals became as important as whether they crossed the threshold. A new generation of

The Microflow Cytometer by J S Kim & F S Ligler

www.panstanford.com
978-981-4267-41-0

electronics was introduced that could accurately measure and record peak intensities, while simultaneously minimizing noise and fluctuations in the baseline.

Modern optical cytometers have placed significant demands on electronics and data processing. Some commercially available models are capable of identifying over 100,000 cells per second. Others have more than ten simultaneous detection channels. Still others boast a dynamic range of nearly seven orders of magnitude. With greatly increased sensitivity and flow speed requirements over the original instruments of the 1950s, the need for sophisticated cutting edge electronics has become greater than ever.

Breadboard and laboratory-built cytometers present special challenges. Functionality on par with commercially produced instruments must be implemented, but stand-alone versions of the specialized circuit boards that companies have developed through years of experience are not available to the scientific community for purchase. In benchtop-sized breadboard cytometers, some of the functionality of commercial instruments can be duplicated through creative use of software, or performed by stand-alone electronics packages that can be chained together to obtain the desired functions. For portable microflow cytometers, especially those designed to run without a personal computer, space is at a premium. Elements that would normally be quite large must instead be implemented at the circuit board level. This is the design challenge that must be addressed.

In general, two different approaches are used in the development of laboratory prototype breadboard microflow cytometers: one must either disassemble a commercial cytometer and use its proprietary electronics train[1] or use commercially available stand-alone packages, such as DAQ cards, supplemented with other custom electronic elements and a computer or microcontroller.[2] The goal of this chapter is to expand on the second approach, specifically in cases where a custom printed circuit board (PCB) is desired. First, a general discussion of electronic parts is presented. Next, we outline the development of a stack of low-power PCBs for a stand-alone, remotely operated device. Finally, we briefly highlight future directions in microflow cytometry electronics.

The intended audience for this book chapter is the laboratory scientist who is interested in adding electronics to his or her work in order to make a portable microfluidic instrument. The chapter is written from a very basic point of view and is not meant to guide experts such as electronics engineers. We would like to gratefully acknowledge that most of this chapter is heavily influenced by the work of two authors, Shapiro[3] and Snow[4]; electronics designers should consider consulting both manuscripts for additional information and an alternate point of view.

13.2 CYTOMETER ELECTRONICS: COMPONENTS, FUNCTIONS, AND DATA COLLECTION

Data acquisition capabilities as well as supporting electronics and software are typically implemented into a cytometer to enable it to perform a number of different functions. For example, the instrument should filter and amplify signals from the detector(s) and compare them to user-set thresholds or other baseline criteria to determine if a significant event has occurred. In cases where data collection is triggered, the instrument should sample each data stream at a sufficient rate, calculate desired quantities such as peak area, and then either store the data, send it to an external display, or transfer it to an attached computer. In addition, the instrument may include electronics to condition power for components such as light sources, detectors, pumps, and valves. In cases of autonomous operation, the cytometer will need to have enough onboard memory to store the expected amount of data, and be capable of sufficient write speeds to avoid buffer overflows.

Collecting and conditioning signals in a flow cytometer is a very delicate operation. Fluorescent emission from objects in the flow stream can be weak, and it is collected on a noisy background. Analog electronic components can add a significant amount of noise to weak signals, especially in instruments with a large dynamic range. When selecting electronic components for use, it is necessary to consider the inherent noise levels in each element in the circuit design so that the desired dynamic range can be realized, as well as making sure that each element can respond to changes fast enough to meet the desired overall bandwidth. Implementing functionality into a flow cytometer, with discussion on component selection, will be the topic of this section.

13.2.1 Electronic Components

In his review article, Snow[4] has laid out a very general description of the tasks required in any flow cytometry data collection system. The components required to accomplish these tasks, along with power conditioning, communications, and data storage, make up the PCB as required for a microflow cytometer. The functional diagram that we use to describe these tasks is slightly different [Fig. 13.1], but still in line with Snow's basic concept for an analog detector. Briefly, objects flow through one or more illumination regions in the cytometer. Fluorescence and scattered light are emitted, filtered, and collected. A photodetector converts photons to an electrical current. A transimpedance amplifier converts this current into a voltage, usually in conjunction with electronics that provides negative feedback to minimize baseline drift. In an analog detector, the signal then moves through a low-pass filter to prevent aliasing effects. Custom analog circuits may be used to measure desired quantities, such as peak area. Analog-to-digital conversion takes place. The signal is then further processed by a microcontroller, and finally either stored in memory or transferred to an external device (such as a computer) for software manipulation and display. In a digital detector, many or all of the components labeled as optional in Fig. 13. 1 are removed, and their corresponding operations are performed in software by a microcontroller, digital signal processor (DSP) or field-programmable gate array (FPGA).

Before selecting components, it will be important to know two basic desired quantities for the instrument: the sampling rate, and the dynamic range. The sampling rate of the instrument is usually determined by considering the desired

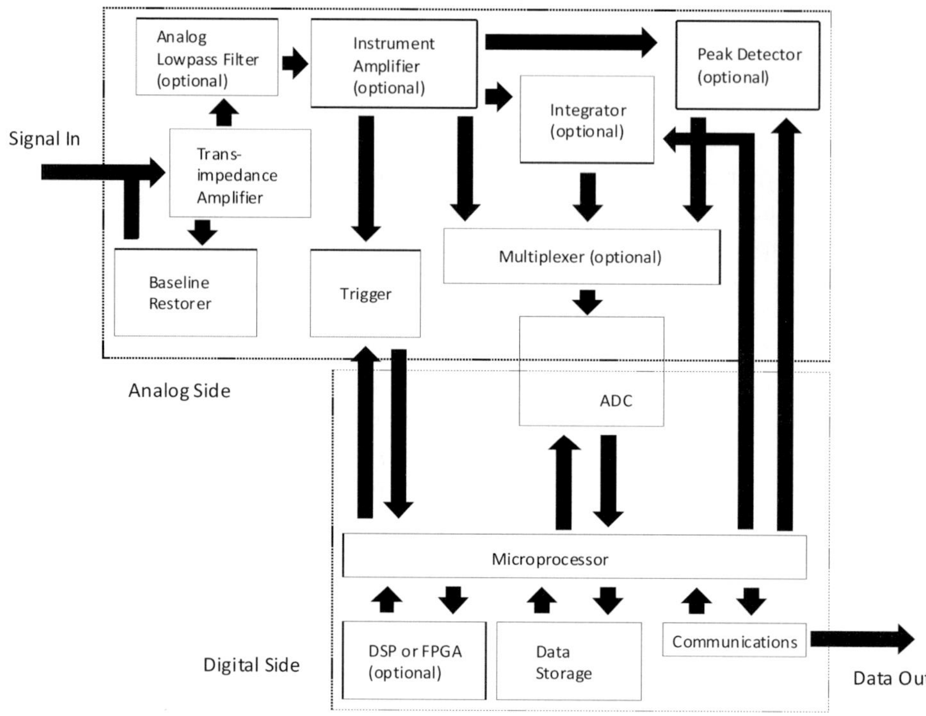

Figure 13.1. A functional diagram for a microflow cytometer electronics board design. Included features will vary depending on the desired board use. Analog detectors will include many or most of the optional features on the analog side of the board. In contrast, all-digital detectors will implement most of the function from the optional analog components in a digital signal detector (DSP) or field-programmable gate array (FPGA).

number of data points per peak when the smallest expected object passes through an illumination region of known width at the maximum expected velocity. In most analog instruments a few data points per peak is sufficient. For more modern, all-digital designs, eight or more data points per peak will be necessary. The result of this calculation is the minimum required sampling speed per channel for the instrument, and will be important for selecting filters, amplifiers, microprocessors, and other components.

The required dynamic range of the conditioned data is usually specified by the designer based on the intended use of the instrument. In order to collect relevant data, some experiments will require a larger range than others. Three or four orders of magnitude are not uncommon in commercial instruments. The number of decades of required dynamic range will influence the acceptable background noise levels, the required quality of the analog-to-digital converters (ADC), microcontroller selection, and the specifications of other components.

Once these parameters are known, the designer is ready to begin selecting specific components.

13.2.1.1 Transimpedance Amplifier and Baseline Restorer

Photodetectors output an electrical current. A transimpedance amplifier converts this current into a voltage. Simple units can be built from operational amplifiers (op-amps), which output a voltage range defined by the value of selected resistors and the power supply to the amplifier.[5] Alternatively, commercially available models are available for purchase. These off-the-shelf units typically use a complex, proprietary design, have preset output ranges, and have well characterized properties such as noise levels. In some cases, it is possible to purchase transipmedance amplifiers bundled together with the photodetector, such as the photomultiplier tube modules available from Hamamatsu.[6]

Regardless of whether the transimpedance amplifier is home-built or commercially obtained, it is important to make sure that it has sufficient bandwidth to capture changes in the signals at the desired frequency. Home-built amplifier designs should be tested with an oscilloscope and signal generator to determine their available bandwidth. Commercially produced amplifiers usually have a frequency cutoff value stated in the specifications. Data collected at rates above this value will be less accurate, weaker, or may be completely removed, and so this cutoff frequency will set a limit for the overall bandwidth of the system.

Signal produced by photodetectors will contain a background that may drift. A number of authors have suggested including a negative feedback component (a baseline restorer) in or directly after the transimpedance amplifier to remove most of the DC component of the background. To the best of our knowledge, commercially built transimpedance amplifiers that come packaged with photodetectors do not contain this kind of circuitry, although it is possible to add a baseline restorer to either a voltage or a current signal. An example of a custom transimpedance amplifier and baseline restorer combination is presented in Shapiro's book[3] on page 191. A slightly different design that was inspired from that work is shown in Fig. 13.2. It is capable of performing baseline restoration on a voltage signal (from a photodetector with a built-in transimpedance amplifier, for example), and uses only basic circuit elements, diodes, and op-amps.

Most transimpedance amplifiers and baseline restorers use operational amplifiers as components. The selection of op-amps for these and other functions on the analog side of the electronics board is an important consideration. The desired maximum frequency of the input to be captured, coupled with the amplification factors in the various elements of the circuit board, will determine the quality of the required op-amps for the entire PCB. At least three different parameters are critical for op-amp selection for microflow cytometry; these are discussed below. There may be other important parameters depending on the specific conditions of use; one example is crosstalk in dual and quad op-amp packages.

First, all op-amps have diminishing gains as a function of input signal frequency. The unity gain bandwidth product identifies the maximum input signal frequency to which an op-amp can respond. At this frequency, only unity gain (i.e., gain = 1) is available. The maximum available gain typically increases by a factor

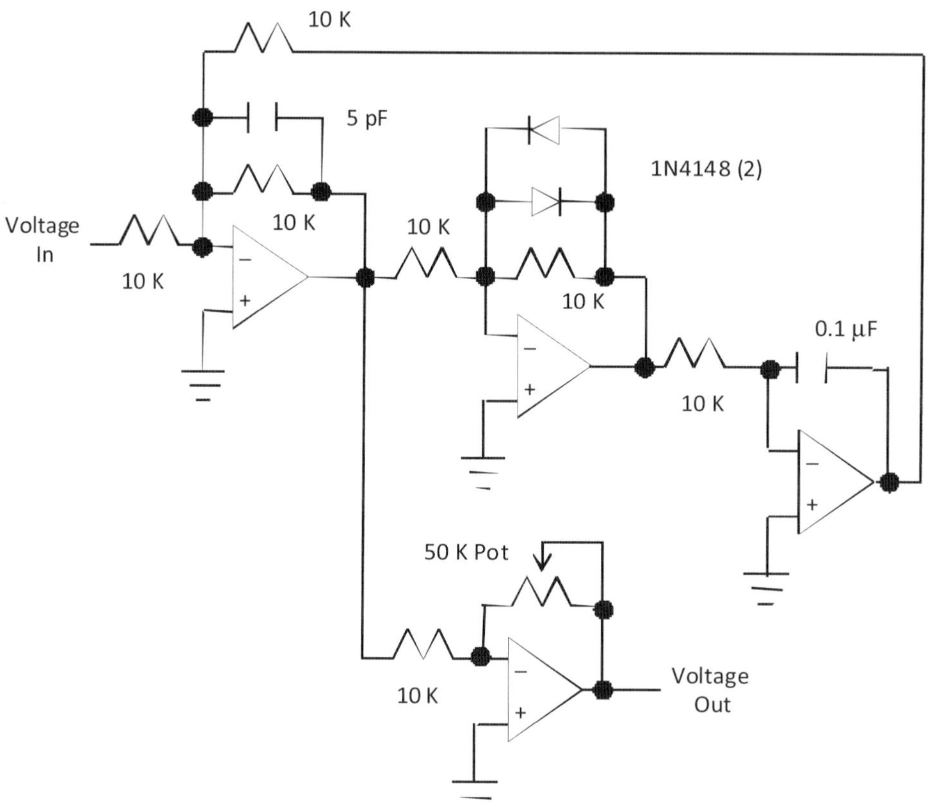

Figure 13.2. An example of a circuit to remove the DC component of the baseline voltage.

of 10 for each decade that the input frequency decreases. Amplifiers will need to be selected based on the gain requirements at maximum data collection speed.

Second, the slew rate of the amplifier must also be considered, which specifies how fast the amplifier can respond to large changes in voltage. In general, the slew rate must be sufficient to capture output voltage swings from zero to maximum for the highest frequency of data expected.

Finally op-amp input bias currents are especially important for the construction of transimpedance amplifiers. An ideal op-amp has inputs that draw no current. In reality, the inputs actually draw a small amount of input bias current during normal operation. Op-amps should be carefully selected in order to prevent signal from getting swamped by stray currents as a result of this input bias.

13.2.1.2 Low-pass Filter

Once a voltage has been produced from the transimpedance amplifier, it will need to be filtered. A digital detector can perform this operation in software. In contrast, analog instruments will require an electronic low-pass filter, which is a device that allows low frequency electrical signals to pass while attenuating those with high frequency. The most important parameter to consider when selecting a filter is the cutoff frequency, which specifies the frequency at which the filter attenuates the signal by 3 decibels (dB). Frequencies marginally above the cutoff will be attenuated at rate proportional to the order of the low-pass filter. For example, a 2nd order filter has a signal attenuation of 2 orders of magnitude (40 dB) per decade in frequency.

The Nyquist sampling criterion suggests that in order to avoid aliasing, incoming data streams must be low-pass filtered at a rate that is no more than half the sampling rate. For example, if the instrument is designed to collect samples at 2 MHz, the low-pass filter should have a cutoff no higher than 1 MHz. If oversampling is desired to help increase the signal-to-noise ratio, the filter cutoff can be even lower.

In cases where there is a high frequency noise source near the cutoff that must be eliminated, a filter with a high-order cutoff will be necessary. Most high-order analog filters tend to include active components such as op-amps; these components will introduce a small amount of noise to the signal. These baseline noise levels limit the dynamic range of the instrument, and therefore need to be considered when selecting components. Dynamic range can be calculated as follows:

$$\textit{dynamic range} = \log_{10}(\textit{maximum voltage}/\textit{minimum distinguishable voltage})$$

The maximum voltage a single component can handle will be specified in its data sheet. The minimum voltage that can be distinguished will depend on the amount of electronic baseline noise generated by that circuit element. This electronic noise, as well as the noise coming from the photodetectors and the rest of the elements in the PCB, set the floor for the weakest signal that can be distinguished.

Noise levels may restrict the type of filter that can be selected for use in the instrument. For example, a wide variety of tuneable switched capacitor filters are available on the market, with very sharp cutoff characteristics. One such vendor is Linear Technology,[7] which has low-pass filters available with cutoffs that are sometimes better than 10th order. However, many of these products also produce a significant amount of electronic noise, often on the order of hundreds of microvolts. Consider that if a filter produces a wideband electronic noise of 250 µV, and the full scale voltage of the signal is 10 V, the theoretical maximum one can expect to obtain is 4.3 orders of magnitude of dynamic range from such an element with a signal-to-noise (S/N) ratio of 2. Obviously, if one desires a system with 5 orders of magnitude dynamic range, including such a filter would pose difficulties.

Alternatives to switched capacitor filters exist. One can easily create an active low-pass filter from op-amps or a passive filter from resistor-capacitor combinations that produce significantly less noise. However, obtaining higher order filtering from such a device can be a design challenge. In cases where higher order filtering is required, there are some packaged low-pass filters available that contain op-amps and passive elements built into a single integrated circuit (IC), with the user required to add only resistors. For example, Maxim[8] makes a continuous time active filter divided into blocks, which can be daisy chained together to obtain arbitrarily high order cutoffs, using resistors and freely available software on their website. Other companies offer similar products.

13.2.1.3 Thresholding

In most experiments, nearly all of the signal coming from a flow cytometer during normal operation will be baseline, with very little of it corresponding to significant events. In cases where the photodetectors are being calibrated, it may be necessary to collect all of this data so that the raw S/N ratio can be calculated. During normal operation of the instrument, it is usually not required or desired to store all of this baseline information, as it takes up an enormous amount of space.

Triggers are used to tell the electronics when an event has started so that it can collect only data corresponding to an actual event. Through the use of comparators and reference voltage inputs (thresholds), either from a user or from a microprocessor, or possibly through the use of a real-time baseline calculation algorithm, the signals in one or more channels are monitored. In cases where a signal rises above the set threshold level, the system will begin to acquire data. When the signal falls below the set threshold level, the system will stop.

In most cases, a single channel is used for thresholding, and it is typically the strongest one (forward scattering in many cases) due to superior signal-to-noise characteristics compared to weak signals like fluorescence. There are exceptions.

13.2.1.4 Integrators, Peak Detectors, and Analog-to-Digital Converters

Several different circuits can be implemented on the analog side of the device for the collection of relevant data. It is not always necessary to gather a large number of data points on each peak in order to make useful measurements. Integrating circuits can be built to determine peak areas. Peak detectors are circuits that follow the voltage from a signal and store the largest value on a capacitor. These circuits can be used to accurately collect amplitude information on a signal regardless of the interval where the analog-to-digital converter (ADC) sampled it. Packaged peak detectors are available, with proprietary design and very good specifications. Both peak detection and integration can also be performed digitally in software, although the accuracy of these operations will depend on the number of data points that are sampled in each peak.

Information from any of these analog circuits, as well as from the signal itself, will need to be digitized. While analog to digital conversion is typically performed through commercially available data acquisition cards, it is also possible to buy ADCs in an IC package and add them to a custom PCB. ADCs are rated in terms of speed (samples per second) and resolution (bit accuracy or bin width). The resolution specifies the least significant bit that can be determined from the ADC. Since the data output will be binary, the dynamic range of the data that can be obtained from the ADC can be directly calculated: the number of channels over the maximum range of collection is given by 2^n, where n is the bit accuracy of the converter.

The $\log_{10}(2^n)$ gives the theoretical dynamic range available for a particular ADC. For example, a 1 MSPS, 14-bit ADC has 4.21 orders of magnitude of dynamic range available and can collect one million samples per second. Dividing the maximum signal value by this number gives the smallest voltage step (the bin width) that can be recorded by the system. For example, in a 14-bit ADC with a maximum signal of 10 V, the width between channels will be 610 μV.

In many cases, an ADC will be used to sample data streams from multiple channels. For example, a user may desire to obtain data from the waveform itself, as well as from a peak detector to obtain an accurate analog value of the peak height. In such cases, it will be necessary to add a switch to the circuit board, which can be controlled through the digital output lines on a microcontroller, microprocessor, or attached computer. Many ADCs contain multiple, switchable input channels to simplify this function.

13.2.1.5 Log Transform

Most flow cytometer data is presented on log-log plots. Obtaining these plots requires a linear-to-log conversion somewhere in the data collection hardware or software. It is possible to simply collect linear data from an autonomous instrument, perform the conversion on a separate desktop computer, and then plot the data. However, the required number of significant bits for log data may be greatly reduced over that required for linear data, which may make a difference when it comes to data storage. In cases where real-time data is required, a log conversion must take place in the instrument as the data is collected.

Older commercial cytometers tend to perform the log transform in hardware using analog electronics. Unfortunately, there are a number of difficulties that must be dealt with: signals may becomes negative (for example, on a fluctuating baseline or due to fluorescence compensation), and the rate at which log amplifiers operate is dependent on the signal strength. As signals become weaker, the bandwidth of the amplifier decreases. More recent commercial cytometers tend to do log conversions in software. Digital log conversions can be computationally expensive and can drastically slow down a microcontroller, especially if real-time data processing is required. Look-up tables can be used for at least part of the transformation to speed up the process. Another advantage of doing such a transform in software is that algorithms can be switched until one is found that has the desired properties. There is much discussion in the literature about alternatives to straight log transforms for better data quality.[9,10]

When performing digital log transformations, the requirement for ADC resolution (for collecting linear data) will depend heavily on the resolution required for the logarithmic data. Digitized data is collected in bins, each with a certain width. When data is collected in linear form, the width of each bin is identical. For example, in a 12-bit system with a 10 V maximum scale, each bin has an absolute width of about 2.4 mV. If this digitized data is then converted to log format, the absolute bin width becomes variable in order to meet the logarithmic requirement of having a constant number of bins per decade. Typically, a designer will truncate the accuracy of the logarithmic data. If the 12-bit linear system mentioned above is used to create 8-bit log data, the upper channels in the log system will each have an absolute width significantly greater than the width of a single linear channel. However, at the low end of the logarithmic scale, the bin width becomes smaller than each linear channel. When data is converted, some channels at the low end of the log scale will be skipped. The result is the famous "picket fence" effect. Picket fences can be avoided by dithering or subranging,[3] although if possible, a better solution is to increase the resolution of the linear data going into the transform. For example, 10-bit log data can be digitally created without a picket fence effect through the use of 20-bit accuracy linear data. ADCs should be selected accordingly.

13.2.2 Evaluation Kits

13.2.2.1 Microcontrollers

Most autonomous instruments will require a component that can duplicate the role of a desktop computer. The element selected to perform these functions is typically a microcontroller. Unfortunately, working with a bare microcontroller or microprocessor can be difficult, because of the delicate soldering work that must be done to incorporate it into a circuit board. Virtually all microprocessors are surface mount components, and many have 128 pins with a very small pitch. Others are attached by ball-grid arrays on the backside of the chip; these ICs require special soldering techniques that are difficult to perform by hand. A viable alternative to using a single PCB and a bare microcontroller is to purchase a small evaluation kit with a microprocessor already soldered on, and use this as a motherboard. Many kits are commercially available for a very reasonable price, usually with software packages to make programming easier as well as customer support. In some cases, this motherboard will provide functionality beyond just the microprocessor (for example, it may be an evaluation kit designed to control flash memory), but ideally it would serve as the heart of the electronics system. Starting from the motherboard, it is possible to add the rest of the functionality, either through an external connection between the motherboard and some daughterboard(s), or by building directly onto the motherboard. Note that in most cases, the second option is only possible if the evaluation kit contains a specific area for the designer to add new electronics.

When selecting a microcontroller or microprocessor, there are a few issues that should be considered. First, it should have enough digital input and output lines to control and read from all of the other electronics on the board. Second, the clock speed must be sufficient to handle data streams and any operations that the microprocessor will perform on it. Microprocessors and microcontrollers are rated in terms of the number of bits per clock cycle operation. An 8-bit microcontroller can handle 8-bits of data per clock cycle, while a 16-bit microcontroller can handle twice the throughput. Another consideration is the instruction cycle time of the microcontroller. Some microcontrollers can process instructions at 1 per clock cycle and others may require 4 or more clock cycles. For example, an 8 MHz chip which can compute most instructions in one clock cycle will run at roughly the same throughput as a chip that runs at 24 MHz but requires 3 clock cycles per instruction. Although there is a trade off, usually microcontrollers and microprocessors that require more cycle time may have an internal pipe-line. This can increase computation when handling very large and complex algorithms.

Different types of software algorithms require different numbers of clock cycles. Averaging, log transforms, integration, peak width determinations, calculation of statistics, threshold adjustments, and software (digital) filtering are some examples of operations that may be desired in real time from data streams. There may be cases where enough run-time digital manipulations are desired that microprocessors become difficult to use. In these cases, field programmable gate arrays (FPGAs) or digital signal processors (DSPs) have been successfully incorporated. Microcontrollers are ubiquitous, easy to use, inexpensive, and relatively simple to code with high level languages such as C. Newer technologies such as FPGAs and DSPs can be much more difficult to work with. One website details some of these difficulties,[11] suggesting that DSP developer software tools including high level coding capabilities lag behind those of microprocessors. In addition, these newer chips tend to be quite expensive compared to microprocessors and microcontrollers.

We certainly expect the cost of FPGAs and DSPs to decrease in the future, and their ease of use to improve. As 24-bit data acquisition speeds up and supporting ICs become faster and quieter, we fully expect that digital signal analysis will completely replace the analog approaches outlined in this document in all but the fastest cytometers. That is, rather than using analog electronics for thresholding, peak detectors, and to integrate areas under data spikes, circuit boards will use ADC and DSP combinations to continuously sample huge amounts of data, store a running stream in its internal buffers, determine when a peak has occurred, and calculate all necessary quantities in real time. Systems with fewer circuit elements produce less noise, and so these all-digital systems will have an inherent advantage over older analog electronic devices. However, all-digital analysis will require at least 8 data points per peak, with 16 or more being recommended for accuracy.[3] Unusually high bandwidth will be required for these types of devices.

13.2.2.2 Communications and Data Storage

Many autonomous instruments will require onboard data storage. There are many ways to do this, including RAM, flash memory, and hard drives. The time required to write data to memory may become important, especially in designs that have a very high duty cycle. For example, while RAM can be quickly accessed, flash memory is much slower and may require more time to write than is available between measurements while actively collecting data.

Communications will be required on any instrument in order to transfer data to the user. In simple cases, this could correspond to a remove-able smart card, flash memory stick, or hard drive. More sophisticated instruments may use wireless communications or a USB connection. In many cases, an instrument will require a live data feed for events such as calibration or setting of thresholds. Before selecting a communication scheme, the maximum sustained and burst data output rates should be considered, as well as ease of implementation. For example, RS-232 communications are slow but tend to be easier to implement than fast protocols such as Ethernet.

For the user that intends to upload data to a computer in real time, and especially in cases like calibration where the background is included, modern communications protocols will be necessary. These include USB, Ethernet, and Firewire. A number of IC manufacturers fabricate chips with the capabilities to run advanced communications schemes. However, all of them can prove challenging to implement directly. For the non-expert electronics designer, this can be a significant hurdle to overcome. Perhaps the easiest solution is to use evaluation boards with included software that is designed to make these transfers easy. Three examples are products made by Rabbit,[12] QuickUSB,[13] and Netburner;[14] these manufacturers design small electronics cards with embedded microprocessors and available software to make high speed communications possible right out of the box. We have found these development boards to be cost effective solutions. In some cases, the embedded microprocessors are fast enough to make the development board a candidate for the system motherboard as well.

Development boards for smart card memory readers are also available to make reading and writing to these elements quick and easy. In cases where the user would prefer to embed their own memory or buffers, they are available in IC form and are relatively cheap, and a microprocessor can be used to directly control read and write operations to them.

13.2.3 Peripheral Operations and Power Conditioning Electronics

13.2.3.1 Amplification and Background Noise Reduction

In many analog instruments, signal amplification will be necessary because of both signal decay as well as the potential difficulty of matching the voltage input ranges of different circuit elements. Amplification will always come with a subsequent increase in background noise, which has an adverse impact on the S/N ratio of the measurement. The only exception to this rule is gain from the photodetector itself. When choosing a photodetector, it is a good idea to make sure that it can perform as much of the required overall signal amplification as possible. In cases where the background is still too high alternate strategies can be employed for improvement. Perhaps the most common method is to use a lock-in amplifier. In such a strategy, the light source is pulsed at a fixed rate. This rate is simultaneously read by the electronics, which can lock onto it and pick out the alternating signal from background noise. Significant gains are possible. In addition to lock-in amplifiers, background noise can be reduced by oversampling or by averaging data. Since the noise should fluctuate randomly while the signal remains relatively constant in strength, averaging several data points will cause some of these baseline fluctuations to cancel out. Finally, there are methods of taking the same data over multiple channels that increase S/N ratios.[15,16]

13.2.3.2 Software Operations

After collecting processed data from the board electronics, it may be necessary to perform further manipulations in software. These operations may include log transforms, fluorescence compensation, gating, and the calculation of statistics. Other software operations, such as automated analysis, are possible.

Fluorescence compensation is important in multi-color flow cytometers when the emission bands of the different channels overlap.[17] Most compensation algorithms involve subtracting a constant value from certain channels to account for overlap with other ones. For example, the emission spectra of fluorescein and phycoerythryn overlap and must be treated accordingly. In cases where there are just two or three overlapping channels, compensation can easily be built into the analog electronics, with computer controls for real-time user adjustments. However, increasing the number of circuits will invariably add noise to the board, affecting sensitivity. Digital or automated software compensation is preferred, especially in cases where there are many channels to correct. Fluorescence compensation should be performed before a log conversion takes place. Unfortunately, compensation can subtract too much value at the weakest end of the signal, especially on a noisy baseline, creating negative values which are difficult to deal with in logarithmic conversion.

Gating, as opposed to thresholding, is a user operation performed on data after it is collected to isolate regions of interest (especially in cases where there is no clear division between positive and negative populations), or to set a region in which subsequent data is to be collected. Gating software has traditionally been used to isolate rectangular regions in the data, although software exists to isolate quadrilateral, circular, or elliptical regions. Changes in fluorescence compensation may change the shape of the populations before gating takes place, although even correctly compensated data will not necessarily fall into rectangular quadrants, making flexibility in gating operations desirable. Gating can be performed either before or after a log transform.

13.2.3.3 Power Conditioning

Care should be taken in the design of the power supply for the electronics, especially in cases where very low noise is required. Traces should be made as small as possible to eliminate stray capacitance. Digital and analog sides of the board should have separate grounds to prevent noise between the two. Ground planes should be used whenever possible to prevent loops. In cases where a ground plane is impossible, star formation is preferred. Commercially available voltage regulators are suggested for deriving steady power sources, rather than voltage divider circuits. In especially sensitive ICs, the power supply should be filtered with capacitors to remove noise, somewhere near that particular circuit element. Most data sheets for electronic components will suggest power coupling guidelines for optimizing performance.

13.2.4 Design and Fabrication Notes

Making electrical connections to evaluation boards is usually easy, and dual-inline package (DIP) electronics can be simply connected by through-hole soldering or wire wrapping techniques. However, at some point it may become necessary to design a board to hold surface mount electronic parts, because many high performance ICs only come in this format. There are a number of commercially available layout packages that can be used to design a circuit board. In our work, we have used a freeware version of Eagle,[18] which is limited to two-layer boards but was sufficient for our purposes. Other freeware solutions exist. In cases where large numbers of layers are desired, commercial packages will become necessary. Especially for the manual soldering of surface mount parts, we highly recommend ordering a board with a solder mask. Solder will not stick to the masked areas, which optimizes the ability to make connections to parts with an extremely small pitch. We suggest using a soldering iron with a small-sized chisel tip. The tip holds a bead of solder, which can be dragged across the mask to fill in the gaps. Solder paste can be used in a reflow system as an alternative to manual attachment with a soldering iron. For example, we used a Pace ST 350 Convective Rework Center[19] for the manual soldering of surface mount components.

13.3 DEVELOPMENT OF THE NRL AUTONOMOUS DATA COLLECTION SYSTEM

In this section, we describe the PCB that we are developing at the Naval Research Laboratory (NRL). This system is intended to analyze and characterize marine algae as a function of ocean depth. It operates with four detector channels: forward scatter, side scatter, and two fluorescent colors. As an alternative, one of the scattering channels could be replaced by a third fluorescent color with no additional difficulties. The information presented here is a work in progress, and should not be viewed as our final product.

13.3.1 NRL Version 1 System

The main requirement for our microflow cytometer is a volume constraint: it must fit into a small, water-tight can (approximately 6″ in diameter and 12″ long), including all of the fluidics, pumps, light sources, and detectors. A secondary requirement is that it must be able to run autonomously for at least 48 hours, and possibly longer. We desire a low power electronics system that has onboard data storage, and is built from stackable cards not much larger than 4″ x 4″ each. An embedded programmable microprocessor is necessary.

The dataflow that we envision is as follows: algae will flow through the beams of two lasers, a 405-nm violet diode laser and a 532-nm green DPSS laser. The sources will simultaneously impinge upon the core stream in the cytometer, resulting in light scattering events as well as fluorescence due to chlorophyll and phycoerthyrin content. These signals will be optically filtered, recorded by the detectors and converted into voltages. The voltages will be conditioned, amplified, and collected with an ADC and peak detector. A microprocessor will control the flow of data and control thresholding operations, and will send data either directly into onboard flash memory, or through a USB line to a connected computer during an upload event.

Specific requirement for the components are derived from the optical train and detectors. Our particular device uses optical fibers to deliver and collect light. Because these fibers have a small numerical aperture compared to a microscope objective, and because it is difficult to include lensing in this type of design, we opted to use photomultiplier tubes as detectors in every channel including forward scattering. In his book, Shapiro[3] has listed a number of commonly used photomultiplier tubes for cytometry work. We opted to use the R6357 tube from Hamamatsu,[6] which is one of two of the smaller (1/2" vs. the older 1-1/8") side-on multi-alkali tubes that Shapiro has evaluated. Specific package options which include a housing, electrical connectors, and high voltage amplifier are the H9306 (1.0 V per microamp conversion and DC to 20 kHz bandwitdth), and H9307 (0.1 V per microamp conversion and DC to 200 kHz bandwidth). In all cases, we are using the -03 models due to their superior optical properties at the wavelengths at which we will be detecting. We are currently evaluating a few of each PMT model. The voltage input is shunted first into a baseline restoring circuit, the design of which is presented in Fig. 13.2. Because we have not yet narrowed down the PMT selection to a single module, the board is being designed to sample at 200 kS/s, which is a modest rate but should be sufficient to cover the different model PMTs that we are considering, while keeping data transfer rates reasonable during calibration or live feed. In the case where we decide to move from a PMT to a photodiode for forward scattering, it would be necessary to build a separate transimpedance amplifier for the photodiode and then place that output into the voltage input of our DAQ board.

The version 1 board was designed with a USB communications option to make burst data transfers possible between the board and a personal computer, either during calibration, or for removal of stored information when the device is not operating underwater. It should also be possible to use the USB interface to collect real-time data if the device is being pulled in a towed vehicle by installing a data feed wire into the tow rope. Due to the USB system requirements, we designed the version 1 board to have two separate microprocessors: one to control the data acquisition, processing, and transfer to memory; and the other to run the USB interface. The bulk of the configuration was designed onto a single board. A separate board was used for power conditioning. Directly after the detector inputs, there is a switched capacitor lowpass filter followed by an amplifier to adjust voltage levels to fit into the ADC. For this particular build, we desire a dynamic range of only 3 orders of magnitude, so the noise from the switched capacitor filter is not an issue. The trigger in version 1 is designed to use only the strongest channel. Since it is unknown at the time of fabrication which channel this will be, the trigger monitors all four channels simultaneously with a single, user-set threshold value. When the voltage from any one of the channels exceeds threshold, a signal is sent to the microprocessor to begin a data collection event.

During a collection event, each detector channel is sampled simultaneously by 4 different multiplexed 12-bit ADCs at 200 kSPS each, and the data is fed into the main microprocessor. Meanwhile, peak detectors monitor each data stream to record maximum intensities. The entire time, the trigger is also monitoring the data streams. As soon as the strongest stream falls below the threshold, the microprocessor switches the multiplexer on the ADCs to the peak detector, and takes a reading of the maximum value on the stream. The microprocessor then packages all of the data from each stream together, and sends it into flash memory. During a live readout event, it is also possible to send the data directly out the USB port and into an attached computer. To do this, the main microprocessor communicates with the USB microprocessor to initiate the transfer.

The firmware of the system works on an interrupt basis instead of a polled basis. An interrupt system sits idle and waits for an event to occur whereas a polled system continuously checks the hardware for an event. An interrupt system frees up the processor to do additional tasks while waiting for an event to occur. In our case the voltage signal from the photo-detectors and threshold voltage level are both wired into a voltage comparator. Once the threshold voltage is achieved the voltage comparator sends a signal to the processor. The processor then immediately responds to this signal and starts data collection. All data is stored into a buffer in the processors internal memory and the firmware continues to read data from the ADC and peak detectors until the voltage comparator outputs a LOW signal to the processor indicating that the desired signal peak from the flow cytometer has passed. At this point the firmware then organizes the data and sends it to the memory device for storage. The processor can do memory manipulation while waiting for another signal from the voltage comparator; this is the advantage of an interrupt driven system. The firmware was written in C programming language using free software from Atmel.[20] Assembly language could be used to achieve massive speed gains but requires a much longer development time since assembly works on an instruction basis and uses only fundamental logic and data flow.

A layout of the electronics and traces on the the NRL version 1 board is shown in Fig. 13.3, which was generated with Eagle,[18] a free electronics layout software package. In Fig. 13.4, we present photos of the bare and soldered boards. We used a two-layer board with electronics elements only on the top side. The final board size was roughly 3.25″ × 4″, with a second stacked board to provide power (not shown).

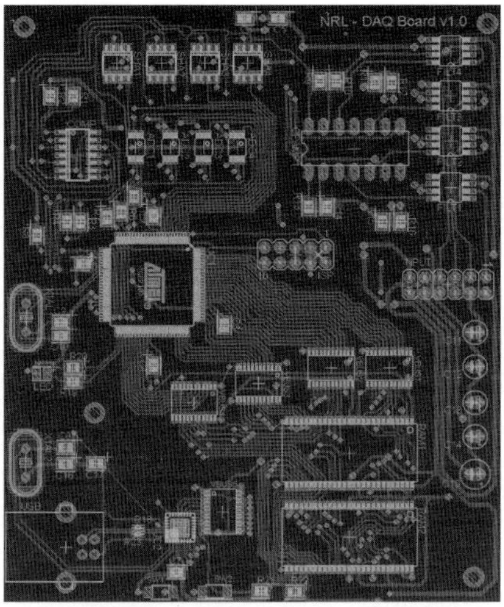

(a)

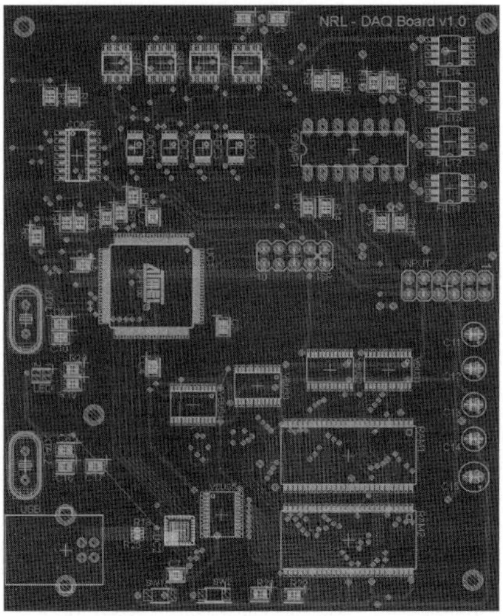

(b)

Figure 13.3. Layouts for the NRL version 1 microflow cytometer DAQ board. (a) Top layer. (b) Bottom layer. The overall board dimensions are 4″ × 3″. Color reference – pg. 351.

13.4 FUTURE OUTLOOK

Along with many others, we believe that digital signal processors and real-time high-throughput data analysis is the key to future microflow cytometers, because it will allow portable and autonomous devices to perform sophisticated analysis without the need for a dedicated computer, and because operations can easily be added or changed by altering the PCB firmware. This new generation of flow cytometer electronics will likely contain few elements other than ADCs, DSPs,

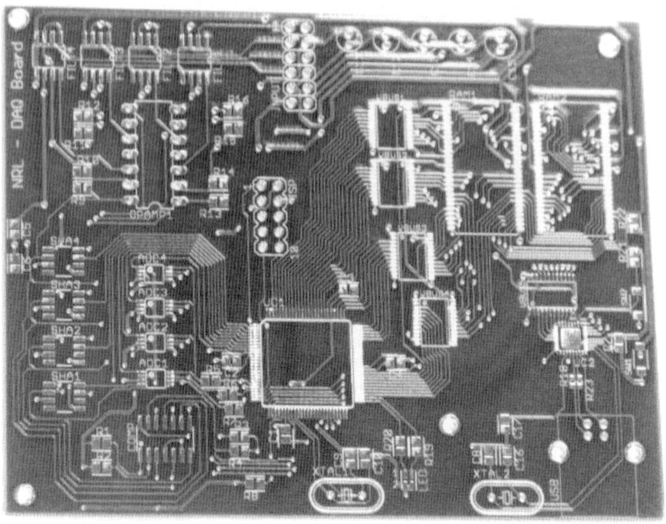

(a)

(b)

Figure 13.4. Photographs of the NRL version 1 microflow cytometer DAQ board (a) without, and (b) with soldered components. Color reference – pg. 352.

memory buffers, and microprocessors. A live stream of data will be continuously digitized and fed into the memory buffers. Detection of an event, lowpass filtering, run-time averaging, baseline compensation, and even some statistical functions may be performed in the DSP in real time on data stored in these buffers. Due to the absence of a large number of analog circuit elements, inherent noise should be low.

We acknowledge the difficulties involved in the implementation of these all-digital electronic systems using currently available technology. However, as other fields such as digital television, optical communications, and software defined radio continue to grow, more powerful specialized electronics parts will become available that can potentially be implemented into the circuit designs of portable microflow cytometers. Perhaps someday soon, it will be possible to obtain the accuracy, power, and functionality of a full benchtop cytometer electronics board in a small, commercially available all-digital PCB, ready to be implemented into your own handheld application.

ACKNOWLEDGMENTS

The work presented here was performed under NIH grant UO1 A1075489 and ONR/NRL 6.2 work unit 6006. The views presented here are those of the authors and do not represent the opinion of the US Navy or the Department of Defense.

References

[1] W. G. Lawrence, G. Varadi, G. Entine, E. Podniesinski and P. K. Wallace, *Cytometry A* **73A**, 767–776 (2008).
[2] H. M. Shapiro, N. G. Perlmutter and P. G. Stein, *Cytometry A* **33**, 280–287 (1998).
[3] H. M. Shapiro, *Practical Flow Cytometry*, 4th *Ed.* John Wiley & Sons: New Jersey (2003).
[4] C. Snow, *Cytometry A* **57A**, 63–69 (2004).
[5] R. F. Coughlin and F. F. Driscoll, *Operational Amplifiers and Linear Integrated Circuits*, 6th *Ed.* Prentice-Hall, Inc.: New Jersey (2001).
[6] http://www.hamamatsu.com/
[7] http://www.linear.com/
[8] http://www.maxim-ic.com/
[9] C. B. Bagwell, *Cytometry A* **64A**, 34–42 (2005).
[10] D. R.Parks, M. Roederer and W. A. Moore, *Cytometry A* **69A**, 541–551 (2006).
[11] http://www.microcontroller.com/Embedded.asp?did=61
[12] http://www.rabbit.com/
[13] http://www.quickusb.com/store/
[14] http://www.netburner.com/
[15] C. H. Chen, F. Tsai, V. Lien, N. Justis and Y. H. Lo, *IEEE Photonics Technology Letters* 19, 441–443 (2007).
[16] P. Kiesel, M. Bassler, M. Beck and N. Johnson, *Applied Physics Letters* **94**, 041107 (2009).
[17] Maecker, H. T. Trotter, *J. Cytometry A* **69A**, 1037–1042 (2006).
[18] http://www.cadsoft.de/
[19] http://www.pacedirect.com/
[20] http://www.atmel.com/products/

Chapter Fourteen

Miniaturized Sorters: Optical Micro Fluorescence Activated Cell Sorter

Kamlesh D. Patel* and Thomas D. Perroud

Department of Biosystems R&D, Sandia National Laboratories P.O. Box 969, Livermore, CA 94551 USA
**kdpatel@sandia.gov*

We present an overview of optical cell sorting, based on optical tweezers, as a promising approach for developing micro fluorescence-activated cell sorting (μFACS). Presented is a description of the principle behind optical trapping and cell sorting in microfluidic devices. We describe the configuration and operation of our group's optical μFACS and compare its performance with other relevant designs. General limitations of optical sorting and novel strategies are presented, followed by a discussion on its future outlook.

14.1 IMPORTANCE OF OPTICAL CELL SORTING TO MICROFLOW CYTOMETRY

Optical tweezers, or single-beam optical traps, have become an increasingly powerful interdisciplinary tool in engineering, physics, and biology. Laser-based optical traps have unprecedented precision and fidelity to manipulate particles at the microscopic scale. In the late 1980s, Arthur Ashkin first showed that laser light focused to a diffraction-limited spot using a large numerical aperture lens could trap and manipulate live bacteria, viruses, yeast cells, and even small organisms like protozoas.[1–4] As a direct result, laser tweezers have led to new discoveries in the dynamics of kinesin molecular motors,[5] the understanding of viscoelastic properties of cell plasma membranes,[6,7] the polarity of T-cell's shape to antigen,[8] and the biophysical nature of single DNA molecules.[9] These works have ushered laser-based optical traps as a new tool in cell biology.

Not surprisingly, the ability of optical tweezers to handle microscopic objects within their microscale environment has led to the coupling of this technology with laboratory-on-a-chip (LOC) platforms. As the name implies, the concept of LOC,

The Microflow Cytometer by J S Kim & F S Ligler

www.panstanford.com
978-981-4267-41-0

also known as micro total analysis system (microTAS),[10] is an emerging approach to integrating multiple bench-top functionalities into a single microfluidic chip only millimeters to a few square centimeters in size. Numerous comprehensive reviews of LOC devices for cell analysis can be found in the literature that highlight the recent advances and accomplishments in this field.[11–18] These microfluidic devices offer a number of advantages over traditional laboratory methods such as low sample consumption (especially valuable when handling human or animal cell samples),[19] decreased reaction times for diffusion-limited processes,[20] portability for point-of-care analysis,[21] high throughput through parallelization,[22] and the capability for high-level integration and automation for multi-parameter analyses.[23,24] Additionally, the small dimensions of LOC devices result in an important physical phenomena, where viscous forces dominate inertial forces, making fluid flow highly deterministic — a characteristic described by a low Reynolds number.[25]

Since the development of fluorescence-activated cell sorting (FACS),[26] scientists have been able to perform multiparametric cell separations at sorting speeds of $10^4 - 10^5$ cells/s.[27] Such rapid and efficient analysis of heterogeneous cell suspensions has positioned benchtop FACS systems as an important therapeutic tool for hematology,[28] gene therapy,[29,30] pancreatic islet cell transplantation,[31] and sperm sorting for gender pre-selection.[32] However, for small samples ($< 100,000$ cells), such sorting efficiency cannot be sustained.[33] Additionally, the large footprint and frequent maintenance of FACS instruments make them impractical for laboratories with limited space (e.g. biosafety-level 3 and 4 facilities).

A microfluidic-based fluorescence-activated cell sorter (μFACS) represents a promising alternative for sorting a limited number of cells from a precious sample at a high yield, while guaranteeing sterility, biosafety, and cleanliness through disposability. A small, portable, and potentially inexpensive μFACS would also enable wide acceptance of this technology throughout the biomedical research field. For these reasons, the microfluidic community has focused their efforts to go beyond microflow cytometry and develop a complete μFACS system. The main challenge in developing a robust μFACS is the integration of a precise and reliable sorting mechanism at this small scale. One attractive sorting strategy is to adapt the same laser-based, gradient forces used to manipulate (or "tweeze") cells as a way to rapidly deflect or reposition cells flowing in a microfluidic channel. Recent efforts by our group[34] and inspiring work by others[35–38] have shown optical cell sorting to be an effective method for sorting cells in a microfluidic device to solve unique and challenging biological problems.

In the following section, we focus on the recent advances in coupling laser-based, optical sorting techniques with microfluidic devices to sort cells. First, we present a brief overview of the fundamental principle of optical forces within the context of its implementation with LOC devices as cell sorters. Key characteristics along with inherent advantages are discussed using metrics to gauge performance, followed by a brief discussion on the current limitations of the technology and ongoing efforts for novel applications.

14.2 CHARACTERISTICS OF OPTICAL CELL SORTING

Numerous proof-of-concept strategies for microfluidic-based cell sorting have been reported in the literature and have been reviewed extensively.[13,18,39–41] Demonstrated proof-of-concept sorting techniques based on fluid handling include electrokinetic flow switching,[35,42] hydrodynamic flow switching using on-chip[43] and off-chip valves,[44,45] MEMS-based micro-T switches,[46] and a thermoreversible gelation polymer.[47,48] Dielectrophoretic approaches for cell sorting include segregation of tagged cells,[49,50] untagged cells,[51,52] and droplets.[53]

The non-invasive nature of optical tweezers is the main motivation behind cell sorting based on optical forces. Firstly, the only requirement needed to couple this sorting mechanism to the microfluidic device is an optical window transparent to the wavelength of the trapping laser. This allows for a closed system during the sorting process — a safer approach compared to conventional droplet-based FACS instruments, which generate aerosols. Secondly, since the sorting mechanism is external, the fabrication of the chip remains simple, making it relatively inexpensive to manufacture and consequently disposable. Thirdly, with a simplified chip, integration of additional cell handling functionalities is also easier to implement, allowing the true potential of optical sorting in LOC devices to be fully realized.

14.2.1 Deflection of Flowing Cells by Optical Forces

The general principles and methods for building an optical trap are discussed in detail in previous reviews.[54–56] Notably, Hunt and Wilkinson reviewed the state-of-the-art in optical techniques applied to microfluidic systems for chemical and biochemical analysis.[57] Perhaps the most relevant review for theoretical consideration on optical forces in the manipulation of cells and colloidal particles is the work by Dholakia and coworkers.[40,58] Their reviews expand on prior work[36] and divide optical sorting into four categories: (1) static fluid with static light pattern to induce flow and thus separation of particles; (2) dynamic fluid with static light patterns to flow particles through an optical lattice; (3) static fluid with dynamic (moving) optical patterns to transport particles; and (4) dynamic fluid with a dynamic optical pattern to utilize motional light with flow.[40] The application of optical forces in μFACS systems falls under this last category and is the primary focus of this chapter.

14.2.1.1 Principle of Laser-Based, Gradient-Forces Optical Trapping

When a particle is located in a focused laser beam, the radiation pressure from the laser imparts a force that can be divided into two components: a *scattering force* and a *gradient force*. The scattering force is in the direction of propagation of light (i.e. Poynting vector) and is proportional to the laser intensity; whereas the gradient force points towards the region of maximum optical intensity — the focal spot of the laser beam.[54] The gradient force can be thought of in an analogous manner to dielectrophoresis where instead of non-uniform electrical field gradients, light gradients impart forces on a cell.[57] An object is said to be in a stable optical trap when the gradient forces overcome the scattering forces in the region beyond the focal spot.

Since eukaryotic cells have dimensions significantly larger than the wavelength of the trapping laser, a condition referred to as the Mie regime, the trapping of such cells by optical tweezers can be described using simple ray optics.[54] When a cell interacts with a light gradient generated by a focused laser beam, the sum of all the light rays passing through the cell results in a reaction force that acts on the cell. More specifically, rays of light carrying momentum are bent by refraction when passing through the cell according to Snell's law since the refractive index of the cell is different than that of its aqueous trapping medium. By conservation of momentum, the rate of change of momentum in the deflected rays creates an opposite rate of change of momentum in the cell. This rate of change of momentum is proportional to the gradient forces that trap the cell near the focal spot (Newton's second law).[54]

The force and stiffness of an optical trap is mainly defined by the refractive index of the trapped object relative to that of its surroundings, the steepness of the light gradients, and the power of the trapping laser.[56] Given its heterogeneous content, the refractive index of a cell is difficult to determine and should be considered as a continuum of refractive-index fluctuations.[59] For mammalian cells, the refractive index is approximately 1.38,[59] which is close to the refractive index of water (1.33) since cells are mainly comprised of water. To make matters worse, mammalian cells should be kept in cell media, which have a slightly higher refractive index ($n = 1.34$). These facts contribute to the cell's small relative refractive index, reducing the overall strength of the optical trap. Since the steepness of the light gradients is defined by the numerical aperture of the objective, high numerical aperture objectives such as oil- or water-immersion objectives (typically, 1.2 – 1.4 NA) are preferred.[56] Finally, the trapping laser should have an incident power ranging from milli-watts (mW) to watts (W) depending on the configuration of the optical system.[54] As a rule of thumb, the magnitude of the trapping force on a 10 μm polystyrene bead ranges from 0.1-to-1 pN per mW of laser power.[60] However, for cells, the maximum laser power is limited by potential optical-induced damage due to long exposures to intense radiation — opticution.[2] As Ashkin found out early on,[4] the best wavelengths for trapping cells while minimizing light-induced heat damage are in the near-infrared region between 780-1330 nm where biological material is quasi-transparent.[54]

14.2.1.2 Optical Deflection in a Microfluidic Channel

Given the laminar nature of fluid flow in a μFACS,[25] the sorting mechanism simply consists of deflecting a cell laterally a few microns (typically, 20 – 60 μm). The deflected cell positioned in a neighboring flow stream is then directed to a different microfluidic channel downstream of the sorting region. In a microfluidic device, the height of the channel physically restricts the vertical position of the cell to a

20-to-50 μm range. Consequently, a stable optical trap, in which gradient forces are greater than scattering forces, is not required for cell deflection. It is actually more important to have a larger optical trap to compensate for minor variations in the lateral and/or vertical position of successive flowing cells. Large optical traps are achieved using low numerical aperture objectives (typically, 0.2 – 0.45 NA)[34,37,38] which is significantly lower than in typical optical tweezers experiments. The resulting loss in optical trap strength is compensated by an increase in laser power to maintain similar gradient forces. Correspondingly, the interaction time between the cell and the trapping laser is decreased to avoid opticution while sustaining a reasonable throughput. As a rule of thumb, the 1 nN Stokes drag generated by a 10 μm diameter cell being laterally displaced at a velocity of 10 mm/s requires 1-10 watts of laser power focused by a 0.2-0.45 NA objective.

14.2.2 Active Sorting Using Optical Forces

Of the two major optical cell sorting mechanisms (active and passive), our discussion for this section will primarily focus on active cell sorting methods since this approach mimics the mechanism and types of applications seen with conventional FACS systems. A detailed discussion of passive sorting strategies, which sort objects based on their intrinsic physical properties (size, shape, optical polarizability) has been previously described by Dholakia *et al.*[36,40] In active sorting schemes, an external decision is made in real time by probing the cell as it passes through the interrogation region of a microfluidic chip. Because of its great sensitivity, most active sorting schemes use laser induced fluorescence similar to conventional flow cytometry where external (e.g. fluorescently labeled antibodies) or internal markers (e.g. green fluorescent protein) are typically used to differentiate cell types. Alternative interrogation approaches such as Raman spectroscopy,[61] conductivity,[62] or optical imaging[63] have also been reported.

14.2.2.1 Development of Optical μFACS

The implementation of active sorting by optical forces in a μFACS is the result of almost two decades of research. In 1987, Buican *et al.*[64] pioneered the use of optical tweezers as a passive sorting mechanism when the authors realized that lymphocytes could be more easily trapped than erythrocytes. In 2003, Ozkan *et al.*[65] showed that an array of vertical-cavity surface-emitting lasers (VCSELs) could be used to sort beads and cells laterally at a T-junction using gradient forces but also vertically in a multilayer microfluidic device using scattering forces. Unfortunately, this approach was hindered by the limited power of VCSELs resulting in a small trapping force ($\sim$ 0.1 pN). In a similar manner, *Applegate et al.*[66] showed that bovine red blood cells would follow the path defined by a diode laser bar and could theoretically be directed into seven different channels. Closer to the principle of active sorting, Enger *et al.*[67] used a Y-type junction to detect and sort yeast cells from a 1/1000 mixture of yeast cells and *E. coli* bacteria. The main intent of these previous reports was to show proof-of-principle of cell sorting using optical forces as a deflection mechanism: these instruments were operated manually on a few cells. It was not until Tu *et al.*[37] and the corresponding follow-on work by Wang *et al.*[38] that an automated optical μFACS was reported. In their work, the authors characterized the performance of their instrument on HeLa cells using metrics commonly accepted by the cell sorting community. More recently, our group (Perroud *et al.*[34]) has successfully used a μFACS instrument to sort *Francisella-tularensis*-infected macrophage cells for studying host-pathogen interactions. To realize the full potential of LOC platforms, our optical μFACS has been integrated with an upstream cell preparation module (pathogen infection, immunostaining) and a downstream single-cell array for high-resolution imaging.[68]

14.2.2.2 Optical μFACS System Configuration

The optical μFACS can be roughly broken into four essential components: a microfluidic chip with its fluidic interface, an interrogation laser with its detection system, a trapping laser with dynamical control, and a decision-making process

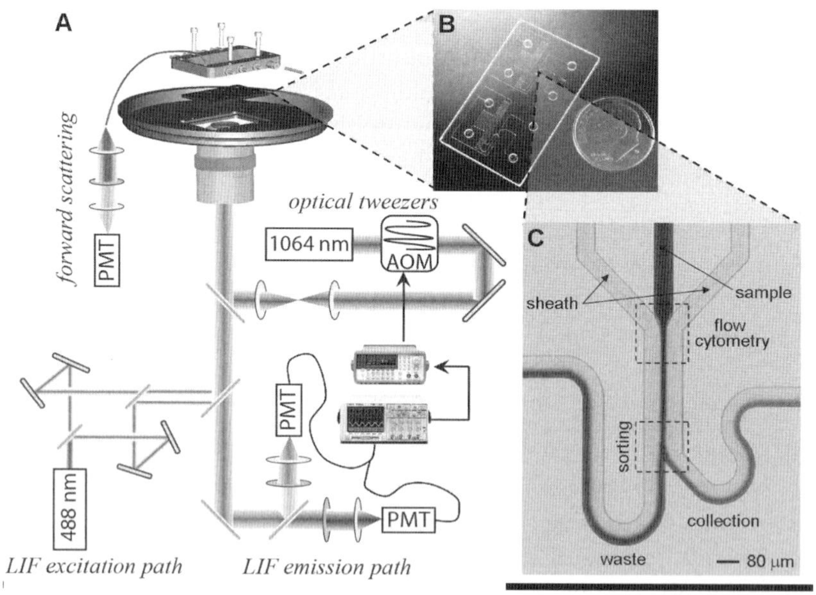

Figure 14.1. Schematic of μFACS based on optical forces. (A) Instrument layout with laser-induced fluorescence (LIF) excitation and emission paths, forward scattering detection, optical tweezers, and world-to-chip interface. (B) Microfluidic chip of two independent sorting modules with inlets on the left and outlets on the right. (C) Close-up view on flow cytometry and sorting regions of the chip where sample (blue dye) is hydrodynamically focused by two neighboring sheath flows (yellow) and split between a waste (left) and collection channel (right). [Reprinted with permission from T. D. Perroud, J. N. Kaiser, J. C. Sy, T. W. Lane, C. S. Branda, A. K. Singh and K. D. Patel, Microfluidic-based cell sorting of Francisella tularensis infected macrophages using optical forces. *Anal. Chem.* **80**(16), 6365–6372 (2008). Copyright 2008 American Chemical Society.] Color reference – pg. 353.

that synchronizes the detection event to the sorting event. Fig. 14.1 shows a general schematic of these basic components adapted to a standard inverted microscope.

Given the high laser power used in an optical μFACS, the microfluidic chip is generally made out of a substrate with a high transmission in the near-infrared region such as fused silica. A rigid substrate also prevents flow instabilities, which are detrimental to the overall efficiency of the cell sorter. To ensure precise control of the hydrodynamic focusing, cell velocity, and cell position, both input and output fluidic reservoirs are pressurized with individual electronic pressure controllers. A manifold connects the fluidic reservoirs to the microfluidic chip using an O-ring compression seal. Note that to maintain a constant throughput over an extended period of time, the buoyancy of the sample reservoir containing the cells must be increased to prevent cell sedimentation.

Following the same approach used on commercial flow cytometry instruments, the interrogation laser consists of solid-state lasers with wavelengths in the visible range for optimal compatibility with biologically relevant fluorophores. The interrogation laser beam is focused into the sample using a cylindrical lens to minimize signal fluctuations caused by lateral variations in the position of hydrodynamically focused cells. In the detection system, the forward-scattering and laser-induced fluorescence signals are isolated by interference optical filters before being detected by photomultiplier tubes. Note that forward scattering does not need to be detected on a photomultiplier tube since the signal is generally large enough for a photodiode. Side scattering is more difficult to achieve in a microfluidic chip given its planar geometry, but novel strategies such as integration of an optical waveguide in the chip have been demonstrated.[69]

For optical cell sorting, the required output of several watts in the near infrared region for the trapping laser limits the number of possible candidates. The conventional solid-state lasers for optical trapping such as neodymium : yttrium-aluminum-garnet (Nd:YAG), the neodymium : yttrium-lithium-fluoride (Nd:YLF), and the neodymium : yttrium-orthovanadate (Nd:YVO_4) lasers[54,56] are being replaced by more efficient and more compact diode-pumped solid state fiber lasers (*e.g.* Ytterbium fiber laser). Alternatively, diode bar lasers offer a low-cost solution in a compact package and do not require beam steering, but are limited by the small optical force generated (typically, 10-100 pN).[70] The rapid scanning of the focused trapping laser can be achieved using galvanometer mirrors, acousto-optical modulators, or electro-optical modulators. While the galvanometer mirrors offer the best transmission efficiency, their limited speed (1-2 KHz) and step response impacts the overall throughput of cell sorting. The acousto-optical modulator probably offers the best compromise between speed (up to GHz) and cost, but suffers from poor optical transmittance ($< 85\%$ per axis), which realistically limits the deflection to one dimension. The electro-optical modulator has unmatched speed coupled to efficient transmission, but is rather costly and has limited deflection range.

To couple the detection event to the sorting event, the analog signals from the photomultiplier tubes are sent into voltage comparators with preset voltage limits. If the signal of a particular channel falls within the preset window of voltages, a digital signal is sent from the voltage comparator to a programmable logic device to enable Boolean operations on several channels. The final output is a transistor-transistor logic (TTL) pulse that triggers a function generator to modulate the acousto-optical modulator, and thus steer the beam accordingly.

14.2.3 Operation of Optical μFACS

Similar to conventional FACS, each cell is focused into a single file, detected and analyzed by an interrogation laser, and then individually sorted. In Fig. 14.2, an illustration of the key steps involved in the sorting process are correlated with a sequence of four bright-field images showing the sorting of a macrophage cell. As illustrated previously in Fig. 14.1, the main microfluidic channel of an optical μFACS is divided into two distinct regions: an upstream flow cytometry region followed by a downstream optical-force cell-deflection region.

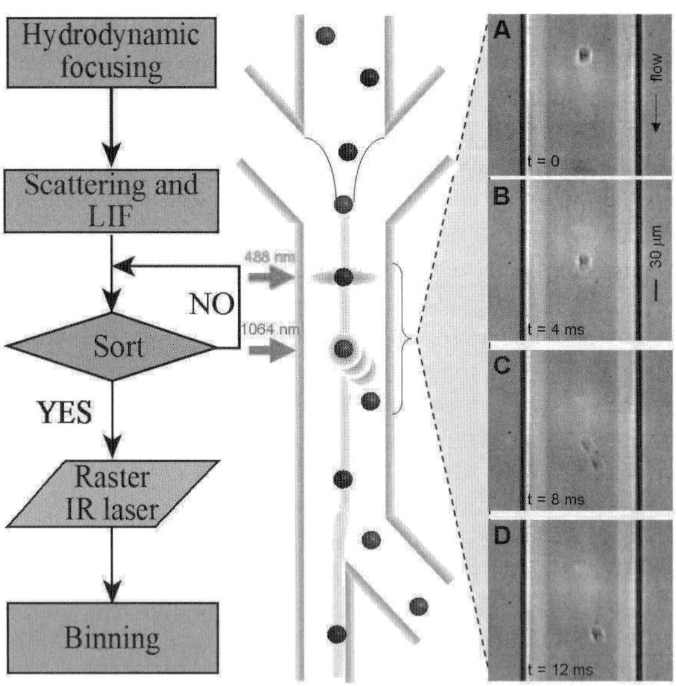

Figure 14.2. Principle and illustration of μFACS based on optical forces. The hydrodynamically focused macrophage cell: (A) is detected by forward scattering; (B) enters the near-infrared laser spot; (C) is deflected by optical gradient forces; (D) and finally is released in a different laminar flow stream. [Reprinted with permission from T. D. Perroud, J. N. Kaiser, J. C. Sy, T. W. Lane, C. S. Branda, A. K. Singh and K. D. Patel, Microfluidic-based cell sorting of Francisella tularensis infected macrophages using optical forces. *Anal. Chem.* **80**(16), 6365–6372 (2008). Copyright 2008 American Chemical Society.] Color reference – pg. 353.

Cells of interest are hydrodynamically focused by two neighboring sheath flows into a narrow 10–15 μm wide vertical plane. Other configurations that enable both in-plane and out-of-plane hydrodynamic focusing have been demonstrated using multi-layer soft lithography[71] or high Reynolds numbers.[72] Once hydrodynamically focused, each cell is detected by forward-scattering and interrogated by laser-induced fluorescence. These signals detected by photomultiplier tubes are compared to a threshold voltage, which in this particular case serves as the decision-making criteria for sorting. Crossing the threshold value, the signal generates a TTL pulse after a predetermined delay, corresponding to the cell's velocity multiplied by the distance between the interrogation and sorting regions. This delayed TTL pulse triggers a function generator sending a negative ramp function to the acousto-optical modulator, which spatially rasters the focused trapping laser at an angle to deflect the cell into a neighboring laminar flow stream. The shape of the function is important to maximize the interaction time between the laser and the cell. The function consists of four events: an immediate rise to turn on the near-infrared laser at the center of the focused-sample stream; a plateau to hold the beam in place until the cell arrives thus compensating for small variations in cell velocity; a negative slope to translate the beam at a defined angle (22°) for a distance of 54 μm; and a rapid fall to turn off the laser, resetting it for the next sorting event. When the cell enters the trapping laser spot (Fig. 14.2(B)), the laser starts translating at the same speed but with a different trajectory than that of the cell. This difference results in gradient forces that deflect the cell laterally (Fig. 14.2(C)). The overall effect is a displacement of a macrophage from the center of the channel to a neighboring streamline (Fig. 14.(2)D). The laminar nature of fluid flows in microfluidics ensures that the cell will stay on this path for downstream binning. The 150-μm-wide central channel is bifurcated into 80-μm-wide waste and 70- μm-wide collection channels. The channel lengths and widths have been designed to ensure that the hydrodynamically focused flow is directed into the waste channel. Only cells deflected by the laser can enter the collection channel. Although the exact split geometry is not critical for sorting, the configuration shown in Fig. 14.1 is preferred to a symmetrical T- or Y-type geometry. The side-split geometry is more forgiving when sorting partially deflected cells as it prevents aggregates from undesirably entering the collection channel. Furthermore, this design has also been shown to minimize the sticking of cells irreversibly at the sorting interface, avoiding biofouling and ultimately increasing the chip lifetime.

14.3 PERFORMANCE METRICS

The vast majority of μFACS reviewed in the literature is proof-of-concept devices and thus, lacks rigorous quantitative evaluation. As μFACS technology matures, researchers in the field have now begun to adopt metrics commonly used to characterize and compare benchtop FACS systems. Throughput, recovery, and purity are three well-accepted criteria for evaluating and comparing FACS performance.

For μFACS systems, criteria to track cell viability and enrichment factors for rare cells and precious samples are also worth considering.

Throughput can be defined as the number of cells that pass through the sorting region per unit of time; recovery is the ratio of successfully isolated cells to the total number of input target cells; purity is the fraction of target cells relative to the total number of cells sorted into the collection well. Lastly, the enrichment factor refers to the relative increase in the number of targeted cells collected from the starting mixture — an important parameter for μFACS systems applied to rare cell sorting. Each metric is interdependent, and as with most FACS parameters, compromises must be made to either balance the parameters for optimal performance or enhance one desired metric at the expense of the others. For example, a large throughput is achieved using high initial cell densities, fast cell velocities, and a small angular deflection (13 degrees). The first two factors result in small distances between successive cells and in large drag forces, increasing the probability of a miss and thus decreasing recovery. A small angular deflection results in a small cell displacement during deflection, thus increasing chances that unwanted cells accidentally enter the collection channel (false positive) thereby decreasing overall purity.

14.3.1 Comparison of Throughput, Recovery, and Purity for Different μFACS Sorting Strategies

Table 14.1 summarizes the performance of optical μFACS and compares this approach to alternative microfluidic-based cell sorting techniques that report metrics for sorting cells. As stated earlier, the efficiency of optical deflection in a microfluidic channel depends on the characteristics of the cell (e.g. diameter, relative refractive index, and shape). This dependence is illustrated in Table 14.1 when comparing the performance of optically sorting HeLa[38] and macrophage cells.[34] As a result, cells grown in suspension are more suitable to optical deflection than adherent cells because of their rigid cytoskeleton.

Compared to pneumatic valve[43] and thermoreversible gelation polymer[48] techniques, optical deflection can potentially achieve much higher throughputs and better detection accuracy. Throughput is ultimately limited by the minimum time needed to deflect a cell. For optical μFACS, the deflection time ($<$ 4 milliseconds) is shorter than the reaction time required for actuating pneumatic valves, forming thermally responsive gels, or overcoming the capacitance of the fluid. Fluidic displacement also induces undesired flow perturbations in the interrogation region, consequently limiting the detection accuracy.

Recently reported magneto- and dielectrophoretic sorting strategies have demonstrated impressive metrics for throughput, purities, and enrichment. These sorting methods are highly specialized techniques for limited applications, and thus, are not as general as optical μFACS. Both MACS (magnetic-activated cell sorting) and DACS (dielectrophoretic-activated cell sorting) strategies require that each target cell be covalently bound to either a magnetic or dielectrophoretic tag through antibodies (typically a 1 μm polystyrene bead). External labeling not

Table 14.1 Performance comparison for microfluidic-based cell sorters.

Ref.	Method	Cell type Sorted	Throughput [cells/s]	Purity [%]	Recovery [%]	Enrichment [fold]
Perroud[34]	Optical	Highly infected GFP-macrophages from uninfected	14–22	75–93	$55 \sim 60$	NA
Wang[38]	Optical	GFP HeLa from non-expressing cells	20–100	82–98	> 85	$63 \sim 71$
Fu[35]	Pneumatic switching	GFP tagged *E. coli* from non-tagged	1–44	NA	16–50	7–83
Sugino[48]	Thermo-reversible polymer	GFP-*E. coli*, YFP-*E. coli*, and DS-Red *E. coli*	0.4	93–99	89–92	NA
Adams[73]	Magnetic (MACS)	Magnetic particle tagged *E. coli* from non-tagged *E. coli*	28,000	NA	> 99	523
Kim[50]	DEP (DACS)	Dielectric particle tagged *E. coli* from non-tagged *E. coli*	$\sim 4,200$	> 99	> 99	260–930

only prevents sorting cells based on intra-cellular events, but also can interfere with the cell's physiology (e.g. phagocytosis of the particles). These particles have either high superparamagnetic or dielectric properties, which interact with the electromagnetic field generated by preprinted electrodes on the surface of the microfluidic channel. These sorters operate passively by guiding tagged cells along transverse field lines, and, as a result, can achieve very high throughputs and enrichment factors. However, caution must be used when comparing their metrics with optical µFACS. The guidelines used to report the metrics with optical µFACS are more stringent since the work is focused on sorting mammalian cells. Indeed, 1 µm *E. coli* bacteria, as reported with these techniques, has a smaller Stokes drag than 10 µm mammalian cells making it easier to sort. It remains to be seen if magnetic or dielectrophoretic forces represent a viable approach for sorting large mammalian cells.

One particularity of optical µFACS is the capability of changing the media surrounding the cell during a deflection;[67] whereas, hydrodynamic-based µFACS strategies deflect packets of fluid containing the cell. This "washing effect" can be particularly desirable if subsequent processing steps require the cell to be separated from contaminants present in the initial sample media (e.g. antibiotics, growth factors, or stabilizing agents).

Overall, optical µFACS mimics many of the protocols and well-accepted approaches used in conventional FACS, making it easy to implement with current methods used in cell biology.

14.3.2 Cell Health and Viability

As discussed in Section 2.1.2, the force necessary to deflect flowing cells (1-10 nN) in an optical µFACS requires several watts of laser power. Despite being quasi-transparent in the near-infrared region, cells can suffer from optically induced damage due to long exposures to intense radiation (opticution). Table 14.2 summarizes experimental parameters used in optical tweezers cell studies, where the key figure of merit is the total energy delivered to the cell. These works have shown that opticution can affect viability in human sperm,[74] reduce clonal growth in CHO cells,[75] inhibit motility in *E. coli* bacteria,[76] and lead to DNA damage in lymphoblasts.[77] Despite using equivalent power density, optical µFACS differs from conventional optical tweezers studies in that short exposure times are used (a few milliseconds). This brief interaction limits the total energy delivered to a cell to several milli-joules during each deflection, consequently avoiding potential damages induced by opticution. Wang *et al.*[38] showed that, following optical deflection, HeLa cells were not only viable, but also not activated by heat stress. Through our work,[34] we have thoroughly shown that viability, proliferation, activation state, and functionality of macrophage cells were not affected by the optical deflection. Additionally, initial observation by Kovac *et al.*[63] showed no effect on the membrane integrity of BA/F3 cells by laser tweezers. Note that the addition of near-infrared light-absorbing materials (e.g. organic dyes, gold nanoparticles) could have deleterious effects on the viability of the cells and should be carefully evaluated prior to use with optical μFACS. Nevertheless, optical μFACS remains a noninvasive approach to microfluidic-based cell sorting.

14.4 CRITICAL ISSUES

14.4.1 New Concepts to Overcome Limitations in Optical μFACS Systems

The previous section has shown optical μFACS as a complete system that has been not only characterized using established metrics, but also applied to biologically relevant experiments. Yet, innovative approaches are constantly being developed to decrease size and cost, improve the detection system, and multiplex the sorting process.

The high power necessary for cell deflection requires a class IV laser with its accompanying opto-mechanical components for beam steering. Such optical system is costly (> $10,000) and bulky. Diode lasers offer a lower-cost alternative in a compact package. For example, Applegate *et al.*[78] used a series of inexpensive fiber-focused diode lasers to create a trapping bar or row of diodes instead of a single-point modulated laser beam for binning particles. Currently, the limited output power of these lasers does not provide a sufficient optical force to deflect flowing cells at a reasonable throughput but future advances in laser technology could circumvent this problem.

Kovac *et al.*[63] reported an alternative to fixed single-point detection system by performing image-based cell sorting. Cells initially placed in microwell arrays were individually imaged, lifted out of their microwell using scattering forces, and swept away by a perpendicular crossflow. This approach is attractive because it allows sorting based on fluorescence localization (e.g. localized protein expression or translocation events) — not possible by traditional FACS instruments. Since this process is rather slow (up to 45 s per cell), this technology has low throughput (1 cell/min) and is realistically limited to dozens of cells.

Lau *et al.*[61] showed that Raman spectroscopy could be combined with laser tweezers to create a Raman-activated cell sorter (RACS). RACS relies on the intrinsic Raman spectrum of the cell instead of a fluorescent signal. Because of the inherently weak signal of Raman spectroscopy, a fundamental limitation of RACS is a long integration time (second to minutes) required for acquiring the spectrum than that of FACS (micro- to milliseconds). Nevertheless, this technique can be useful for cell samples that cannot be labeled fluorescently.

Most cellular functions are heterogeneous in nature and far from a simple "yes or no" answer. New μFACS strategies with multiple output bins are emerging to enhance the multiplexing capability of cell sorting. An implementation of optical sorting with multiple outputs was demonstrated by Applegate *et al.*[69] and illustrated in Fig. 14.3. Target particles were sorted into specific bins by modulating the amplitude of the deflection. A similar strategy was implemented using a thermoreversible gel polymer and fully characterized using *E. coli* cells expressing different fluorescent proteins.[48]

Table 14.2 Laser parameters at onset of opticution (adapted from Wang *et al.*[38] and Kovac *et al.*[63]).

Ref.	Sample	λ[nm]	Power [W]	Exposure time [s]	Spot size [μm]	Power Density [W/cm^2]	Energy density [J/cm^2]	Energy [J]	Damage
Liu[74]	Human sperm	1064	0.3	120	0.75	6.8e + 07	8.1e + 09	36	viability
Liang[75]	CHO	990	0.088	180	0.70	2.0e + 07	1.2e + 09	16	clonal growth
Neuman[76]	*E. coli*	1064	0.1	600	0.81	1.9e + 07	1.2e + 10	60	cell motility
Mohanty[77]	NC37 lymphoblasts	1064	0.12	30	0.75	2.7e + 07	8.1e + 08	3.6	DNA damage
Kovac[63]	BA/F3	980	0.125	20	8.6	2.2e + 05	4.3e + 06	2.5	none obvious
Wang[38]	HeLa	1070	13.2	0.004	4.9	7.0e + 07	2.8e + 05	0.053	none
Perroud[34]	Macrophage cells	1064	9.6	0.004	3.8	6.6e + 07	2.7e + 05	0.038	none

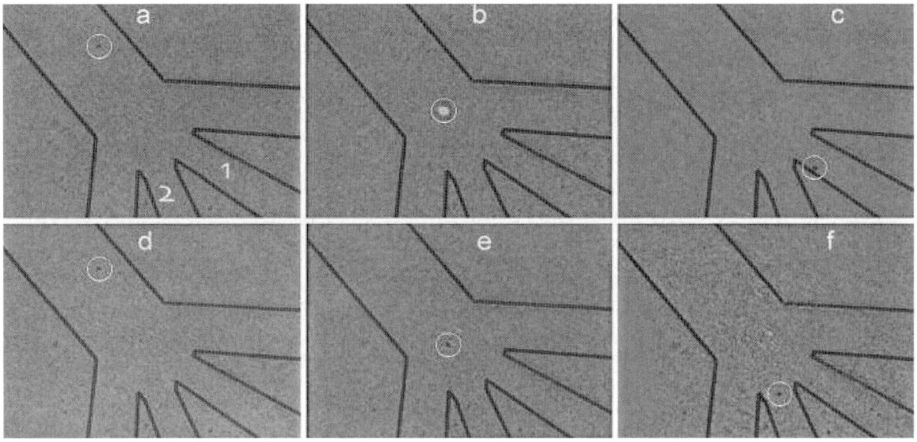

Figure 14.3. Images showing non-fluorescing colloid and fluorescing colloids being sorted into different output channels. Frames (a-c) show the path of the fluorescing colloid going into the outlet labeled 1; whereas (d-f) show the non-fluorescing colloid towards outlet 2. [R. W. Applegate, J. Squier, T. Vestad, J. Oakey, D. W. M. Marr, P. Bado, M. A. Dugan and A. A. Said, *Lab Chip* **6**(3), 422 (2006). Reproduced by permission of The Royal Society of Chemistry.]

The microfluidic substrates used with optical sorting are mainly made from glass because of its transparency in the near-infrared region. Commonly used alternatives for LOC devices such as polydimethylsiloxane[79] and plastics do not have optimal transmission properties and are generally not compatible. However, new strategies that move away from traditional materials have been introduced. Just recently, Lin *et al.*[80] reported on an optically induced dielectrophoretic (ODEP) sorter. In this interesting work, light patterns are used to make "virtual electrodes"

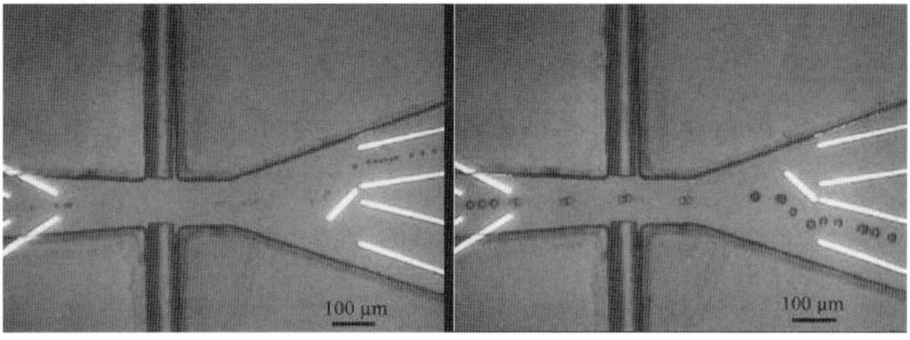

Figure 14.4. The focusing and sorting of microparticles with ODEP forces. A virtual electrode induced by illuminating the photoconductive layer generates a negative ODEP force. Particles are focused with virtual electrodes and the smaller (left) microparticles are sorted from the larger (right). [Reproduced with permission from Lin *et al. Biosensors & Bioelectronics* **24**, 572–578 (2008).] Color reference – pg. 354.

in an amorphous silicon layer inside the microfluidic channel. When an external projected light source interacts with the amorphous silicon layer, electron-hole pairs are excited, thus decreasing the impedance in the silicon. The applied AC voltage decreases in the liquid layer directly above the illuminated area such that it creates non-uniform electric fields. This concept is shown in Fig. 14.4, where virtual electrodes focus and sort particles into different "virtual channels".

In the subfield of droplet-based microfluidics, Cordero *et al.*[81] reported on the use of reconfigurable holographic optical beam patterns to actively manipulate droplets in a two-phase flow microfluidic system. The authors demonstrate that optical forces can be used to redirect, slow, stop, and merge droplets with diameters as large as 200 μm. Such techniques could be applied to optical sorting of whole organisms (e.g. *C. elegans*[82,83]).

14.4.2 Outlook on the Future of Optical μFACS

The most promising potential of optical μFACS is its integration with other microfluidic-based functionalities such as upstream cell sample preparation[84] and downstream single-cell imaging. Such novel instrumentation capable of multiplexed measurements would help decipher complex biological events in a quantitative, dynamic way at a single-cell level.

The Microscale Immune and Cell Analysis (MICA) platform in development at Sandia National Laboratories offers the ability to understand cell behavior at the molecular and cellular levels with unprecedented speed, resolution, sensitivity, and multiplexing.[85] The goal of the MICA platform is to provide a valuable tool to biologists for making correlated measurements not feasible with conventional instrumentation. To illustrate this concept, we coupled flow cytometry to single-cell imaging through optical cell sorting. On-chip flow cytometry is capable of rapid, multi-parametric measurements on large cell populations; whereas, microscopy imaging can provide spatial information at high-resolution, but only on a limited number of cells. These two fundamentally different, but complementary, techniques provide correlated measurements at the single-cell level. A picture of this integrated microfluidic platform is shown in Fig. 14.5, where the 12-trap single-cell array is occupied by individual cells.[86] In this integrated microfluidic platform, a large population of cells is analyzed using flow cytometry and particular cells of interest are optically sorted into a single-cell array, where each cell is trapped in a micropore by hydrodynamic confinement and imaged at high-resolution.[87]

14.5 CONCLUSIONS

The coupling of laser-based, optical traps into microfluidic platforms has led to the development of optical microfluidic-based fluorescence activated cell sorting. Optical μFACS is a proven non-invasive approach for the rapid sorting of live, unstressed, and functional cells, while guaranteeing sterility, biosafety, and

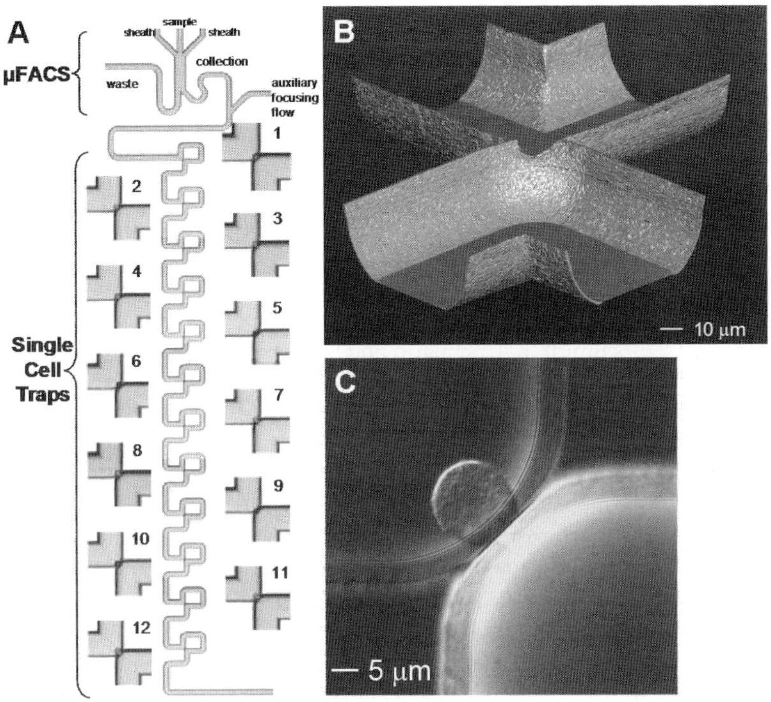

Figure 14.5. Tandem integration of µFACS with a single-cell array. (A) Upstream optical µFACS with downstream 12-trap single-cell array. Sorted cells enter the collection channel and are trapped sequentially against the micropores. Inset figures show fluorescently labeled macrophage cells trapped at each pore. (B) 3D reconstruction of a micropore from laser-scanning confocal microscopy. (C) Single macrophage cell trapped by a micropore and imaged by differential interference microscopy. Color reference – pg. 354.

cleanliness through disposability. The true potential of this technique will be fully realized through the monolithic integration of additional functionalities for multiplexed measurements. Continued development in this field will result in a portable and inexpensive instrument, thereby enabling the wide acceptance of this technology throughout the biomedical research field.

ACKNOWLEDGMENTS

Sandia is a multiprogram laboratory operated by Sandia Corporation, a Lockheed Martin Company, for the United States Department of Energy's National Nuclear Security Administration under contract DE-AC04-94Al85000. This work was supported by the Sandia LDRD program through the Microscale Immune Studies Laboratory project.

References

[1] A. Ashkin and J. M. Dziedzic, Optical trapping and manipulation of viruses and bacteria. *Science* **235**(4795), 1517–1520 (1987).

[2] A. Ashkin and J. M. Dziedzic, Internal cell manipulation using infrared-laser traps. *Proceedings of the National Academy of Sciences of the United States of America* **86**(20), 7914–7918 (1989).

[3] A. Ashkin, J. M. Dziedzic, J. E. Bjorkholm and S. Chu, Observation of a single-beam gradient force optical trap for dielectric particles *Optics Letters*, **11**(5), 288–290 (1986).

[4] A. Ashkin, J. M. Dziedzic and T. Yamane, Optical trapping and manipulation of single cells using infrared-laser beams. *Nature* **330**, 6150, 769–771 (1987).

[5] K. Svoboda, C. F. Schmidt, B. J. Schnapp and S. M. Block, Direct observation of kinesin stepping by optical trapping interferometry. *Nature* **365**(6448), 721–727 (1993).

[6] J. W. Dai and M. P. Sheetz, Mechanical-properties of neuronal growth cone membranes studied by tether formation with laser optical tweezers. *Biophysical Journal* **68**(3), 988–996 (1995).

[7] D. Raucher, T. Stauffer, W. Chen, K. Shen, S. L. Guo, J. D. York, Sheetz, M. P. and Meyer, T., Phosphatidylinositol 4,5-bisphoshate functions as a second messenger that regulates cytoskeleton-plasma membrane adhesion. *Cell* **100**(2), 221–228 (2000).

[8] P. A. Negulescu, T. B. Krasieva, A. Khan, H. H. Kerschbaum, Cahalan and M. D. Polarity of T cell shape, motility, and sensitivity to antigen. *Immunity* **4**(5), 421–430 (1996).

[9] A. D. Mehta, M. Rief, J. A. Spudich, D. A. Smith and R. M. Simmons, Single-molecule biomechanics with optical methods. *Science* **283**(5408), 1689–1695 (1999).

[10] A. Manz, N. Graber and H. M. Widmer, Miniaturized total chemical-analysis systems — a novel concept for chemical sensing. *Sensors and Actuators B: Chemical* **1**, (1-6), 244–248 (1990).

[11] D. N. Breslauer, P. J. Lee and L. P. Lee, Microfluidics-based systems biology. *Molecular Biosystems* **2**, 97–112 (2006).

[12] P. S. Dittrich, K. Tachikawa and A. Manz, Micro total analysis systems. Latest advancements and trends. *Analytical Chemistry* **78**(12), 3887–3907 (2006).

[13] J. El-Ali, P. K. Sorger and K. F. Jensen, Cells on chips. *Nature* **442**, (7101),403–411 (2006).

[14] W. H. Huang, F. Ai, Z. L. Wang and J. K. Cheng, Recent advances in single-cell analysis using capillary electrophoresis and microfluidic devices. *Journal of Chromatography B-Analytical Technologies in the Biomedical and Life Sciences* **866**, (1-2), 104–122 (2008).

[15] C. E. Sims and N. L. Allbritton, Analysis of single mammalian cells on-chip. *Lab Chip* **7**(4), 423–440 (2007).

[16] Y. Tanaka, K. Sato, T. Shimizu, M. Yamato, T. Okano and T. Kitamori, Biological cells on microchips: New technologies and applications. *Biosensors & Bioelectronics* **23**(4), 449–458 (2007).

[17] J. West, M. Becker, S. Tombrink and A. Manz, Micro total analysis systems: Latest achievements. *Analytical Chemistry* **80**(12), 4403–4419 (2008).

[18] P. Chen, X. J. Feng, W. Du and B. F. Liu, Microfluidic chips for cell sorting. *Frontiers in Bioscience* **13** 2464–2483 (2008).

[19] S. Sjolander and C. Urbaniczky, Integrated fluid handling-system for biomolecular interaction analysis. *Analytical Chemistry* **63**(20), 2338–2345 (1991).

[20] E. T. Lagally, P. C. Simpson and R. A. Mathies, Monolithic integrated microfluidic DNA amplification and capillary electrophoresis analysis system. *Sensors and Actuators B: Chemical* **63**(3), 138–146 (2000).

[21] A. E. Herr, A. V. Hatch, D. J. Throckmorton, H. M. Tran, J. S. Brennan, W. V. Giannobile and A. K. Singh, Microfluidic immunoassays as rapid saliva-based clinical diagnostics. *Proceedings of the National Academy of Sciences of the United States of America* **104**(13), 5268–5273 (2007).

[22] A. Khademhosseini, R. Langer, J. Borenstein and J. P. Vacanti, Microscale technologies for tissue engineering and biology. *Proceedings of the National Academy of Sciences of the United States of America* **103**(8), 2480–2487 (2006).

[23] N. Y. Zhang, H. D. Tan and E. S. Yeung, Automated and integrated system for high-throughput DNA genotyping directly from blood. *Analytical Chemistry* **71**(6), 1138–1145 (1999).

[24] H. D. Tan and E. S. Yeung, Automation and integration of multiplexed on-line sample preparation with capillary electrophoresis for high-throughput DNA sequencing. *Analytical Chemistry* **70**(19), 4044–4053 (1998).

[25] T. M. Squires and S. R. Quake, Microfluidics: Fluid physics at the nanoliter scale. *Reviews of Modern Physics* **77**(3), 977–1026 (2005).

[26] L. A. Herzenberg, R. G. Sweet and L. A. Herzenberg, Fluorescence-activated cell sorting. *Scientific American* **234**(3), 108–116 (1976).

[27] S. F. Ibrahim and G. van den Engh, High-speed cell sorting: Fundamentals and recent advances. *Current Opinion in Biotechnology* **14**(1), 5–12 (2003).

[28] L. Murray, B. Chen, A. Galy, S. Chen, R. Tushinski, N. Uchida, R. Negrin, G. Tricot, S. Jagannath, D. Vesole, B. Barlogie, R. Hoffman, and A. Tsukamoto, Enrichment of human hematopoietic stem-cell activity in the Cd34(+)thy-1(+)lin(-) subpopulation from mobilized peripheral-blood. *Blood* **85**(2), 368–378 (1995).

[29] N. Eliopoulos, A. Al-Khaldi, C. M. Beausejour, R. L. Momparler, L. F. Momparler and J. Galipeau, Human cytidine deaminase as an *ex vivo* drug selectable marker in gene-modified primary bone marrow stromal cells. *Gene Therapy* **9**(7), 452–462 (2002).

[30] C. M. F. Dirven, J. Grill, M. L. M. Lamfers, P. Van der Valk, A. M. Leonhart, V. W. Van Beusechem, H. J. Haisma, H. M. Pinedo, D. T. Curiel, W. P. Vandertop and W. R. Gerritsen, Gene therapy for meningioma: Improved gene delivery with targeted adenoviruses. *Journal of Neurosurgery* **97**(2), 441–449 (2002).

[31] R. M. Jindal, P. McShane, D. W. R. Gray and P. J. Morris, Isolation and purification of pancreatic-islets by fluorescence-activated islet sorter. *Transplantation Proceedings* **26**(2), 653–653 (1994).

[32] L. A. Johnson, G. R. Welch, K. Keyvanfar, A. Dorfmann, E. F. Fugger and J. D. Schulman, Gender preselection in humans — Flow cytometric separation of X-spermatozoa and Y-spermatozoa for the prevention of X-linked diseases. *Human Reproduction* **8**(10), 1733–1739 (1993).

[33] H. M. Shapiro, *Practical Flow Cytometry*. Wiley-Liss: New York, (2003).

[34] T. D. Perroud, J. N. Kaiser, J. C. Sy, T. W. Lane, C. S. Branda, A. K. Singh and K. D. Patel, Microfluidic-based cell sorting of Francisella tularensis infected macrophages using optical forces. *Analytical Chemistry* **80**(16), 6365–6372 (2008).

[35] A. Y. Fu, C. Spence, A. Scherer, F. H. Arnold and S. R. Quake, A microfabricated fluorescence-activated cell sorter. *Nature Biotechnology* **17**(11), 1109–1111 (1999).

[36] M. P. MacDonald, G. C. Spalding and K. Dholakia, Microfluidic sorting in an optical lattice. *Nature* **426**(6965), 421–424 (2003).

[37] E. Tu, H. C. Zhang, M. M. Wang, D. E. Raymond, J. M. Yang, N. Hagen, B. Dees, E. M. Mercer, A. H. Forster, S. Attari, C. Richardson, I. B. Kariv, M. Chachisvillis, M. Hara, P. J. Marchand and W. F. Butler, Microfluidic cell analysis and sorting using photonic forces. In *Optical Trapping and Optical Micromanipulation*, Dholakia, K. Spalding, G. C., Eds. **5514**, 774–785, (2004).

[38] M. M. Wang, E. Tu, D. E. Raymond, J. M. Yang, H. C. Zhang, N.Hagen, B. Dees, E. M. Mercer, A. H. Forster, I. Kariv, P. J. Marchand, and W. F. Butler, Microfluidic sorting of mammalian cells by optical force switching. *Nature Biotechnology* **23**(1), 83–87 (2005).

[39] H. Andersson and A. van den Berg, Microfluidic devices for cellomics: A review. *Sensors and Actuators B-Chemical* **92**(3), 315–325 (2003).

[40] K. Dholakia, M. P. MacDonald, P. Zemanek and T. Cizmar, Cellular and colloidal separation using optical forces. *Laser Manipulation of Cells and Tissues* **82** 467–495 (2007).

[41] C. Q. Yi, C. W. Li, S. L. Ji and M. S. Yang, Microfluidics technology for manipulation and analysis of biological cells. *Analytica Chimica Acta* **560**(1–2), 1–23 (2006).

[42] P. S. Dittrich and P. Schwille, An integrated microfluidic system for reaction, high-sensitivity detection, and sorting of fluorescent cells and particles. *Analytical Chemistry* **75**(21), 5767–5774 (2003).

[43] A. Y. Fu, H. P. Chou, C. Spence, F. H. Arnold and S. R. Quake, An integrated microfabricated cell sorter. *Analytical Chemistry* **74**(11), 2451–2457 (2002).

[44] J. Kruger, K. Singh, A. O'Neill, C. Jackson, A. Morrison and P. O'Brien, Development of a microfluidic device for fluorescence activated cell sorting. *Journal of Micromechanics and Microengineering* **12**(4), 486–94 (2002).

[45] A. Wolff, I. R. Perch-Nielsen, U. D. Larsen, P. Friis, G. Goranovic, C. R. Poulsen, J. P. Kuttera and P. Telleman, Integrating advanced functionality in a microfabricated high-throughput fluorescent-activated cell sorter. *Lab Chip* **3**(1), 22–7 (2003).

[46] C. T. Ho, R. Z. Lin, H. Y. Chang and C. H. Liu, Micromachined electrochemical T-switches for cell sorting applications. *Lab Chip* **5**(11), 1248–1258 (2005).

[47] Y. Shirasaki, J. Tanaka, H. Makazu, K. Tashiro, S. Shoji, S. Tsukita, and Funatsu, T., On-chip cell sorting system using laser-induced heating of a thermoreversible gelation polymer to control flow. *Analytical Chemistry* **78**(3), 695–701 (2006).

[48] H. Sugino, K. Ozaki, Y. Shirasaki, T. Arakawa, S. Shoji and T. Funatsu, On-chip microfluidic sorting with fluorescence spectrum detection and multiway separation. *Lab Chip* (2009).

[49] X. Y. Hu, P. H. Bessette, J. R. Qian, C. D. Meinhart, P. S. Daugherty and H. T. Soh, Marker-specific sorting of rare cells using dielectrophoresis. *Proceedings of the National Academy of Sciences of the United States of America* **102**(44), 15757–15761 (2005).

[50] U. Kim, J. R. Qian, S. A. Kenrick, P. S. Daugherty and H. T. Soh, Multitarget dielectrophoresis activated cell sorter. *Analytical Chemistry* **80**(22), 8656–8661 (2008).

[51] U. Kim, C. W. Shu, K. Y. Dane, P. S. Daugherty, J. Y. J. Wang and H. T. Soh, Selection of mammalian cells based on their cell-cycle phase using dielectrophoresis. *Proceedings of the National Academy of Sciences of the United States of America* **104**(52), 20708–20712 (2007).

[52] B. H. Lapizco-Encinas, B. A. Simmons, E. B. Cummings and Y. Fintschenko, Dielectrophoretic concentration and separation of live and dead bacteria in an array of insulators. *Analytical Chemistry* **76**(6), 1571–1579 (2004).

[53] K. Ahn, C. Kerbage, T. P. Hunt, R. M. Westervelt, D. R. Link, and D. A. Weitz, Dielectrophoretic manipulation of drops for high-speed microfluidic sorting devices. *Applied Physics Letters* **88**(2), 24104–1 (2006).

[54] K. Svoboda and S. M. Block, Biological applications of optical forces. *Annual Review of Biophysics and Biomolecular Structure* **23**, 247–285 (1994).

[55] M. J. Lang and S. M. Block, Resource letter: LBOT-1: Laser-based optical tweezers. *American Journal of Physics* **71**(3), 201–215 (2003).

[56] K. C. Neuman and S. M. Block, Optical trapping. *Review of Scientific Instruments* **75**(9), 2787–2809 (2004).

[57] H. C. Hunt and J. S. Wilkinson, Optofluidic integration for microanalysis. *Microfluidics and Nanofluidics* **4**(1–2), 53–79 (2008).

[58] K. Dholakia, W. M. Lee, L. Paterson, M. P. MacDonald, R. McDonald, I. Andreev, P. Mthunzi, C. T. A. Brown, R. F. Marchington and A. C. Riches, Optical separation of cells on potential energy landscapes: Enhancement with dielectric tagging. *IEEE Journal of Selected Topics in Quantum Electronics* **13**(6), 1646–1654 (2007).

[59] R. Drezek, A. Dunn and R. Richards-Kortum, Light scattering from cells: Finite-difference time-domain simulations and goniometric measurements. *Applied Optics* **38**(16), 3651–3661 (1999).

[60] W. H. Wright, G. J. Sonek and M. W. Berns, Radiation trapping forces on microspheres with optical tweezers. *Applied Physics Letters* **63**(6), 715–717 (1993).

[61] A. Y. Lau, L. P. Lee and J. W. Chan, An integrated optofluidic platform for Raman-activated cell sorting. *Lab Chip* **8**(7), 1116–1120 (2008).

[62] M. D. Vahey and J. Voldman, An equilibrium method for continuous-flow cell sorting using dielectrophoresis. *Analytical Chemistry* **80**(9), 3135–3143 (2008).

[63] J. R. Kovac and J. Voldman, Intuitive, image-based cell sorting using optofluidic cell sorting. *Analytical Chemistry* **79**(24), 9321–9330 (2007).

[64] T. N. Buican, M. J. Smyth, H. A. Crissman, G. C. Salzman, C. C. Stewart and J. C. Martin, Automated single-cell manipulation and sorting by light trapping. *Applied Optics* **26**(24), 5311–5316 (1987).

[65] M. Ozkan, M. Wang, C. Ozkan, R. Flynn, A. Birkbeck and S. Esener, Optical manipulation of objects and biological cells in microfluidic devices. *Biomedical Microdevices* **5**(1), 61–67 (2003).

[66] R. W. Applegate, J. Squier, T. Vestad, J. Oakey and D. W. M. Marr, Optical trapping, manipulation, and sorting of cells and colloids in microfluidic systems with diode laser bars. *Optics Express* **12**(19), 4390–4398 (2004).

[67] J. Enger, M. Goksor, K. Ramser, P. Hagberg and D. Hanstorp, Optical tweezers applied to a microfluidic system. *Lab Chip* **4**(3), 196–200 (2004).

[68] K. D. Patel, T. D. Perroud, J. N. Kaiser, C. S. Branda, T. W. Lane and A. K. Singh, In On-chip Flow Cytometry and Single-Cell Imaging in tandem: Integration of a μFACS with a Single-Cell Array. *The 12th International Conference on Miniaturized Systems for Chemistry and Life Sciences*, San Diego (CA), October 12–1 (2008)

[69] R. W. Applegate, J. Squier, T. Vestad, J. Oakey, D. W. M. Marr, P. Bado, M. A. Dugan and A. A. Said, Microfluidic sorting system based on optical waveguide integration and diode laser bar trapping. *Lab Chip* **6**(3), 422–426 (2006).

[70] R. W. Applegate, J. Squier, T. Vestad, J. Oakey and D. W. M. Marr, Fiber-focused diode bar optical trapping for microfluidic flow manipulation. *Applied Physics Letters* **92**(1), 3 (2008).

[71] C. Simonnet and A. Groisman, Two-dimensional hydrodynamic focusing in a simple microfluidic device. *Applied Physics Letters* **87**(11) (2005).

[72] X. L. Mao, J. R. Waldeisen and T. J. Huang, Microfluidic drifting — Implementing three-dimensional hydrodynamic focusing with a single-layer planar microfluidic device *Lab Chip* **2007**, 7, 1260–1262.

[73] J. D. Adams, U. Kim and H. T. Soh, Multitarget magnetic activated cell sorter. *Proceedings of the National Academy of Sciences of the United States of America* **105**(47), 18165–18170 (2008).

[74] Y. Liu, G. J. Sonek and M. W. Berns, B. J. Tromberg, Physiological monitoring of optically trapped cells: Assessing the effects of confinement by 1064-nm laser tweezers using microfluorometry. *Biophysical Journal* **71**(4), 2158–2167 (1996).

[75] H. Liang, K. T. Vu, P. Krishnan, T. C. Trang, D. Shin, S. Kimel and M. W. Berns, Wavelength dependence of cell cloning efficiency after optical trapping. *Biophysical Journal* **70**(3), 1529–1533 (1996).

[76] K. C. Neuman, E. H. Chadd, G. F. Liou, K. Bergman and S. M. Block, Characterization of photodamage to Escherichia coli in optical traps. *Biophysical Journal* **77**(5), 2856–2863 (1999).

[77] S. K. Mohanty, A. Rapp, S. Monajembashi, P. K. Gupta and K. O. Greulich, Comet assay measurements of DNA damage in cells by laser microbeams and trapping beams with wavelengths spanning a range of 308 nm to 1064 nm. *Radiation Research* **157**(4), 378–385 (2002).

[78] R. W. Applegate, J. Squier, T. Vestad, J. Oakey and D. W. Marr, Fiber-focused diode bar optical trapping for microfluidic flow manipulation. *Applied Physics Letters* 013904–1–3 (2008).

[79] D. C. Duffy, J. C. McDonald, O. J. A. Schueller and G. M. Whitesides, Rapid prototyping of microfluidic systems in poly(dimethylsiloxane). *Analytical Chemistry* **70**(23), 4974–4984 (1998).

[80] Y. H. Lin, and B., L. G., Optically induced flow cytometry for continous microparticle counting and sorting. *Biosensors & Bioelectronics* **24**, 572–578 (2008).

[81] M. L. Cordero, D. R. Burnham, C. N. Baroud and D. McGloin, Thermocapillary manipulation of droplets using holographic beam shaping: Microfluidic pin ball. *Applied Physics Letters* **93**(3) (2008).

[82] K. H. Chung, M. M. Crane and H. Lu, Automated on-chip rapid microscopy, phenotyping and sorting of *C. Elegans*. *Nature Methods* **5**(7), 637–643 (2008).

[83] C. B. Rohde, F. Zeng, R. Gonzalez-Rubio, M. Angel and M. F. Yanik, Microfluidic system for on-chip high-throughput whole-animal sorting and screening at subcellular resolution. *Proceedings of the National Academy of Sciences of the United States of America* **104**(35), 13891–13895 (2007).

[84] N. Srivastava, J. S. Brennan, R. F. Renzi, M. Wu, S. S. Branda, A. K. Singh and A. E. Herr, Fully integrated microfluidic platform enabling automated phosphoprofiling of macrophage response. *Analytical Chemistry* DOI: 10.1021/ac8024224 (2009).

[85] Microscale Immune and Cell Analysis. http://www.ca.sandia.gov/mica/ (Dec 15, 2008).

[86] K. D. Patel, T. D. Perroud, C. S. Branda, T. W. Lane and A. K. Singh, In on-chip flow cytometry and single-cell imaging in tandem: integration of a uFACS with a single-cell array. *The 12th International Conference on Miniaturized Systems for Chemistry and Life Sciences*, San Diego, CA, (2008). Locascio, L. E. Gaitan, M. Paegel, B. M. Ross, D. J. Vreeland, W. N., Eds. The Chemical and Biolgoical Microsystems Society: San Diego, CA, 1867–1869 (2008).

[87] T. D. Perroud, R. J. Meagher, M. P. Kanouff, R. F. Renzi, M. Y. Wu, A. K. Singh and K. D. Patel, Isotropically etched radial micropore for cell concentration, immobilization, and picodroplet generation. *Lab Chip* **9**(4), 507–515 (2009).

Chapter Fifteen

Raman Spectroscopy: Label-Free Cell Analysis and Sorting

James W. Chan

NSF Center for Biophotonics Science and Technology, University of California, Davis, 2700 Stockton Blvd Suite 1400, Sacramento, CA, USA
Physics Division, Lawrence Livermore National Laboratory
7000 East Ave, L-211, Livermore, California, USA
chan19@llnl.gov

Quantitative, label-free biochemical analysis of single, living cells is a highly sought after modality in the biological sciences. While flow cytometry provides a variety of parameters for characterizing and sorting cells, it currently lacks a method for noninvasive and nondestructive chemical analysis. This chapter introduces Raman spectroscopy as a new component in flow cytometry for label-free analysis and sorting of living cells. While similar in principle to current fluorescence-based sorting systems in its design layout and operation, key differences between the Raman and fluorescence process necessitate research focused on developing novel optical and microfluidic schemes applicable for Raman analysis of cells. A particular emphasis must be placed on addressing the weak Raman signal intensity, which is a major obstacle to realizing a Raman sorting system. Although unlikely to rival current cytometers in its throughput, a Raman system could offer new capabilities such as sorting smaller populations for broad biological applications, especially when noninvasive, label-free analysis of living cells is a key requirement. This chapter will introduce the basic principles of Raman spectroscopy, the basic instrumentation, and both seminal and recent studies that have established the foundation for achieving Raman-activated flow cytometry.

15.1 NOVEL RAMAN MARKERS FOR MICROFLOW CYTOMETRY

Flow cytometry has become a powerful technique for single-cell analysis in fields such as molecular biology, immunology, pathology, and medicine.[1] An important area in flow cytometry is the search for new parameters that can be measured from a single cell to improve the identification and sorting of cell populations. New parameters are needed to improve the discrimination of cell populations, especially

The Microflow Cytometer by J S Kim & F S Ligler

www.panstanford.com
978-981-4267-41-0

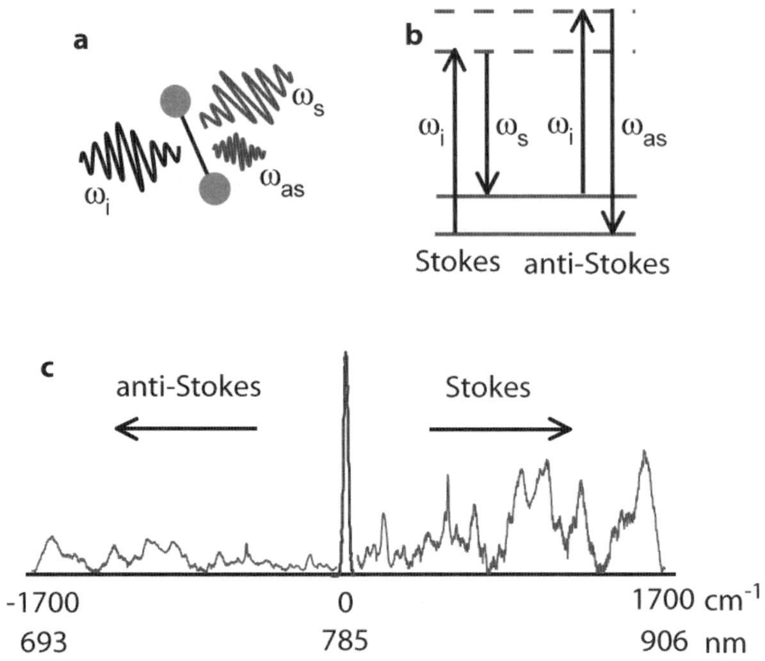

Figure 15.1. (a) Raman scattering is the inelastic scattering of photons at frequency ω_i by molecular bonds, generating new frequencies at ω_s or ω_{as}. (b) Energy diagram depicting the Stokes and anti-Stokes vibrational transitions (solid and dashed lines denote vibrational levels and virtual states, respectively) (c) Stokes shifted signals are more intense than anti-Stokes signals. In spontaneous Raman spectroscopy, Stokes signals are detected to create a Raman 'molecular fingerprint' spectrum of the sample. Peak positions are represented in wavenumber units (cm^{-1}). Corresponding wavelength units are shown for the representative case of excitation with 785 nm light. Color reference – pg. 355.

when no known definitive markers exist for certain cell types, such as cancer and stem cells. Currently, the three most common parameters used in flow cytometry for quantitative analysis and sorting of cells are fluorescence, light scattering, and impedance. Of these variables, fluorescence is the only parameter that provides chemical selectivity. Fluorescence-based cell analysis and sorting rely on the use of exogenous fluorescent dyes to label specific biomolecules. These dyes may often be toxic. In addition, cell permeabilization and fixation may be required for intracellular staining. In either case, the cells may be rendered unviable and unusable after the analysis, which is undesirable for applications such as clinical transplantation of the sorted cells. In these cases, a label-free method for cell classification that minimizes cell perturbation would be more suitable. Electrical impedance and light scattering, which measure cell size and granularity, respectively, are the only two available label-free parameters for cell sorting. However, because both do not provide specific biochemical information of the cell, it is usually insufficient to rely on only these two parameters to achieve accurate sorting of cell populations.

Raman spectroscopy[2] is one of the few optical techniques that can provide optical signatures related to the intrinsic molecular composition of tissues and cells without needing to introduce labels. Acquisition of the Raman signals can be achieved noninvasively and nondestructively without cell perturbation, thus potentially allowing living biological samples to remain viable after the interrogation. As illustrated in Fig. 15.1, Raman spectroscopy depends upon the inelastic scattering of incident photons by intrinsic molecular bonds of the sample, which results in scattered photons that are shifted in energy from that of the incident light. These photons either gain (anti-Stokes shifted) or lose (Stokes shifted) energy from this process by an amount equivalent to the characteristic vibrational energy of the particular bond. At room temperature, the Stokes-shifted photons generate a stronger signal than the anti-Stokes signal due to the majority of the molecules being in the ground vibrational state instead of the excited state, as defined by the Boltzmann's distribution. Since the signal intensity is proportional to the number of bonds present, Raman spectroscopy can be used for quantitative analysis. Detection of all scattered Stokes signals results in a Raman fingerprint spectrum of the sample, with Raman bands identified by their shift, in wave number (cm^{-1}) units, from the excitation wavelength.

The potential advantages of Raman spectroscopy can be best appreciated by comparing it to the more familiar fluorescence process used in flow cytometry. Unlike fluorescence, Raman spectroscopy yields *multiple* narrow spectral lines simultaneously. Each is assignable to specific vibrational modes of different biomolecules that, collectively, reflect the overall biochemical inventory of the cell. Multiplexed detection of these Raman signatures can potentially be used to improve discrimination of cell populations. Also, Raman signatures can be very sensitive to the biological state of the cell[3–5] (e.g. apoptosis, stage of cell cycle) and therefore may better reflect the inherent biology of the cell as well as differences between cells. Because Raman is a scattering process, any wavelength of light can theoretically be used to generate a Raman spectrum. This is because the Raman process involves transitions to virtual states (as denoted by the dashed lines in the energy diagram of Fig. 15.1) as opposed to real electronic transitions such as in a fluorescence process. Consequently, longer wavelength light in the near infrared can be used, which reduces light absorption by the sample that can lead to photodamage. As a label-free approach that induces minimal sample perturbation on a living cell while providing novel 'optical markers', Raman spectroscopy has the potential to offer unique capabilities to address current limitations in single cell analysis and sorting for clinical and biomedical applications.

15.2 CHARACTERISTICS OF A RAMAN-BASED CYTOMETER

The overall design layout of a Raman-based flow cytometer is, in principle, equivalent to standard fluorescence-based flow cytometers. A schematic of the general layout is shown in Fig. 15.2. It is comprised of similar hardware components (excitation source, optical filters, detectors, microfluidics, data processors) with comparable functionality arranged in a similar configuration. However, the key fundamental differences between the Raman and fluorescence process, as discussed in the previous section, dictate the unique parameters and design requirements for a Raman system. The following discussion will highlight the major differences between a Raman-based and fluorescence-based system.

Unlike fluorescence, which requires multiple lasers at specific wavelengths to excite different dyes, Raman spectroscopy only requires a single laser. Since Raman spectroscopy does not require a specific wavelength to generate Raman signals, any wavelength can theoretically be used. However, a near-infrared (NIR) laser source is typically chosen because this wavelength significantly reduces the generation of autofluorescence signals from biological materials that can overwhelm the weaker Raman signals. In addition, the optical window for

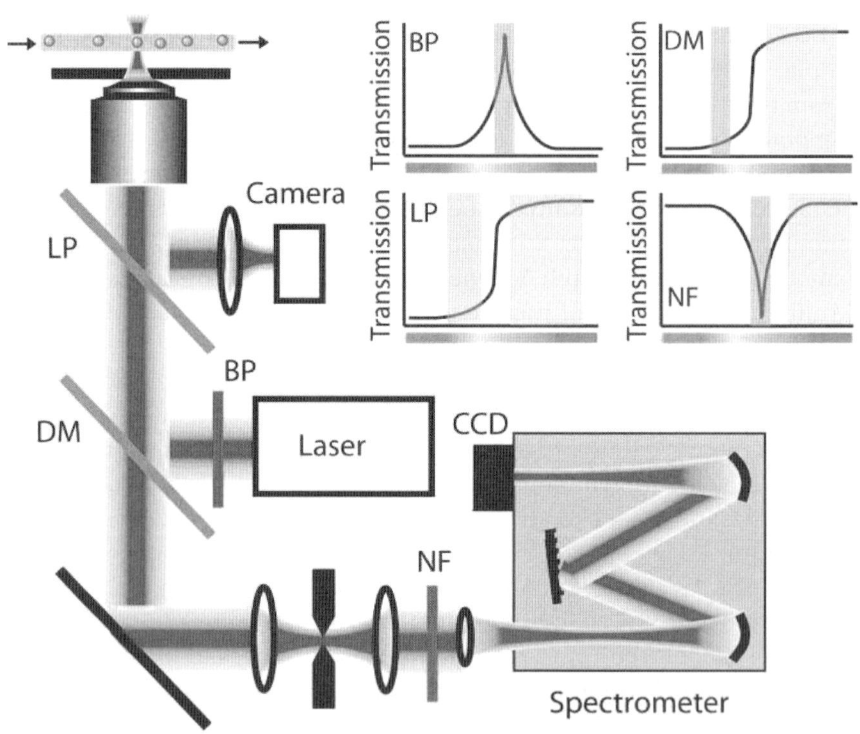

Figure 15.2. A schematic diagram of a single cell Raman analysis system. A 785 nm CW laser is delivered into a high numerical aperture (1.2-1.4) objective focused into a microfluidic device. Raman signals detected in the epi-direction are sent through a confocal pinhole, and into a spectrograph equipped with a back-illuminated, deep depletion thermoelectrically cooled CCD camera. An additional camera captures white light images of the cells. Inset shows the transmission profiles of the optics in the system (BP–785 nm bandpass filter, DM –785 nm dichroic longpass mirror, LP–longpass dichroic mirror, NF–785 nm notch filter). Red, brown, and green shading represents laser, Raman signal, and visible white light respectively. Color reference – pg. 355.

most biological materials, where there is minimal light absorption, occurs in the near-infrared region; hence, the use of light at this wavelength for Raman spectroscopy minimizes potential cell damage due to light absorption. The excitation laser must be capable of delivering monochromatic light with a narrow linewidth (e.g. 1×10^{-4} nm). This narrow line shape is needed in order to generate spectra with spectrally narrow Raman peaks that are typically of order 10 cm^{-1} in width. A narrow bandpass (BP) filter is typically placed in front of the laser to suppress additional laser lines from interfering with the Raman signal. A dichroic longpass mirror (DM) with a sharp cut-on wavelength is needed because Raman shifts of $\sim 400\,cm^{-1}$ need to be detected. An additional notch filter (NF) is usually needed to suppress residual Rayleigh scattered light in front of the detector. A Raman spectrum consists of multiple Raman peaks within a ~100 nm spectral window. In order to detect all signals, a spectrometer equipped with a sensitive CCD array detector simultaneously records the intensity of the multiple spectral lines. The spectrometer should have a typical resolution of $5 - 8\,cm^{-1}$. This detection system is used in lieu of a single-channel point detector (e.g. photomultiplier tube) and filters that are commonly used to discriminate fluorescence at defined wavelengths. A full system can sit on a small benchtop footprint (e.g. 2′×2′), with the potential for further miniaturization. The one major limitation of Raman spectroscopy that challenges its practical use in flow cytometry is the inherently low Raman cross sections. While fluorescence cross sections can reach as high as $10^{-11}\,cm^2$, typical Raman scattering cross sections are $10^{-26}\,cm^2$, some 15 orders of magnitude weaker! This means that longer interrogation times for a single cell are necessary, resulting in a low throughput system. Therefore, the weak Raman scattering is probably the single most important issue that needs to be addressed in order to realize a Raman-activated flow cytometer.

One solution is to achieve high laser intensities using tight focusing conditions with high numerical aperture microscope objectives to illuminate a single cell with a diffraction-limited spot. These focusing conditions are much higher than in current flow cytometer configurations. Focal volumes of $\sim$ 0.5–1 μm^3 can be achieved with 1.2–1.4 NA microscope objectives. Typical laser powers ranging from 5–20 mW are used. This assists in increasing the Raman signal and therefore reducing the acquisition time for each individual cell. This tight focusing condition also reduces the generation of background signals from the solution or any materials in close proximity to the cell, which can often be strong enough to interfere with the weak Raman signal from the biological cell. Background is typically not a concern in fluorescence flow cytometry due to the stronger fluorescence signals. Even with ideal focusing conditions, the optimal Raman sorting system still requires acquisition times in the tens of seconds (typically at least 30 seconds). Therefore, unlike fluorescence-based systems in which cells can rapidly flow past the laser beam and still generate a detectable signal, the Raman-based system would require each cell to remain within the excitation beam path for a longer time.

The requirement for long signal acquisition times necessitates novel optical and microfluidic approaches that deviate from conventional flow cytometry designs. Unique optical schemes are needed that can immobilize a cell within the laser beam as well as appropriate microfluidic designs that are compatible with the Raman analysis and sorting of cells. Another important research focus is the development of alternative Raman methods, such as coherent anti-Stokes Raman scattering (CARS)[6] spectroscopy, that can be used in place of Raman spectroscopy to significantly amplify the weak Raman signals, which can lead to higher throughput Raman systems. These research topics will be discussed in greater detail in the next section.

In its current state, Raman cytometry cannot compete with the speed of existing cytometers, which can reach as high as 70,000 cells/sec. Even with research developments to improve the system, achieving comparable throughput rates in a Raman-based system will be a major challenge. However, a Raman system has the unique advantage of being able to analyze small cell populations, something that is not possible with current flow cytometers, which depend on large numbers of cells. Therefore, even with lower throughput rates (for example, it is anticipated that analysis of several hundred to several thousand cells per hour can be achieved for optimized Raman and CARS systems), Raman flow cytometry may be ideally suited for specific applications where small cell populations are available for analysis and sorting, such as for isolating and purifying small stem cell populations or diagnosing disease (e.g. cancer, infectious disease). Such analyses could be performed without any extensive sample preparation that raises toxicity issues or that adversely affects the cells' biological functions.

15.3 REVIEW OF PAST AND CURRENT DEVELOPMENTS

A series of pioneering studies in the early 1990s and more recent works in the past ten years are helping to define Raman spectroscopy as a viable label-free technique for single cell analysis. These studies have mainly served to (1) improve the performance of the method and (2) to demonstrate its biological, biomedical, and clinical relevance through various application studies. In this section, a brief summary of these studies will be presented in chronological order to review the progress towards achieving Raman-based cytometry.

15.3.1 Single Cell Raman Spectroscopy

The seminal work by Puppels *et al.* in 1990[7] was the first to demonstrate the ability to analyze *single* living cells with nonresonant Raman spectroscopy using a novel, highly sensitive confocal Raman microspectrometer capable of high spatial resolution ($<1\ \mu m^3$). These studies opened up the possibility of studying single-cell behavior and obtaining Raman information about distributions of cell populations, rather than providing only bulk-averaged information. Prior to this, single-cell Raman analysis was only feasible on sperm cells (due to the high density of condensed DNA that gave strong Raman signals), and when using ultraviolet-resonance Raman spectroscopy, which is not a viable technique for analyzing live samples because they can be photodamaged by the low wavelength light that is used to probe the sample. The key enabling features, many of which are now routinely implemented in Raman systems, as illustrated in Fig. 15.2, were the use of a confocal arrangement to detect the Raman signals, the use of highly sensitive detectors, and the use of longer wavelength 660 nm light to minimize background interference from autofluorescence and to avoid sample degradation. This combination helped counter the low Raman scattering cross-sections of biological macromolecules, reducing the integration time sufficiently to enable live cells to be probed. Acquisition times ranged from 150–900 seconds, which was fast for vibrational spectroscopy at the time, but would be unacceptably long for flow cytometer applications now. In addition, samples had to be immobilized on a glass surface to ensure that the cell remained motionless and perfectly overlapped in three dimensions with the probe beam during this long acquisition time (Fig. 15.3(A)). For sampling large populations, cells would have to be adhered to a glass surface, each cell would be individually probed, and glass slides would have to be continuously replaced to probe new cells. Although feasible, this method would not be practical nor easily automatable for handling larger cell populations or for cell sorting requirements. For over a decade, this approach had been the generally accepted method for single cell Raman analysis.

15.3.2 Laser Tweezers with Raman Spectroscopy

In the late 1990s, a novel scheme combining laser tweezers (optical trapping) with Raman spectroscopy was introduced for studying single emulsion droplets and polystyrene beads in water.[8,9] These studies expanded the scope of single particle Raman studies to small micron-sized objects suspended in solution. The seminal work by Xie *et al.*[10–12] further extended this technique for analyzing biological samples by using a low powered near-infrared (NIR) laser beam to avoid potential photodamage. Laser tweezer Raman spectroscopy (LTRS) greatly simplified the analysis of biological particles and single living cells, particularly those that are naturally non-adherent suspension cells (Fig. 15.3(B)). Optical tweezers use a focused laser beam to generate forces strong enough (typically picoNewtons) to manipulate microscopic objects. The underlying physical principle,[13] which can be explained using a ray optics approach, is the transfer of momentum from photons to the dielectric object as the light rays are refracted by the object, resulting in forces being imparted to the trapped object. For a laser beam with a Gaussian intensity profile, the net force acts to pull the dielectric object towards the center of the beam and the region with the highest intensity. The symmetry of the intensity profile at the beam center results in no net lateral force at this equilibrium position, resulting in stable trapping of the particle. In the axial direction, gradient forces counter gravitational and scattering forces to achieve stable trapping along the beam axis. The simplest optical tweezers configuration uses a single tightly focused beam to trap particles. Such an arrangement is straightforward to implement with single-cell Raman spectroscopy, which already utilizes a single focused beam to acquire Raman spectra from single cells. There are several advantages of performing Raman using this optical arrangement: (1) cells suspended in solution can be analyzed, allowing many cells to be randomly sampled in solution without adhesion of cells to a surface, (2) cells are probed away from surfaces, which have strong signals that interfere with the sample spectra, (3) maximum overlap of the laser focus with the cell ensures optimal Raman signal generation (4) cells can be physically immobilized, manipulated, and sorted into different regions, and (5) analysis of submicron particles is possible.

Examples of more complex optical trapping arrangements implemented with LTRS have been demonstrated to improve system performance and function. Two collinear laser beams at two different wavelengths can be used, one at a long (e.g. 1064 nm) wavelength to trap the object and a second laser (e.g. 785 nm) for Raman excitation.[14] This arrangement provides additional flexibility for trapping larger objects with higher powers at longer wavelengths to minimize photodamage. Also, by having independent trapping and excitation beams, different regions within a cell can be positioned within the focal volume of the Raman exci-

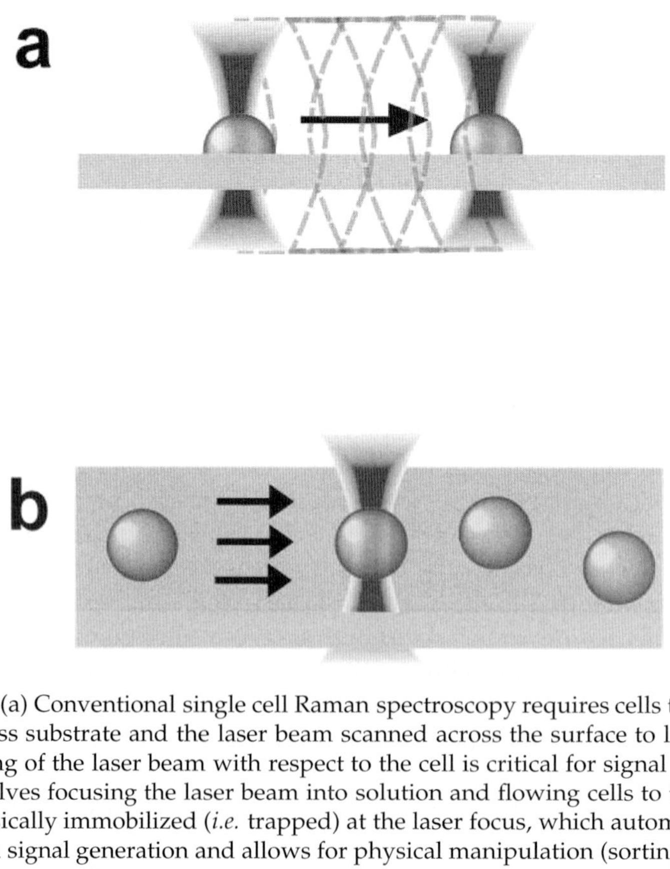

Figure 15.3. (a) Conventional single cell Raman spectroscopy requires cells to be immobilized on a glass substrate and the laser beam scanned across the surface to locate the cell. 3-D positioning of the laser beam with respect to the cell is critical for signal optimization. (b) LTRS involves focusing the laser beam into solution and flowing cells to the laser trap. Cells are physically immobilized (*i.e.* trapped) at the laser focus, which automatically maximizes Raman signal generation and allows for physical manipulation (sorting) of the cells. Color reference – pg. 356.

tation beam for analysis by scanning the trap beam. This is particularly important when analyzing larger cells. Yet another configuration[15] utilizes two optical fibers in a counter-propagating geometry to trap a cell with two divergent light fields. A third orthogonal beam probes the Raman spectra of the trapped cell. By using divergent fields, this design allows for trapping large cells while reducing the possibility of photodamage. Raman spectra of local parts of a trapped cell can be recorded by independently moving the cell position relative to the probe beam. A third example[16] is the use of a spatial phase modulator (SPM) to generate multiple trapping points placed around the cell periphery to trap the cell, while a separate probe beam acquires the Raman spectrum of the cell at various locations. This setup distributes the trapping laser power more efficiently to reduce possible photodamage and provides a more stable trapping configuration via multiple 'anchor' points.

15.3.3 Integration of LTRS with Microfluidic Systems

The earliest single-cell LTRS experiments were performed on standard optical microscopes with a drop of solution containing single cells and particles placed on a glass substrate for analysis. Although a simple and useful research approach for sampling a small number of cells, this arrangement was not applicable for sampling an increasing number of cells and for sorting them into different populations after the analysis. Such an application would require microfluidic devices for cell delivery and handling. The merger of LTRS and microfluidic devices first occurred in 2005, when Ramser *et al.*[17] used a device consisting of three reservoirs connected in series by two microchannels for the purpose of precisely controlling the environmental conditions surrounding the cells during the Raman measurements of single cells. Cells placed in the center reservoir for spectral analysis could be flushed with specific types of buffers flowing in from the side reservoirs via electro-osmotic flow. Xie *et al.*[18] demonstrated a similar device for Raman sorting and identification of micro-organisms. A small sample holder consisting of two sample chambers connected together by a microchannel was used in their study. A mixed population of cells was introduced into one chamber. A laser beam was used to trap and analyze individual cells, which were then sorted by using the trap beam to physically transport the cell into the second chamber. By confining the analysis to batch cultures, this system limits the volume and number of cells that can be examined before a fresh sample solution needs to be prepared for analysis.

A more practical automated approach for Raman sorting is needed for continuous sampling of a larger number of cells. Our group has devised an alternative scheme that uses multi-channel microfluidic devices to enable continuous Raman sampling of cells.[19] The concept and basic operation of a hydrodynamic focusing device are illustrated in Fig. 15.4(a). Buffer solution flowing in the two side channels of the device have the primary function of hydrodynamically focusing and delivering a stream of cells in the center channel to the laser beam. A single cell trapped at the focus by the laser tweezers is physically manipulated out of the cell stream into an adjacent buffer stream that has no flowing cells; this is done to avoid the possibility of flowing cells from displacing the trapped cell that is being probed. Once the cell spectrum is acquired (typically tens of seconds), the laser tweezers can physically move the cell into either buffer stream and release the cell towards the outlet downstream for collecting and sorting the cells.

Although hydrodynamic focusing (Fig. 15.4(b)) is one design that is applicable for Raman sorting, other schemes such as pinch-flow fractionation[20] (Fig. 15.4(c)) have been evaluated that are also suitable for such purposes. The pinch-flow fractionation system maintains a stream of cells aligned to one sidewall of the device by a second liquid buffer stream. The choice of the appropriate device largely depends on the maximum particle velocity that still allows for the laser tweezers to optically trap a particle, at a given laser power. For example, at the current laser powers (10–20 mW) that are used in our LTRS system, the maximum flow velocity at which particles can still be trapped is 400 μm/s. Therefore, the pinch-flow scheme is more appropriate in this case, because it can still maintain a linear profile of cells at these lower flow rates.

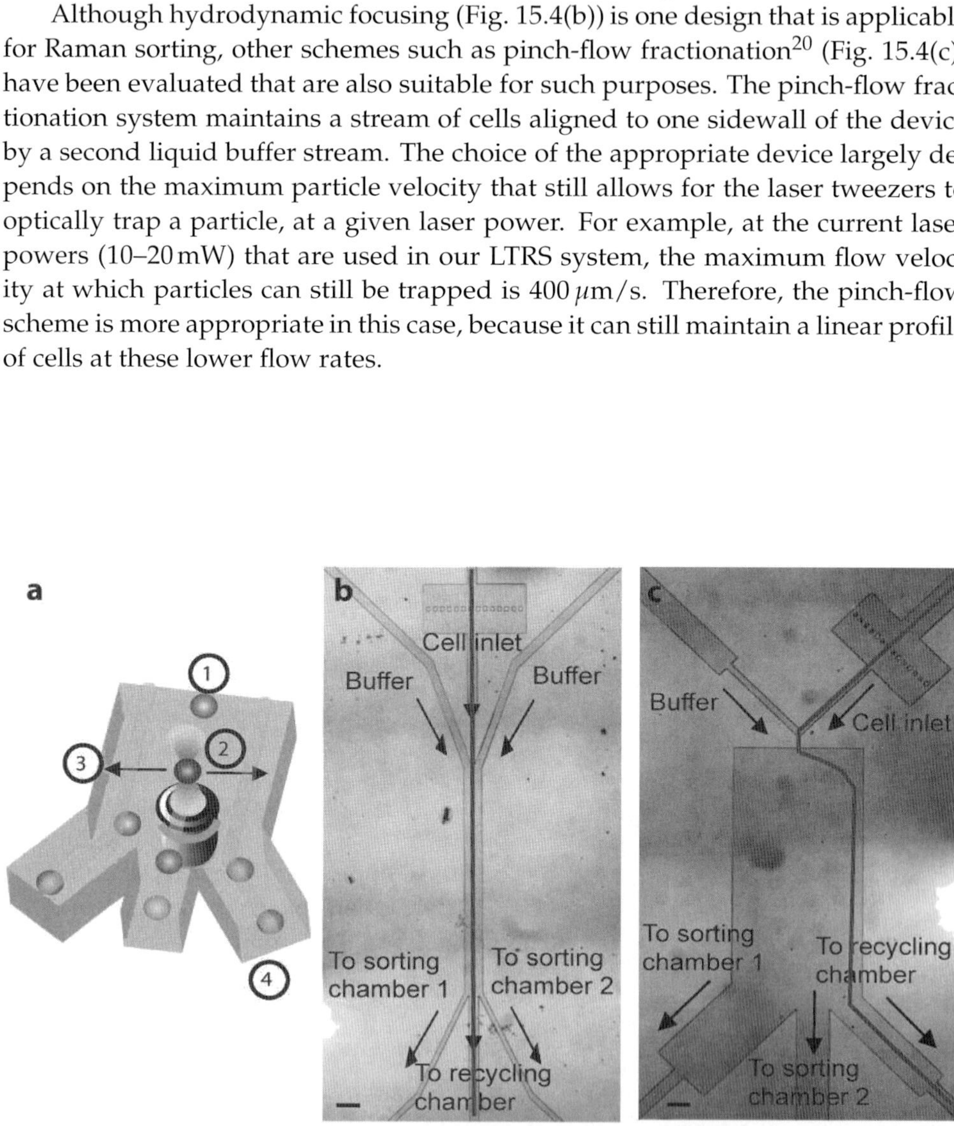

Figure 15.4. (a) Example of the general operating principle of a LTRS sorting system. (1) A focused stream delivers cells to the laser trap (2) A single cell is trapped and (3) is moved to a cell-free buffer stream for Raman acquisition (4) The laser trap releases cells into adjacent side channels that flow downstream into sorting chambers. Microfluidic devices based on (b) hydrodynamic focusing and (c) pinch flow fractionation that have multiple channels and sorting chambers have been implemented with LTRS for delivery, analysis, and sorting of single cells. Red line denotes flow profile of cell stream. Color reference – pg. 356.

15.3.4 Biomedical Applications of LTRS

LTRS has been applied to a broad range of biomedical and biological systems. It has proven to be valuable for characterizing and identifying cell populations and for studying cellular composition and dynamics based purely on the unique intrinsic Raman signatures of the samples. This section briefly reviews some important studies to give a sense for the important information that Raman signatures can provide and the potential biomedical and clinical utility of Raman-based cytometry.

Cancer — Our group was the first to apply LTRS for cancer detection at the single-cell level.[21,22] The studies focused on identifying unique Raman markers of leukemia, a cancer of nonadherent blood cells, that could improve the diagnosis and detection of the disease. Normal and cancer cells isolated from both cell lines and human subjects were individually probed by LTRS. Direct comparison of the spectra and data analysis using multivariate statistical methods showed highly

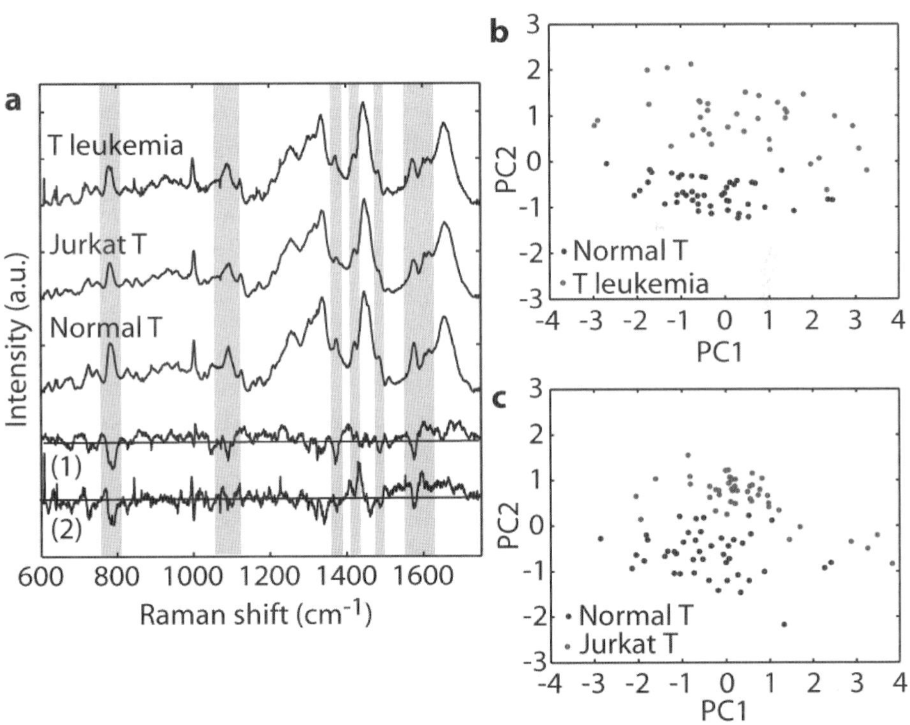

Figure 15.5. (a) Mean Raman spectra of normal T, Jurkat T, and patient T leukemia cell groups. Also shown are the difference spectra comparing (1) Jurkat T and normal T and (2) T leukemia and normal T spectra to highlight the spectral differences. Gray bars indicate the major Raman markers that differ between normal and cancerous cells. (b), (c) 2-D principal component analysis (PCA) graphs demonstrate identification and visualization of distinct clusters corresponding to different cell types, based purely on their intrinsic Raman signatures. Color reference – pg. 357.

reproducible Raman spectral differences between normal and cancer cells that enabled accurate detection and classification of cell populations. Figure 15.5(A) shows representative mean spectra of a healthy T cell population, a transformed T cell line, and cancer T cells from a pediatric leukemia patient. Each spectrum is an average of 20–30 individual cell spectra. Highlighted in gray bars are the regions of significant spectral differences, which denote the major Raman markers that can be used to detect and identify the disease. These Raman peaks, located at e.g. 785, 1093, 1126, 1575, and 1615 cm^{-1}, are related to intrinsic differences in the DNA and protein content of normal and cancer cells. A common approach for the direct comparison of all individual spectra is to apply principal component analysis (PCA) to analyze the Raman data. PCA is a method to reduce the multidimensionality of the Raman data set to a few new variables ('principal components' or PCs) that still reflect the maximum variance in the data. Two-dimensional scatter plots that are generated based on these PCs reveal the relative positions of the individual spectra (represented as single points on the plot). Such a scatter plot is comparable to the results obtained in conventional flow cytometry for identifying and visualizing cell populations. Figures 15.5(B) and 15.5(C) are the scatter plot results following PCA analysis of the Raman data, which show that cancer and normal cell groups can be accurately identified and separated, based purely on their intrinsic Raman signatures. The sensitivity and specificity for discriminating cell populations were routinely greater than 90% for both cancer cell lines and clinical samples. LTRS has also been applied to the detection and diagnosis of colorectal cancer.[23] Single epithelial cell suspensions were prepared from surgically removed human colorectal tissues and examined with LTRS. Observed spectral differences at 788, 853, 938, 1004, 1095, 1257, 1304, 1446, and 1657 cm^{-1} were indicative of differences in the nuclear acids and proteins between normal and cancerous cells. These markers were used to discriminate between normal and cancer cells with a sensitivity and specificity of 82.5% and 92.5%, respectively. When coupled with the microfluidic technology discussed in the previous section, it is anticipated that these cell populations can be identified and sorted for disease diagnosis.

Bacteria — Our early studies[24] with LTRS demonstrated its usefulness in trapping and characterizing single bacterial spores in solution, with short integration times due to improved signal-to-noise ratios as a result of optimal positioning of the spore within the laser focus. Furthermore, we demonstrated the ability to discriminate biological spore particles from nonbiological particles of similar sizes, such as polystyrene and silica beads and to determine the composition of a mixed particle solution. The use of LTRS was a significant improvement over methods used in a prior Raman study[25] in which bacterial spores had to be prepared on glass substrates. This made the analysis time-consuming and yielded poor signal to noise due to difficulties in overlapping the laser focus with the single particles. Xie *et al.*[26] then demonstrated LTRS as a rapid sensing technique for identifying different bacterial species. It was determined that the Raman signatures of *Bacillus cereus, Enterbacter aerogenes, Escheria coli, Streptococcus Pyrogenes, Enterococcus faecalis,* and *Streptococcus salivarius* were sufficiently distinct for each species to be accurately identified purely by their spectra.

In addition to its suitability as a detection platform, LTRS also has the added capability of monitoring dynamic cellular processes of a single cell in real time. A single trapped cell can be stimulated by external perturbation of its local environment and continuously monitored by Raman spectroscopy. Two examples[27,28] that demonstrate this function involved monitoring the expression of proteins in single transfected host bacterial cells upon induction with a chemical. Our group was able to monitor, over several hours, the overexpression of a myelin oligodendrocyte glycoprotein (MOG) from transfected *E. coli* cells upon activation with isopropyl thiogalactoside (IPTG) by continuously monitoring increases in the intensity of several Raman bands that are known protein markers (e.g. 1257, 1340, 1453, 1660 cm^{-1}). A similar experiment was performed on transfected *E. coli* bacteria and *Pichia pastoris* yeast cells induced to express somatolactin protein with isopropyl β-D-1-thiogalactopyranoside. Increases in the protein Raman markers at e.g. 1344, 1451, and 1665 cm^{-1} were observed in the cells at increasing culture times up to 24 hours after induction. Assessing the dynamic response of cells to external stimuli holds promise for biomedical applications such as determining drug susceptibility of bacteria or therapy monitoring in cancer.

Biological particles — Optical trapping of particles in the sub-micron size range is feasible, opening up the possibility of studying subcellular biological particles of biomedical relevance. Biological particles such as individual mitochondria[29] isolated from cells, synaptosomes from brain tissue,[30] and triglyceride-rich lipoproteins[31] isolated from blood are three examples of nanometer-sized particles that have been characterized with LTRS. Ajito *et al.*[30] demonstrated that single synaptosomes could be trapped and probed by Raman spectroscopy, and the quantity of glutamate released upon addition of a K^+ channel blocker could be quantified by comparison of the intensity of the Raman bands of glutamate to those from a bulk standard. Amino acids such as glutamate are neurotransmitters that are utilized at neuronal synapses. Detection of glutamate is of particular interest for revealing its role in neural cells, but measuring it at the single synaptosome level had previously been difficult. Our group has applied LTRS for the characterization of the Raman signatures of individual triglyceride-rich lipoproteins (TGRL),[31] Particles, 100 nm in diameter, consisting of lipids and proteins are found in the bloodstream and are responsible for transporting triacylglycerol and cholesterol to cells and tissues throughout the body by direct particle interaction with proteins on the cell surface. This interaction is not well understood, but is believed to be an important mechanism in the buildup of plaque in the arteries leading to atherosclerosis. *In vitro* chemical characterization of these particles and their interaction with the endothelium may shed light on this mechanism. LTRS has been able to reveal that particles isolated from human subjects before and after meal consumption change their intrinsic chemical composition from largely unsaturated to fully saturated lipid structures during hydrolysis of triglycerides to free fatty acids, based on Raman changes in the peak intensities at the 1000-1200 cm^{-1} region. These same Raman changes have been observed when single trapped lipoprotein particles are exposed to lipoprotein lipase, an enzyme that is responsible for breaking down triglyceride in capillaries,. These results suggest that we have identified

Raman markers reflecting the level of particle-tissue interaction that could be used to monitor the mechanism.

Blood disease — Thalassemia is a blood disease caused by a genetic defect that results in reduced synthesis of the α- or β-globin chains that make up hemoglobin (Hb), leading to chronic anemia. A pilot study[32] has applied LTRS for detecting red blood cells from healthy individuals, α-thalassemia (α-thal) patients, and β-thalassemia (β-thal) patients. Results have shown significant differences in the Raman spectra of the three cell populations, such as decreases in the peaks at 1545, 1602, and 1620 cm^{-1} of the thalassemic cells relative to the normal cells. When PCA was applied to the data, the three cell populations were discernable in a 3-D PCA plot based on the first three components, thus showing the potential for Raman spectroscopy as a screening and pre-diagnosis tool. A more recent study[33] has shown that Hb Raman bands of these cells can be used to detect the reduced oxygenation capability of Hb in β-thal cells, and that normal and thalassemic cells have different responses to photo-induced oxidative stress. For exposure times up to 80 seconds with a 532 nm beam, the Raman signals of thalassemic cells were reduced by 80%, as opposed to only 50% for normal cells, indicating that thalassemic cells are more susceptible to photo-oxidation.

15.3.5 Coherent Anti-Stokes Raman Scattering (CARS) Spectroscopy

The late 1990 s[34–37] saw a revival of coherent anti-Stokes Raman scattering (CARS) spectroscopy as a new optical technique for biological imaging and spectroscopy. A four-wave mixing process, CARS is a nonlinear Raman technique that uses two laser beams at a pump frequency, ω_p, and a Stokes frequency, ω_s, to generate a beat frequency ($\omega_p - \omega_s$). When this beat frequency matches a particular Raman vibration, the Raman resonant vibration oscillators are coherently driven. This results in a very strong anti-Stokes signal at frequency $\omega_{as} = 2\omega_p - \omega_s$, when a second pump photon at ω_p interacts with the oscillators. The primary advantage of CARS is the much stronger vibrational signals that can be achieved compared to spontaneous Raman spectroscopy that has been discussed so far. This opens up new possibilities for using CARS for rapid Raman-based imaging and spectral analysis. For Raman-activated cell sorting, CARS potentially addresses the low throughput issue described above; i.e. the generation of stronger Raman signals enables the dwell time of a cell within the laser beam to be significantly shortened. The implementation of CARS into the Raman system depicted in Fig. 15.2 is fairly straightforward. The key differences are the use of two pulsed laser beams at ω_p and ω_s that are temporally and spatially overlapped and delivered into the objective in lieu of a single CW laser beam. Also, since CARS signals are generated at wavelengths shorter than the excitation laser, different filters are needed than in Raman spectroscopy for separating the excitation light from the Raman signal.

Multiple excitation schemes[6] are available that can yield different CARS spectra. Depending on the specific parameters of the laser excitation source that are chosen, the CARS signal that is generated can either be a narrow spectral signal generated from one specific molecular vibration, or a broad spectral signature reflecting multiple molecular vibrations. When ~7 picosecond (ps) pulses are used for both the pump and Stokes beams, the spectral width of the excitation sources is confined to the width of a typical Raman molecular vibration (~10 cm^{-1}), resulting in a narrow CARS signal from one molecular vibration (Fig. 15.6(A)). However, when femtosecond laser pulses, which have a broad spectral width of ~100 cm^{-1} (for ~100 fs pulses) are used for the Stokes ω_S beam, the CARS signal can span tens of wavenumbers as multiple molecular vibrations are excited (Fig. 15.6(B)). When even broader spectral light sources are used, CARS spectra covering several thousand wavenumbers can be achieved (Fig. 15.6(C)). For example, a highly nonlinear photonic crystal fiber pumped by ultrashort femtosecond laser pulses can generate a high spatial coherence and broad bandwidth pulsed white light supercontinuum source for CARS spectroscopy.[38,39] By combining this broad Stokes beam with a spectrally narrow probe beam, a broad CARS molecular fingerprint spectrum can be obtained for chemical characterization analogous to a Raman fingerprint spectrum. Single channel point detectors can be used to detect single peak CARS signals, while broad CARS spectra require the use of a spectrometer and a CCD camera.

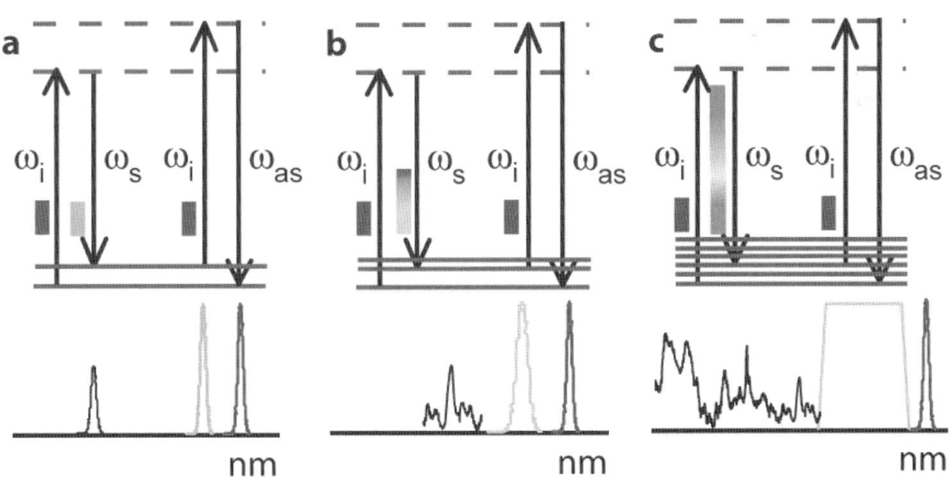

Figure 15.6. llustration of the CARS process using (a) ps pulses (b) fs pulses and (c) broadband supercontinuum sources for the Stokes beam. The narrow spectral width of ps pulses results in excitation of one molecular vibration, generating a narrow CARS signal. Excitation with spectrally broad fs pulses enables multiplexed CARS detection of multiple vibrations. The use of a supercontinuum white-light source as the Stokes beam enables CARS spectral acquisition of multiple molecular vibrations over a ~2000 cm^{-1} range. Shown below each energy diagram is a conceptualized illustration (not real data) of the different spectral signals that can be obtained for each condition (red – probe beam, yellow – Stokes beam, black – CARS signal). Color reference – pg. 357.

Our group[40] was the first to demonstrate the use of CARS, in combination with optical tweezers, for the spectroscopic analysis of sub-micron particles suspended in solution. A proof-of-principle experiment used two collinear ps pulsed laser beams to optically trap a single polystyrene bead. A strong CARS signal was generated when the laser wavelengths were tuned to the 1001.5 cm^{-1} molecular vibration. Equivalent experiments were performed on synthetic lipid vesicles. Following this work, Shi *et al.*[39] demonstrated the first broadband CARS in a supercontinuum optical trap by optically trapping polystyrene beads with the focused white light source, enabling the acquisition of a full spectrum of the particle spanning over $\sim$3000 cm^{-1}. This broadband CARS tweezers approach has the potential benefit of providing more detailed chemical information of a sample in a single acquisition compared to a single CARS peak from ps pulsed excitation, and at much greater speeds than can be provided by LTRS.

Recent research advancements in the CARS field have resulted in the first microfluidic CARS cytometer.[41] In this study, a ps-pulsed CARS system was combined with a microfluidic hydrodynamic focusing system capable of generating a 6 μm wide core stream in which particles flowed to the excitation beams. By line scanning the focused CARS beam across the channel and adjusting the ratio of this scan speed to the mean flow speed, individual particles were able to be detected and visualized, and the particle size determined based purely on its intrinsic CARS signature at 2840 cm^{-1} due to the CH_2 stretch vibration. Flow speeds between 1–30 mm/s and millisecond level CARS acquisition times were demonstrated, thus showing the potential improvement in speed of this system compared to the LTRS system discussed earlier. It should be noted that optical trapping was not implemented in this system due to the strong scattering signals of their samples, which allowed the particles to flow non-stop through the beam. However, the analysis of weaker biological scatterers (e.g. DNA, protein) may require the cell to stay within the probe beams for a longer duration, thus necessitating the use of optical trapping schemes as discussed earlier.

15.4 CONCLUSIONS AND OUTLOOK

Raman flow cytometry is a novel concept still in its early stages of development. This article has focused on the different research areas that are being carried out in parallel in order to translate this technology from the research lab to practical use by end users in the biological and biomedical institutions. As discussed throughout this chapter, the technical hurdles center primarily on the issue of the intrinsically weak signals of the Raman process. This dictates the need to develop novel laser tweezers Raman and CARS methods (both ps and broadband schemes) as potentially viable solutions to improve the Raman signal-to-noise ratio, which in turn will improve particle throughput of the system. The design of appropriate microfluidic systems that are compatible with the tight focusing and trapping requirements of the Raman/CARS technique is another essential component. At this point, it is important to discuss several issues within these areas that remain to be resolved. For example, the CARS technique, whether in ps or broadband modes, suffers from a nonresonant electronic signal that is present even in the absence of the sample.[6] This undesired background contribution currently is a factor that limits the detection of CARS signals primarily to the strong 2840 cm^{-1} vibration of cells rich in lipid structures. Suppression of this nonresonant signal, such as with polarization schemes,[42] needs to be developed for detection of weaker scatterers. Improvements will need to be made to the LTRS microfluidic systems as well. Since the system will not be able to sample every cell through the channel on first pass, suitable designs for recirculating and recycling the cells back into the device for Raman analysis will need to be explored. In addition, the current LTRS systems are based primarily on single beam traps. More complex trapping configurations, such as with dual collinear beam traps and traps with multiple foci would enable analysis of larger trapped objects and multiple objects simultaneously to increase throughput, respectively. A complete design and integration of the optical and microfluidic components into an automated, compact benchtop system is also required.

It remains to be seen which biomedical applications seek to benefit the most from this technology, although it is expected that they would take advantage of the unique features of Raman spectroscopy (e.g. unique markers, label-free, non-destructive analysis/sorting of viable cells). The specific application will largely be defined by the discovery of useful Raman markers of the particular biological system for cell sorting and/or monitoring biological processes. Anticipated areas in which Raman flow cytometry can make an impact include cancer detection and diagnosis, stem cell identification[43–45] for regenerative medicine, and bacterial identification for infectious disease, to name a few.

ACKNOWLEDGMENTS

This work has been supported by funding from the National Science Foundation. The Center for Biophotonics, an NSF Science and Technology Center, is managed by the University of California, Davis, under Cooperative Agreement No. PHY 0120999. This work was also supported, in part, by the Keaton-Raphael Memorial Fund.

References

[1] H. M. Shapiro, *Practical Flow Cytometry, 4th ed..*, Wiley-Liss: New York (2003).

[2] R. Manoharan, Y. Wang and M. S. Feld, *Spectrochimica Acta Part a-Molecular and Biomolecular Spectroscopy* **52**, 215–249 (1996).

[3] N. Uzunbajakava, A. Lenferink, Y. Kraan, B. Willekens, G. Vrensen, J. Greve and C. Otto, *Biopolymers* **72**, 1–9 (2003).

[4] K. W. Short, S. Carpenter, J. P. Freyer and J. R. Mourant, *Biophysical Journal* **88**, 4274–4288 (2005).

[5] G. P. Singh, G. Volpe, C. M. Creely, H. Grotsch, I. M. Geli and D. Petrov, *Journal of Raman Spectroscopy* **37**, 858–864 (2006).

[6] J. X. Cheng and X. S. Xie, *Journal of Physical Chemistry B* **108**, 827-840.(2004).

[7] G. J. Puppels, F. F. M. Demul, C. Otto, J. Greve and M. Robertnicoud, D. J. Arndtjovin and T. M. Jovin, *Nature* **347**, 301–303 (1990).

[8] K. Ajito, *Applied Spectroscopy* **52**, 339–342 (1998).

[9] K. Ajito and K. Torimitsu, *Trac-Trends in Analytical Chemistry* **20**, 255–262 (2001).

[10] C. G. Xie, M. A. Dinno and Y. Q. Li, *Optics Letters* **27**, 249–251 (2002).

[11] C. G. Xie and Y. Q. Li, *Applied Physics Letters* **81**, 951–953 (2002).

[12] C. G. Xie, Y. Q. Li, W. Tang and R. J. Newton, *Journal of Applied Physics* **94**, 6138–6142 (2003).

[13] A. Ashkin, *Biophysical Journal* **61**, 569–582 (1992).

[14] C. M. Creely, G. P. Singh and D. Petrov, *Optics Communications* **245**, 465–470 (2005).

[15] P. R. T. Jess, V. Garces-Chavez, D. Smith, M. Mazilu, L. Paterson, A. Riches, C. S. Herrington, W. Sibbett and K. Dholakia, *Optics Express***14**, 5779–5791 (2006).

[16] C. M. Creely, G. Volpe, G. P. Singh, M. Soler and D. V. Petrov, *Optics Express* **13**, 6105–6110 (2005).

[17] K. Ramser, J. Enger, M. Goksor, D. Hanstorp, K. Logg and M. Kall, *Lab Chip* **5**, 431–436 (2005).

[18] C. G. Xie, D. Chen and Y. Q. Li, *Optics Letters* **30**, 1800–1802 (2005).

[19] A. Y. Lau, L. P. Lee and J. W. Chan, *Lab Chip* **8**, 1116-1120 (2008).

[20] M. Yamada, M. Nakashima and M. Seki, *Analytical Chemistry* **76**, 5465–5471 (2004).

[21] J. W. Chan, D. S. Taylor, S. M. Lane, T. Zwerdling, J. Tuscano and T. Huser, *Analytical Chemistry* **80**, 2180–2187 (2008).

[22] J. W. Chan, D. S. Taylor, T. Zwerdling, S. M. Lane, K. Ihara and T. Huser, *Biophysical Journal* **90**, 648–656 (2006).

[23] K. Chen, Y. J. Qin, F. Zheng, M. H. Sun and D. R. Shi, *Optics Letters* **31**, 2015–2017 (2006).

[24] J. W. Chan, A. P. Esposito, C. E. Talley, C. W. Hollars, S. M. Lane and T. Huser, *Analytical Chemistry* **76**, 599–603 (2004).

[25] A. P. Esposito, C. E. Talley, T. Huser, C. W. Hollars, C. M. Schaldach and S. M. Lane, *Applied Spectroscopy* **57**, 868–871 (2003).

[26] C. Xie, J. Mace, M. A. Dinno, Y. Q. Li, W. Tang, R. J. Newton and P. J. Gemperline, *Analytical Chemistry* **77**, 4390–4397 (2005).

[27] J. W. Chan, H. Winhold, M. H. Corzett, J. M. Ulloa, M. Cosman, R. Balhorn and T. Huser, *Cytometry Part A* **71A**, 468–474 (2007).

[28] C. G. Xie, N. Nguyen, Y. Zhu and Y. Q. Li, *Analytical Chemistry* **79**, 9269–9275 (2007).

[29] H. Tang, H. Yao, G. Wang, Y. Wang, Y. Q. Li and M. Feng, *Optics Express* **15**, 12708–12716 (2007).

[30] K. Ajito, C. X. Han and K. Torimitsu, *Analytical Chemistry* **76**, 2506–2510 (2004).

[31] J. W. Chan, D. Motton, J. C. Rutledge, N. L. Keim and T. Huser, *Analytical Chemistry* **77**, 5870–5876 (2005).

[32] G. Wang, H. Yao, S. Huang, P. Chen and Y. Li, *Biomedical Optics, Technical Digest (CD) (Optical Society of America)* **2006**, *paper ME77.*

[33] A. C. De Luca, G. Rusciano, R. Ciancia, V. Martinelli, G. Pesce, B. Rotoli, L. Selvaggi and A. Sasso, *Optics Express* **16**, 7943–7957 (2008).

[34] A. Zumbusch, G. R. Holtom and X. S. Xie, *Physical Review Letters* **82**, 4142–4145 (1999).

[35] J. X. Cheng, A. Volkmer, L. D. Book and X. S. Xie, *Journal of Physical Chemistry B* **105**, 1277–1280 (2001).

[36] A. Volkmer, J. X. Cheng, L. D. Book and X. S. Xie, *Biophysical Journal* **80**, 164a–164a (2001).

[37] A. Volkmer, J. X. Cheng and X. S. Xie, *Physical Review Letters* **8702** (2001).

[38] K. B. Shi, P. Li and Z. W. Liu, *Journal of Nonlinear Optical Physics & Materials* **16**, 457–470

[39] K. B. Shi, P. Li and Z. W. Liu, *Applied Physics Letters* **90**, (2007).

[40] J. W. Chan, H. Winhold, S. M. Lane and T. Huser, *IEEE Journal of Selected Topics in Quantum Electronics* **11**, 858–863 (2005).

[41] H. W. Wang, N. Bao, T. T. Le, C. Lu and J. X. Cheng, *Optics Express* **16**, 5782–5789 (2008).

[42] J. X. Cheng, L. D. Book and X. S. Xie, *Optics Letters* **26**, 1341–1343 (2001).

[43] B. S. Kim, C. C. I. Lee, J. E. Christensen, T. R. Huser, J. W. Chan and A. F. Tarantal, *Stem Cells and Development* **17**, 185–198 (2008).

[44] L. Notingher, I. Bisson, J. M. Polak and L. L. Hench, *Vibrational Spectroscopy* **35**, 199–203 (2004).

[45] I. Notingher, I. Bisson, A. E. Bishop, W. L. Randle, J. M.P. Polak and L. L. Hench, *Analytical Chemistry* **76**, 3185–3193 (2004).

Chapter Sixteen

The Autonomous Pathogen Detection System

John M. Dzenitis* and Anthony J. Makarewicz

Lawrence Livermore National Laboratory, 7000 East Avenue, Livermore California, 94550, USA

**john.m.dzenitis@llnl.gov*

We developed, tested, and now operate a civilian biological defense capability that continuously monitors the air for biological threat agents. The Autonomous Pathogen Detection System (APDS) collects, prepares, reads, analyzes, and reports results of multiplexed immunoassays and multiplexed PCR assays using Luminex© xMAP technology and flow cytometer. The mission we conduct is particularly demanding: continuous monitoring, multiple threat agents, high sensitivity, challenging environments, and ultimately extremely low false positive rates. Here, we introduce the mission requirements and metrics, show the system engineering and analysis framework, and describe the progress to date including early development and current status.

16.1 IMPORTANCE

Biological terrorism is an increasing concern: biological technology is advancing in capability while also becoming more accessible, and terrorist activities are increasing in number, scale, and diversity. Dissemination of biological threat agents as aerosols is a particular menace because of the ease of widespread dispersal and the effectiveness of infection by inhalation. Lawrence Livermore National Laboratory (LLNL) developed the Autonomous Pathogen Detection System (APDS) for the U.S. Department of Energy (DOE) and Department of Homeland Security (DHS) as a means of detecting biological threat agents in the air. The APDS was developed for high-risk locations to reach the best-achievable combination of competing characteristics such as speed, selectivity, sensitivity, numbers of agents, and cost. For example, while fast detection can prevent exposures and allow the most effective medical treatment of people already exposed,[1] false positive results must also be extremely low in a civilian setting. The APDS is the first actionable autonomous detector[2] component of the DHS's BioWatch Program.[3]

The Microflow Cytometer by J S Kim & F S Ligler

www.panstanford.com
978-981-4267-41-0

The APDS instrument will be of interest to readers of this book because the analytical core of the instrument is a flow cytometer. In the initial stages of development, compact and advanced flow cytometers provided fast and sensitive detection of biological agents. In subsequent development, the more conventional cytometry approach was replaced with a Luminex assay platform and flow cytometer to allow detection of many agent signatures at once. The compact flow cytometer is one subsystem in the larger system which includes aerosol collection, biological reagents, sample preparation, result analysis, field packaging, communications, maintenance support, and remote monitoring. This chapter does not focus on the flow cytometer itself, but instead provides an overview of the characteristics, progress, and critical issues of an advanced, fielded system utilizing an automated flow cytometry system.

16.2 CHARACTERISTICS OF PATHOGEN DETECTION SYSTEMS

16.2.1 *Mission and Metrics*

The fundamental mission of pathogen detection systems is to provide information that saves lives. For DHS's BioWatch Program, the system needs to continuously sample the air at a location over many months and produce regular reports that either the air is uncontaminated or that there is potentially a biological threat agent present. This is true of both the manual system, where dry filters are transported daily to Laboratory Response Network laboratories for analysis and reporting,[3] and the APDS, where the analysis is performed in the field and reported via a network. The response to positive signals could include facility operators closing a facility to limit further exposure, response crews in personal protective equipment searching for sources and performing manual sampling, public health laboratories performing characterization testing, medical testing and countermeasures, and other actions. Other operational systems with similar missions include the U.S. Postal Service's Biohazard Detection System[4] and the Department of Defense's (DoD's) Joint Biological Point Detection System (JBPDS).[5] However, the APDS mission differs in requirements including monitoring for the public, continuous operations, number of agents, and the stakeholders involved.

The overall metrics for the detection system depend a great deal on the stakeholders. Here, the term "stakeholders" is used instead of "users" to focus on those who act on the system's biological information. From this perspective, the personnel who actually keep the equipment running are viewed as part of the system. The DHS commissioned an expert panel in 2006 to examine metrics for biological detection systems from the stakeholder perspectives, not just technical perspectives. The panel included local authorities (facility operators, public health officials, and law enforcement personnel) and national authorities (DHS, Centers for Disease Control and Prevention [CDC], DoD, and the Postal Service). The details of the study are not public information, but some qualitative aspects can be described here.

Ultimately, the utility of the system to the stakeholders should be described in terms of cost *versus* benefit, where cost includes direct, indirect, and opportunity costs, and benefit includes saving lives. More accurately, and perhaps more coldly, the system should be analyzed in terms of Net Present Value, including the probability of an attack and the financial value of a human life. As a step towards the larger picture, cost and benefit categories were used to organize the quantifiable characteristics of the system as follows:

Benefit metrics

- Detect agents of interest: agent types, agent panel, number of signatures, time to add an assay, probability of detection, and operational availability.
- Detect in the required environments: environments of interest, temperature range, humidity range, particle tolerance, battery capability, and additional factors such as electromagnetic interference.
- Detect at effective levels: sensitivity as limit of detection.
- Enable effective response: probability of false alert, probability of actionable false positive, selectivity, data accessibility, remote system diagnostics, data security, physical security, biohazard security, data time-stamping capability, data archiving duration, and sample archiving capability.
- Report on an effective timescale: sampling period and time to results.

Cost metrics

- Direct acquisition cost.
- Direct operation cost.
- Indirect operation cost to the local stakeholders: local interoperability, facility labor requirement, biological monitor labor requirement, maintenance interval, mean time between failure, power, size, visual impact, and noise.

Some of the metrics affect multiple areas (e.g., probability of false positive affects effective response and indirect operation cost), but were placed in what was judged to be their primary category. In a full cost-benefit analysis, the effects would just apply quantitatively as appropriate. Subsequent work with the stakeholders for the particular APDS application established specific threshold values, and more stringent goal values, for these metrics. Although this limited effort fell short of a cost-benefit or net-present-value calculation, having quantified values with stakeholder agreement was an important step. We will point out some of the less obvious results.

Of all of the parameters, probability of false actionable results was the most stringent: the critical locations that would be monitored would also have a tremendous impact if shut down, and clearing a location to reopen after a biological alert would take hours because of the sampling and testing required. There is a relationship between sensitivity, probability of detection, and probability of a false positive, often discussed as the Receiver Operating Characteristic (ROC).[6] In the present mission, the stakeholders felt that the false-positive rate needs be set to a very low level that they can tolerate in operations, and the resulting sensitivity is either acceptable or not. They do not view it as a trade-off.

The ability to detect numerous different agent signatures in every sample was important for several reasons. First, several different agents could be used in the attack, so of all the likely possibilities need to be represented to detect the agents of interest. Second, the false-positive rate must be extremely low, and having multiple independent signatures per agent helps greatly. However, having many signatures can quickly drive up the acquisition and operation costs if many separate analyses need to be done.

Some of the traditional parameters had little meaning by themselves, but primarily affected other metrics. For example, Mean Time Between Failures (MTBF) is a key characteristic, but in itself is not a cost. Its effects are in operational availability, in direct cost for labor and materials, and potentially in indirect cost of inconvenience to the facility operations.

Finally, there is often a focus on direct acquisition cost of complex instrumentation. However, for this type of system and mission, the yearly operation costs will be on the same scale as the acquisition cost, so they need to be considered together.

16.2.2 System Engineering and Analysis

The steps in a detection process can be organized in a number of ways. We will use the following steps for the first level:

- Collect the sample
- Prepare the sample for detection
- Read the prepared sample
- Analyze the results
- Report the result

An example schematic diagram of the process flow over time for a pathogen detection system is shown in Fig. 16.1. The system repeatedly executes the steps of *collect*, *prepare*, *read*, *analyze*, and *report*. In this example, the collection is continuous but split into segments. After the step Collect 1 ends, Prepare 1 begins on that collected sample, while the next collection step begins in parallel. The other processing steps are assumed to be sequential here, so in this case a release that occurs during Collect 2 is reported near the end of Collect 3 and leads to a response action.

16.2.2.1 Process Analysis Approach

It is useful to have a basic mathematical framework to describe the functions of the various steps in the process train. This will allow us to illustrate the important characteristics of the steps, although we will not go into the quantitative parameters here. The mass balance equation that we use is written for rate of change of a

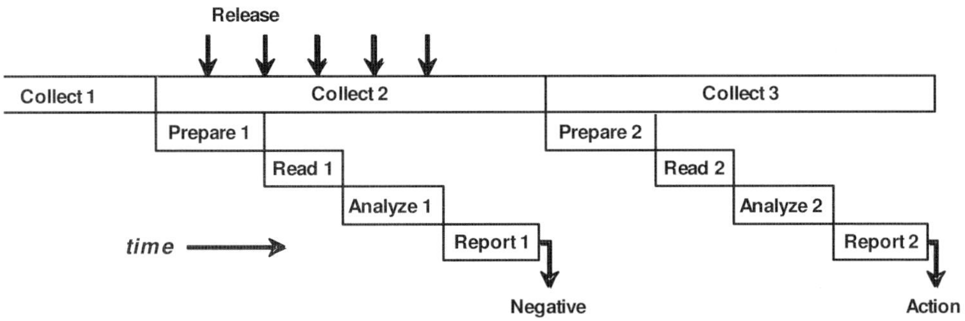

Figure 16.1. A schematic diagram of the overall process flow over time for a pathogen detection system. In this example mode of operation, the collection is continuous but split into distinct segments, and all of the other processing steps are sequential. After the step Collect 1 ends, Prepare 1 begins on that collected sample, while the next collection step begins in parallel. A release that occurs during Collect 2 is reported during Collect 3. The relative times are not to scale.

number of detectable targets, n, in a control volume as

$$\frac{dn_{CV}}{dt} = \dot{n}_{in,CS} - \dot{n}_{out,CS} + \dot{n}_{form,CV} \tag{16.1}$$

where $\dot{n}$ is a flow or formation rate of number of targets, and CV and CS indicate the conceptual control volume and surface enclosing the process step. The targets could be cells, spores, protein epitopes, DNA copies, colony forming units (CFUs) *et cetera*, depending on the detection technique. The last term in Eq. 16.1 represents formation (or if negative, loss) of detectable targets. This is not creation or destruction of mass but rather transformation to or from a detectable form. The usefulness of the concept will be evident later.

Often the processes are carried out batch-wise instead of continuously so we use an integrated form of Eq. 16.1. Also, in order to avoid losses through aliquoting or sub-sampling, it is as desirable to maintain a high concentration as it is to maintain copies of targets. Therefore, we can evaluate the performance of the processing steps as

$$\frac{c_{out}}{c_{in}} = \frac{n_{out}/V_{out}}{n_{in}/V_{in}} \tag{16.2}$$

so that the volumes in and out of the step are part of the rating of the process, along with the number of targets in and out. In the unlikely case that there is no downstream volume limitation, then it may be appropriate not to consider the volume changes and only consider number of targets or efficiency.

16.2.2.2 Collect

The aerosol collector converts the environmental material into an internal sample that can be manipulated. The *net* rate of collection that comes from Eq. 16.1 is

$$\frac{dn_{coll}}{dt} = Q_{coll}\eta_{coll}c_{in} - k_{loss}n_{coll} \tag{16.3}$$

where n_{coll} is the number of targets in the collector, Q_{coll} is the collector flow rate, η_{coll} is the collection efficiency, c_{in} is the target concentration in the air into the collector, and k_{loss} is a loss rate constant.

The first term on the right of Eq. 16.3 is the collection rate. Collection efficiency is important in system design, but the maximum efficiency for any collector is unity while the flow rate for some collectors can be much larger than others. For this reason we emphasize that the collector flow rate is potentially more important than its efficiency. The APDS instrument uses a liquid-based collection, and we value small collection volumes because we want to minimize the material loss from aliquoting downstream. The integrated version of Eq. 16.3 for output liquid concentration considering only the collection term is

$$c_{out} = \left(\frac{Q_{coll}\eta_{coll}}{V_{coll}}\right) \int c_{in}dt = \dot{X}_{coll} \int c_{in}dt \qquad (16.4)$$

if everything is constant over time except for the air concentration. The $\dot{X}$ term is the concentration rate and has units such as (targets/L liquid)/ (targets/L air)/ min, more casually written min^{-1}. This parameter is the best single measure of collection performance.

The last term on the right of Eq. 16.3 is the loss rate, written here with the assumption of first-order decay. This term accounts for the fact that collected targets may be lost through mechanisms like re-aerosolization or degradation, and decreases the output concentration to be less than Eq. 16.4. The loss rate may be significant for long collection times or when the air concentration is low.

16.2.2.3 Prepare: Lyse

Lysis is a sample preparation step that is useful in some cases to convert bound, inaccessible target material to detectable target material. Here, Eq. 16.2 becomes

$$\frac{c_{out}}{c_{in}} = \frac{n_{in} + \eta_{lyse}n_{bound}}{n_{in}}\frac{V_{in}}{V_{out}} = \left(1 + \eta_{lyse}X_{bound}\right)\frac{V_{in}}{V_{out}} \qquad (16.5)$$

where n_{bound} is number of bound or inaccessible targets in, X_{bound} is the ratio of bound to free targets in, and η_{lyse} is the lysis efficiency. We write the performance in this way to point out that the utility of the lysis step depends on not just the efficiency but also on the sample type. Since efficiency is at most unity, the lysis cannot have much effect if the bound target ratio is significantly less than unity. We have tested bacterial spore preparations with X_{bound} between 0.0001 and 100 in our laboratory, but wider ranges are possible.

16.2.2.4 Prepare: Extract

Extraction is a sample preparation step with the goals of purifying and, in some cases, concentrating the material. From the perspective of the downstream analysis, purification creates more detectable target in the sample by removing interfering material. Thus, the measure of the utility of the purification should depend on the analysis. We can use a factor improvement in sensitivity X_{ext} to attribute the benefit from purification to the extraction step, and an extraction efficiency η_{ext} to account for potential loss of material. This gives

$$\frac{c_{out}}{c_{in}} = \eta_{ext} X_{ext} \frac{V_{in}}{V_{out}}. \tag{16.6}$$

We use this form to consider the combined effects of purification and concentration (or dilution), tempered by the efficiency. If the sample is completely undetectable without the purification, c_{in} by the definition here as detectable material would be zero, and a different interpretation should be used instead. Extraction is then a requirement and not a choice, and it is characterized just by volume ratio and efficiency.

16.2.2.5 Prepare: Amplify

PCR, reverse-transcriptase PCR, and other nucleic acid amplification methods form more copies of detectable target. The efficiency should be written such that the number of targets would optimally double with each cycle. At the earlier stages of the reaction the efficiency can be constant over a number of thermal cycles, but more generally for PCR with efficiency η_i for a cycle number i the increase is

$$\frac{c_{out}}{c_{in}} = \frac{V_{in}}{V_{out}} \prod_i (1 + \eta_i). \tag{16.7}$$

The efficiency rolls off for later steps as primers, nucleotides, competitive binding, or enzyme activity becomes limiting. In the extreme case, the concentration of targets out is not strongly dependent on the number of cycles or the concentration of targets in because the reaction becomes limited by the other factors. Different PCR signatures can have different sets of efficiencies and different detection limits. Those differences can be important operationally when multiple signatures target a single agent.

The amplification possible in the PCR is tremendous and detection from a single copy into the PCR can be approached. The enzyme and primers for a PCR can be expensive, and there is a drive to make the reaction and input volumes as small as possible. For small volumes or low concentrations, the implicit assumption of a continuum in Eq. 16.7 hits a statistical limit of a Poisson process.[7] If the average number of targets per aliquot is unity, there is a 37 % probability that any given aliquot has no targets at all. To drive that probability down to 1 % for a more acceptable probability of detection, the input concentration and volume must average at least five targets into the reaction, or

$$n_{in} = c_{in} V_{in} > 5 \text{ targets}. \tag{16.8}$$

Equation 16.8 shows that as the input volume is decreased, the required input concentration must be correspondingly increased. For example, a 6 μL input sample volume requires a concentration greater than 833 targets/mL. This is not necessarily CFUs, as there are generally more nucleic acid sequence copies than CFUs.

The volumetric sampling effect on sensitivity is one of the reasons that "millifluidics" (millimeter length scales) of the type described here are more appropriate than true microfluidics: a 10 pL input volume[8] requires 5×10^8 targets/mL. Preconcentration steps or analyzing many representative subsamples can decrease the impact of this requirement.

16.2.2.6 Prepare: Bind

In the multiplexed antibody and nucleic-acid assays on the Luminex platform, binding of the biological targets to the bead and subsequent labeling with fluorescent targets must occur so the material can be detected. For the antibody assays we use, the steps are (16.1) bind the antigen to the bead conjugated with the corresponding antibody, (16.2) bind the biotinylated labeling antibody to the antigen-bead complex, and (16.3) label the antibody-antigen-bead complex with streptavidin-phycoerythrin.[9] For the DNA assays we use, the steps are (16.1) bind the biotinylated PCR amplicon to the bead conjugated with the corresponding DNA probe and (16.2) label the amplicon-bead complex with streptavidin-phycoerythrin.[10] In both cases, there is a fundamental change in medium from aqueous solution to bead surface. For each signature in the multiplexed assay, we can express the average output concentration per bead as

$$c'_b = \eta_{bind} c_{in} V_{in} / n_b, \tag{16.9}$$

where c'_b is the bead concentration of fluorophore (e.g., molecules of phycoerythrin per bead), η_{bind} is the combined efficiency of all of the binding steps, c_{in} and V_{in} are the target concentration and volume in, and n_b is the number of labeled Luminex beads that result. The prime notation is used to note the shift from volumetric concentration to bead concentration.

In reality, a distribution of concentrations for the bead population results instead of a single concentration, but the average concentration in Eq. 16.9 serves to illustrate a point: In addition to maximizing the number of targets in and the binding efficiency, there is also an impetus to keep the number of beads low in order to maximize the fluorescence per bead and enhance sensitivity. However, the number of beads needs to be kept high enough for good statistics, as will be discussed below. On the other end of the spectrum, when the input amount is very high, Eq. 16.9 is less useful because the bead concentration asymptotes towards saturation.

16.2.2.7 Read

The labeled beads for each signature are read by the Luminex 100 flow cytometer which converts each signature's fluorophore concentration c'_b into a signal s in Median Fluorescence Intensity (MFI). The dependence of MFI on number of beads in the sample has been analyzed in some detail,[11] but a general rule-of-thumb is to have at least 50 to 100 beads of each type in an analysis in order to have good reproducibility. If there are approximately 200 beads in a sample, and if the limit of detection is approximately 500 fluorophores per bead, then with perfect binding Eq. 16.9 requires that at least 1×10^4 biological targets would have to enter the binding step. This is readily achieved for PCR due to the high target amplification (see Eqs. 16.7 and 16.8) from a concentration of about 1×10^3 targets/mL. This is more of a restriction for immunoassays, where a 100 μL sample would need to provide 1×10^5 targets/mL. However, the targets do not have to be CFUs: the requirement is mitigated by the fact that there can be multiple biotin targets per antibody, and multiple antibody epitopes per CFU.

16.2.2.8 Analyze

After the multiplexed signals have been acquired, the first step of the numerical analysis is determining whether the results meet the criteria for validity. In addition to a requirement for a minimum number of beads of each type, there is also a set of positive and negative controls that must satisfy predetermined thresholds.[2,18] For multiplexed PCR, the most stringent of these is the positive amplification control (PC), which requires all of the amplification, binding, and reading processes to be executed properly. This prevents false-negative results.

The next step of the analysis is a comparison of the signal results against predetermined thresholds for each signature. For these thresholds to be at a minimum, the values are required to be below a maximum probability of false result, given a signature's baseline and noise behavior, but may also be set higher if a higher LOD is desired for operational reasons.

Finally, there is an analysis of the signature positives against a decision algorithm for alerts. For the multiplexed assays of interest here, a high-level alert requires that multiple signature positives for an agent occur on the same sample. To the extent that the false-signature results are independent, the probability of a false alert will be much less than a single false-signature reaction. If the noise is independent, the probability of a false result from a signature combination is the product of the individual probabilities, for example

$$p_{f,123} = p_{f,1} p_{f,2} p_{f,3}. \tag{16.10}$$

This form would be modified if there were three of four signatures required or if there is correlation between the signatures' noise.

Equation 16.10 shows how the very low false-positive rates required for the mission can be achieved. Three signatures with individual probability of 2×10^{-3} would have a combined probability of 8×10^{-9}. This is the order of stringency that needs to be approached for this application.

16.2.2.9 Report

An autonomous detection system must communicate its results quickly so appropriate response actions can be taken. For the BioWatch Program, the local public health officials are the critical link converting technical data into human decisions leading to response, independent of whether the data is from manual or autonomous systems. The autonomous detection instruments must be networked, biological alerts must be reported immediately to mobile messaging devices for response, and biological and maintenance raw data must always be available for remote review.

16.2.2.10 Associated Functions

In addition to the core process steps of *collect*, *prepare*, *read*, *analyze*, and *report*, the system engineering must include functions to sustain these processes. One significant part of this is packaging and hardening against the environment, which includes controlling internal temperature in wide ambient temperature variations, handling dust and precipitation, and operating through temporary power loss and voltage spikes.

Internally, the system needs to perform automatic cleaning, adjustments, and self-monitoring of performance. Externally, there need to be manned support functions of preventative maintenance, corrective maintenance, and monitoring of performance. These, in turn, are part of a larger set of integrated logistics and supply functions for the system on the larger scale.

16.2.2.11 Timing

For the biological process steps described above and shown in Fig. 16 1, the time required for the *prepare* step dominates *read*, *analyze*, and *report*, which together require only a minute or two. The preparation of a multiplexed immunoassay can take from about 15 to 60 minutes, and the preparation of a multiplexed PCR assay takes from 90 to 150 minutes, depending on the extent of the sample preparation steps. Faster processing of these assays leads to lower sensitivity.

Some tests that do not require such intricate molecular biology can be executed on significantly shorter time frames. However, these other tests have not been able to meet the high sensitivity and extremely low false-positive rate required for the mission.

16.3 REVIEW OF PROGRESS

16.3.1 Early Development

The reality of the threat of modern chemical and biological terrorism became increasingly apparent at the beginning of the 1990s. The most publicized event was the Aum Shinryko sarin attack in the Tokyo subway in 1995, preceded by an attempted botulinum toxin attack in 1990 and an attempted anthrax attack in 1993.[12] The APDS project was initiated in 1996 by LLNL as an internally funded project, but in 1997 transitioned to be part of the DOE's Chemical and Biological Nonproliferation Program as that program advanced. The APDS development efforts included a high-performance LLNL flow cytometer called the miniFlo,[13,14] successfully demonstrated at the DoD's 1996 Joint Field Trials III (see Fig. 16.2). This early success contributed to the establishment of the DOE program and of the APDS project.

The original version of the APDS instrument included the miniFlo flow cytometer and a solenoid valve-based fluidics system (see Fig. 16.3). LLNL began collaborating with Research International Inc. (Monroe, WA) and using an early version of their SASS 2000 wetted-wall cyclone aerosol collector. The initial advances in automation of environmental biological detection using immunoassays were achieved in that period. The basic philosophy of using automated fluidics instead of robotics and disposables was established at that time and continues on today.

In 1999, the DOE funded LLNL to undertake revisions to the APDS (then termed APDS II and later APDS100) to extend the capabilities of the system. This incorporated related work at LLNL in this area funded by the DoD. The core change was in the assay preparation and reading format from direct staining of cells and the miniFlo to Luminex Corporation's (Austin, TX) bead sandwich assay platform including the Luminex 100 microflow cytometer reader. The Luminex technology is described in more detail by Roth in the chapter "Luminex System" in this book. There were two key system-level drivers for making this change to the Luminex platform: (16.1) being able to multiplex assays up to 100 channels per analysis, and (16.2) having the capability to perform both immunoassays and nucleic-acid assays on the same platform. The multiplexing capability is important because, as noted in the metrics description, many different threat agents could be used and thus many assays should be run at once. Multiplexing gives the ability to perform the analysis without the cost and reliability problems of running many

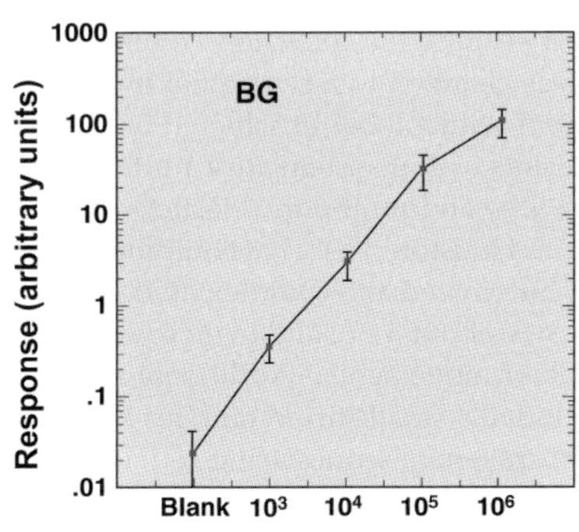

Figure 16.2. Response of the miniFlo flow cytometer to a biowarfare simulant, *Bacillus subtilis* var. *niger* ("BG"), using direct detection of cells by fluorescein-labeled antibodies in the DoD's 1996 Joint Field Trials III at Dugway Proving Ground. The samples were successfully analyzed as unknowns during the trial and later displayed in this format after the samples were unblinded.

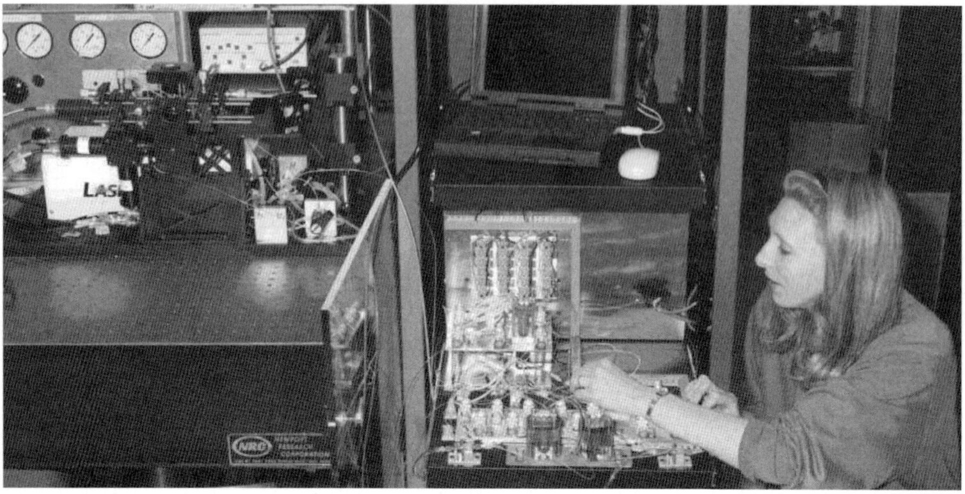

Figure 16.3. An LLNL engineer works on the fluidics for the original version of the APDS in 1998. The miniFlo microflow cytometer is on the optics table in the left of the photograph.

separate analyses. The ability to run both immunoassays and nucleic-acid assays is important for orthogonal detection and the most sensitive and specific assay combination for different threat agents.

The other core subsystems set at that time, and still in use now in upgraded versions, were the aerosol collector and the fluidics module. We changed to a custom two-stage aerosol collector with an LLNL-designed virtual impactor preconcentrator followed by the SASS 2000 wetted-wall cyclone collector. This arrangement was selected to maximize the collector concentration rate by increasing the collection flow rate and keeping the collection volume small (see Eq. 16.4 above). The fluidics module was changed to a sequential injection analysis[15,16] platform and processes developed with Global FIA Inc. (Fox Island, WA) based on their FloPro-4P. Key components of that system are a 1 mL syringe pump (Cavro, Tecan Systems Inc., San Jose, CA) and multiport selection valves (Cheminert, Valco Instruments Company Inc., Houston, TX). That equipment was selected for its ability to perform completely automated and complex fluid manipulations in the volume ranges of interest (down to about 5 μL) and to do so with very high reliability over hundreds of thousands of movements. Additional fluidics hardware specific to the Luminex assay included a small stirred tank for keeping the beads suspended and a bead trap (or "microsphere sequestering cell") invented by Global FIA and used for sample preparation binding and labeling.[17] Fig. 16.4 shows the resulting APDS100 instrument in 2001. At this time, LLNL, in collaboration with Tetracore Inc. (Gaithersburg, MD), had also succeeded in developing multiplexed Luminex immunoassays for biological threat agents.[9]

Later in 2001, there were shifts in the execution of the DOE program due to the September 11 plane attacks and the October anthrax attacks, with increased focus on transitioning to field use. The APDS team conducted a brief off-site operational

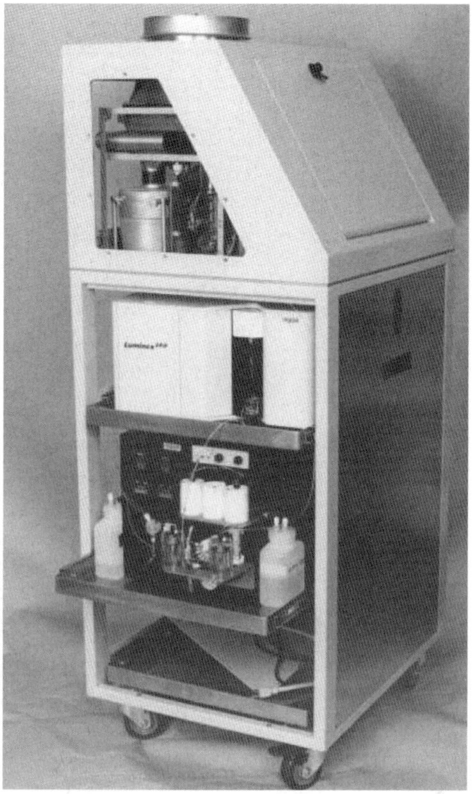

Figure 16.4. The APDS100 instrument in a stand-alone arrangement in 2001. The major subsystems of aerosol collector (top shelf), Luminex reader (second shelf), and fluidics module (third shelf) can be seen. Color reference – pg. 358.

exercise in late 2001; ran an integrated test with chamber releases of biological threat agent simulants in early 2002 at the University of Nevada, Las Vegas; and then demonstrated detection of aerosolized, viable threat agents in late 2002 at the Dugway Proving Ground.[18] Figure 16.5 shows the APDS100 instrument in the foreground rearranged for the live-agent aerosol releases, where the collector was removed from the top of the chassis and placed in the Biological Safety Level 3 chamber. These chamber tests showed the end-to-end functionality of the system and its major subsystems: aerosol collector, fluidics module, Luminex reader, and multiplexed immunoassays.

In 2002, the Department of Homeland Security (DHS) was established, and the responsibility for civilian defense against biological terrorism was transferred from the DOE to the DHS. The DHS established the BioWatch Program in 2003 for environmental monitoring against terrorist attacks using biological threat agents.[3] Subsequently, the APDS program transferred from the DOE to the DHS's Science and Technology Directorate, and later to the Systems Engineering Directorate and Office of Health Affairs for transition and operation. There were two major thrusts

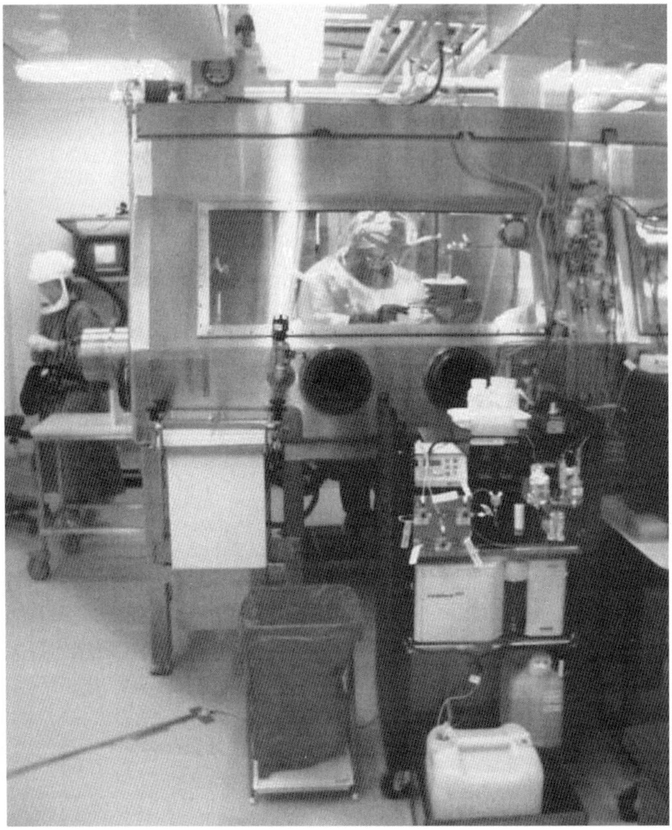

Figure 16.5. The APDS100 instrument running multiplexed immunoassays, attached to a Biological Safety Level (BSL) 3 aerosol chamber at Dugway Proving Ground for testing against live, aerosolized *Bacillus Anthracis* and *Yersinia Pestis*. The APDS instrument can be seen in the foreground in the BSL-2 laboratory; the aerosol collector was placed inside the BSL-3 chamber and pumped samples out for analysis. These chamber tests showed the end-to-end functionality of the system and its major subsystems: aerosol collector, fluidics module, Luminex reader, and multiplexed immunoassays. [Reprinted with permission from M. T. McBride, D. Masquelier, B. J. Hindson, A. J. Makarewicz, S. B. Brown, K. Burris, T. R.Metz, R. G. Langlois, K.W. Tsang, R. Bryan, D. A. Anderson, K. S. Venkateswaran, F. P. Milanovich and B.W. Colston Jr., Microfluidic-based cell sorting of Francisella tularensis infected macrophages using optical forces. *Anal. Chem.* **75**(20), 5293–5299 (2003). Copyright 2003 American Chemical Society.]

in the system development during this time: incorporating nucleic acid detection in addition to the multiplexed immunoassay, and establishing the required operational capabilities for field use. Nucleic-acid detection was integrated using a Flow-Through PCR module based on earlier PCR devices developed at LLNL with DoD and DOE funding.[19,20] The flow-through format is ideally suited to the APDS's fluidics platform; the PCR module integrates essentially as a clamp-on reactor and detector on the existing valves and tubing. This next version of the

Figure 16.6. The APDS150 instrument is shown running multiplexed immunoassays with individual PCR confirmation, field testing in a ventilation system of a major transportation hub. This was part of a series of operational field tests and evaluations during which over 20,000 field samples were autonomously collected, analyzed with multiplexed immunoassays, secondarily analyzed with PCR when necessary, and reported over an encrypted network.

instrument, dubbed the APDS150, used the multiplexed immunoassay as the primary test and real-time TaqMan PCR as a secondary, orthogonal test. This dual assay approach enabled a broad-spectrum but cost-effective test for initial detection of pathogens and proteins followed by a highly specific test for pathogen genetic sequences when necessary. The instrument was tested at Dugway Proving Ground in an aerosol chamber, where the instrument demonstrated autonomous detection of biological threat agents by multiplexed immunoassay with PCR confirmation.[21] The APDS150 also began a series of extended field tests in this phase.[22]

The operational capabilities required for extended field use were developed and refined over a few years and several field tests and evaluations. One of the most important technical developments was establishing the APDS network and data viewer to enable remote, secure monitoring of biological and maintenance signals. Stable networking is necessary for fast and decisive response to biological alerts and is exercised constantly for tracking maintenance data by the minute. Other technical capabilities in the APDS150 period included: changing to a sealed, climate-controlled enclosure for dust, rain, and temperature tolerance; adding a heater to the aerosol collector for operation down to −20°C; adding a pre-separation stage to the aerosol collector inlet to cut down on dust; and incorporating battery backup to handle external power interruptions. The field operations were conducted in two airports, three subway stations, and other high-traffic and critical facilities. Over 20,000 field samples were autonomously collected, analyzed with multiplexed immunoassays, and reported. Figure 16.6 shows an APDS150 instrument during field testing.

In addition to the technical advances in this period, there was a significant and sustained effort with the local stakeholders including public health, law enforcement, and facility representatives on operational aspects. This close collaboration was invaluable in figuring out how autonomous biological detection should fit in the daily operations and emergency responses of a city. The efforts included concepts of operations, response plans, usability assessments, feedback on features, and table-top exercises with key decision-makers.

16.3.2 Recent Development

The final development steps to the current operational system were taken in 2006. The major change for this phase was integration of a multiplexed PCR assay for detection of nucleic acid using the Luminex platform. We extended earlier instrumentation work and assay development on multiplexed PCR and combined that with the field-ready APDS150 platform, thus developing the APDS300 instrument. The hardware for the assay (PCR module, Luminex reader, bead reservoir, and bead trap), and the reagent types (enzyme, primers, beads, and streptavidin-phycoerythrin) were already present in the instrument; what was required was a change in reagent composition and rigorous fluidics process development. The final result was an intricate process that required over 1,200 commands per report cycle, but that was robust enough to run unattended in the field. Figure 16.7 shows the schematic diagram for the subset of the fluidics that conducts multiplexed PCR with Luminex detection.

The APDS300 was tested running a multiplexed PCR assay panel that had already been developed by LLNL for the DHS's BioWatch Program, tested at the CDC, and piloted in multiple Laboratory Response Network laboratories. One phase of the APDS testing was performed in a BSL-2 laboratory, where viable, unlysed, and cleaned *B. anthracis* spores and *Y. pestis* cells in liquid were used to compare the APDS multiplexed results to the manual process with TaqMan PCR. This was not an assay comparison but an integrated process comparison starting with the preparation step (Fig. 16.1); the APDS process used the viable sample directly while the manual process included bead beating and ultrafiltration for sample preparation.

Figure 16.8 shows a comparison between the results for viable biological threat agent. The interpretation is complicated by the thresholds and detection algorithm described earlier, but generally speaking, samples with TaqMan Ct values less than 32 fully satisfied the detection algorithm, whereas samples with Ct values between 32 and 34 were fairly reliably detected by multi-reactive signatures, but not always. In contrast, samples with Ct values above 34 were not completely detected, although reactive signatures were present in all but the least concentrated sample. The cutoff for reliable TaqMan detection is a Ct of about 36, depending on

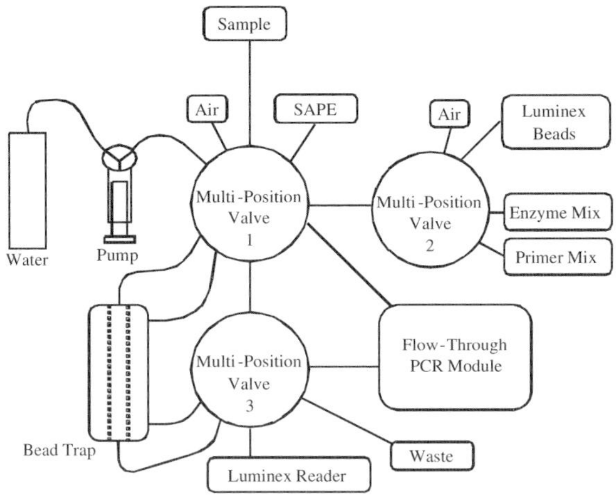

Figure 16.7. A schematic diagram of the APDS300 fluidics manifold that conducts multiplexed PCR, including sample preparation, by amplifying specific sequences using PCR, hybridizing amplicons to specific probes coupled to Luminex beads, labeling hybridized beads with streptavidin-phycoerythrin, and reading the resulting prepared sample in the Luminex microflow cytometer. The amplification and hybridization occur in the flow-through PCR module, and the labeling with washing occurs in the bead trap. [Adapted with permission from J. F. Regan, A. J.Makarewicz, B. J. Hindson, T. R. Metz, D.M. Gutierrez, T. H. Corzett, D. R. Hadley, R. C. Mahnke, B. D. Henderer, J. W. Breneman IV, T. H. Weisgraber and J. M. Dzenitis, *Anal. Chem.* **80**(19), 7422–7429 (2008). Copyright 2008 American Chemical Society.]

the signature. There is room for improvement in the APDS process, and it was demonstrated that the addition of lysis to sample preparation would improve the sensitivity.

One of the multiple real-world testing environments for the APDS300 was a high-traffic subway station. The system remained in continuous operation, sampling 1700 L of air per minute for over two months, during which time 493 aerosol samples were analyzed. Figure 16.9 shows the results for all of the samples, with biological agent signals in Panels A, B, and C, and positive and negative control assays in Panel D. All of the assays yielded control MFI values within the specified ranges, with the exception of sample #91 whose amplification positive control (PC) signal dropped to 135, which is below the acceptable lower limit of 200 for this control (Fig. 16.9, Panel D). The assay on the next sample returned an acceptable PC MFI value without intervention. The signatures shown in the other panels (A, B, and C) have low baseline noise with the primary exceptions being BA-1 and YP-4 signatures, which are two of the more sensitive signatures in this multiplexed PCR assay. These results are expected and are consistent with the analysis thresholds and detection algorithms incorporating multiple signatures per agent.

The APDS300 instrument became the first actionable autonomous detector component of the U.S. Department of Homeland Security's BioWatch program. Figure 16.10 shows the APDS300 readied for operational use running multiplexed PCR assays. Compared to the early integration for proof of concept (Fig. 16.4), the instrument is now less attractive but considerably hardened and more capable. The APDS300 as autonomous BioWatch "Generation 2.5" is operating at small scale. The DHS is entering a formal, full and open "Generation 3" competitive acquisition program for larger-scale autonomous BioWatch with an extensive test and evaluation process.[24]

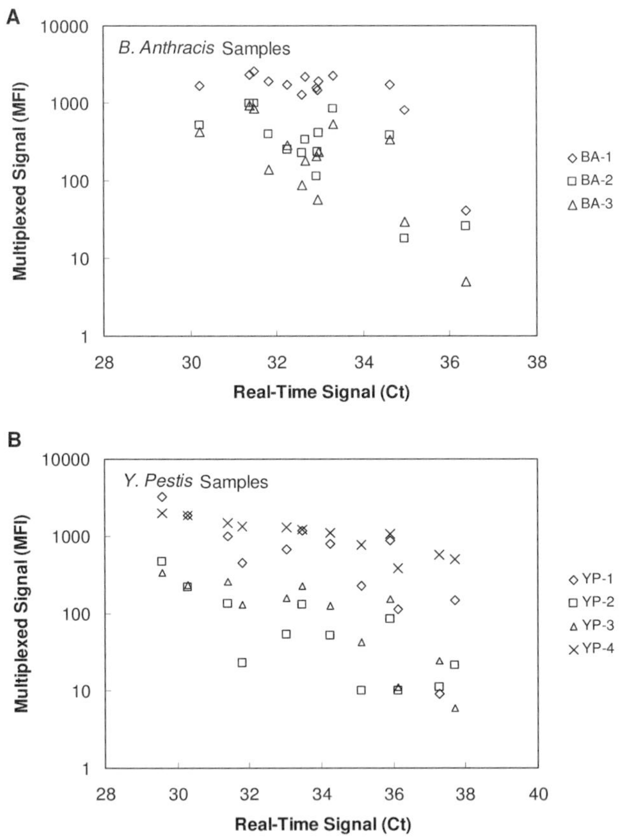

Figure 16.8. Comparison of the performance of APDS multiplexed analysis to a manual laboratory analysis for the detection of concentrations of *B. anthracis* (Panel A) and *Y. pestis* (Panel B). The APDS process introduced unlysed liquid sample directly into a multiplexed PCR assay, whereas the manual laboratory process introduces extracted and purified nucleic acids into single-plex TaqMan assays. The MFI values for each of the *B. anthracis* and *Y. pestis* signatures are shown for each sample concentration tested, represented as Ct values from real-time PCR analysis. [Reprinted with permission from J. F. Regan, A. J.Makarewicz, B. J. Hindson, T. R. Metz, D.M. Gutierrez, T. H. Corzett, D. R. Hadley, R. C. Mahnke, B. D. Henderer, J. W. Breneman IV, T. H. Weisgraber and J. M. Dzenitis, *Anal. Chem.* **80**(19), 7422–7429 (2008). Copyright 2008 American Chemical Society.]

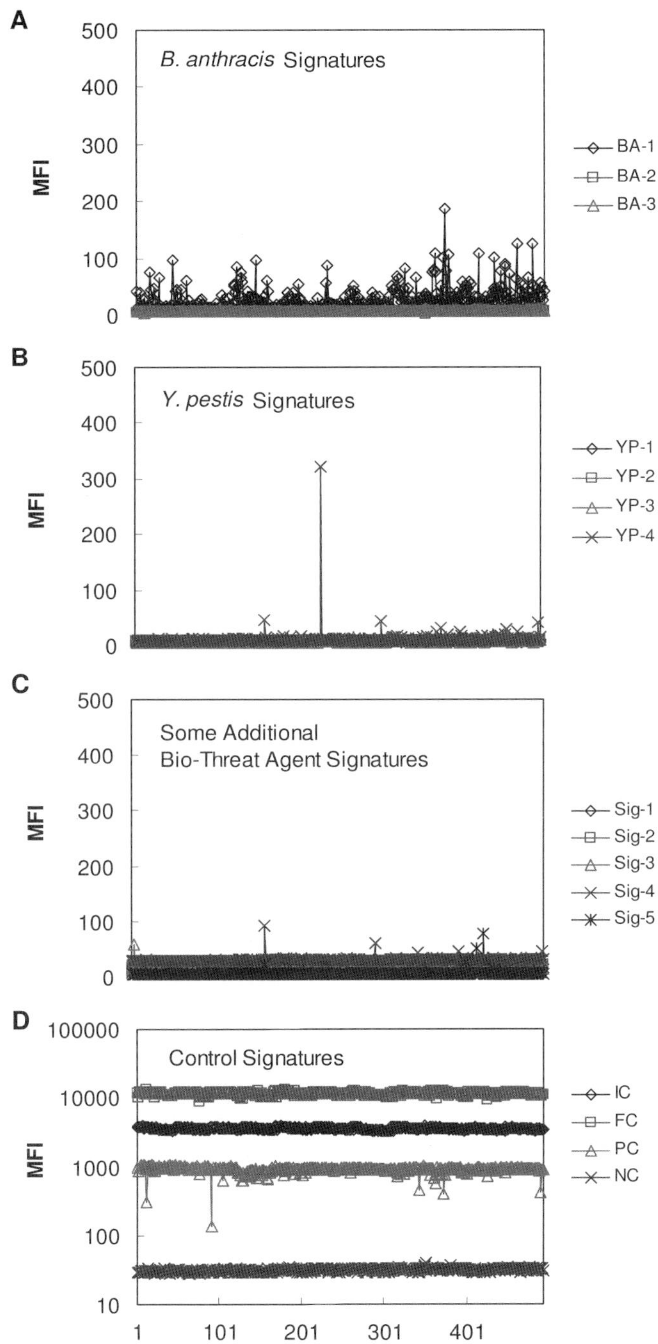

Figure 16.9. Performance of APDS in a high-traffic subway station. The APDS unit performed 493 multiplexed assays over a 2-month period. The MFI values for *B. Anthracis, Y. Pestis*, five selected additional bio-threat agent signatures, and the control signatures are shown in panels A, B, C, and D, respectively. [Reprinted with permission from J. F. Regan, A. J. Makarewicz, B. J. Hindson, T. R. Metz, D.M. Gutierrez, T. H. Corzett, D. R. Hadley, R. C. Mahnke, B. D. Henderer, J. W. Breneman IV, T. H. Weisgraber and J. M. Dzenitis, *Anal. Chem.* **80**(19), 7422–7429 (2008). Copyright 2008 American Chemical Society.] Color reference – pg. 359.

Figure 16.10. The APDS300 instrument running multiplexed PCR assays, readied for deployment for operational field use, with door open to show the internal equipment. The major subsystems of aerosol collector, uninterruptible power supply, fluidics module, and Luminex reader are visible from top to bottom, respectively. Compared to the early integration for proof of concept (Fig. 16.4), the instrument is now less attractive but considerably hardened and more capable. Color reference – pg. 358.

16.4 CRITICAL ISSUES

16.4.1 Problems to be Resolved

As discussed above, the autonomous pathogen detection mission is uniquely demanding: continuous monitoring, multiple threat agents, high sensitivity, extremely low false-positive rates, and challenging environments. Years of field testing followed by a year of operational use have shown that APDS can perform the mission. The main areas for improvement are cost, speed, and sensitivity.

The current APDS acquisition cost is around $100,000 per instrument, depending on the scale on which it is produced and which optional features are included. The true operation cost is less clear because the system is currently being transitioned to industrial operation, but it is probably over $100,000 per instrument per year at small scale. This cost is acceptable for the highest-impact locations where many thousands of people can be protected, but limits the widespread use of the system.

The APDS multiplexed PCR report is issued about every two hours; the report cycle is about an order of magnitude faster than the standard manual process with daily filter collections. This speed is fast enough to help identify and effectively treat those who were exposed, and to prevent unexposed people from entering a contaminated area. However, if the speed could be increased by another order of magnitude so the delay was only a few minutes, this could protect people who are in the area while the release is occurring by getting them to evacuate before they inhaled an infectious dose.

Sensitivity is another area that can always use improvement. Especially for detection outdoors, where the air volumes are great and the area to monitor is large, better sensitivity gives the chance of detecting a release far from its source. However, there may be a limit to how far down the LOD should go: many of the biological threat agents are naturally present in the environment at low levels, and detection at that level should not lead to response actions.

16.4.2 Future Outlook for Progress

LLNL and our industrial partner are engaged in several developmental efforts to improve the cost, speed, and sensitivity of the APDS process. For cost, we are streamlining the operation of the system and reducing the subsystem costs. This includes working with Luminex on a new reader platform tailored for field use in applications like the APDS. For speed and sensitivity, we are pursuing improvements in most of the processing steps covered earlier. Since the preparation time dominates, performing preparation in parallel for two samples can nearly halve the processing time.

Many other approaches to detecting biological threat agents have been described in the technical literature, but few of them make it to integrated field testing. Sometimes the approach is an improvement in one of the attributes of cost, speed, or sensitivity, but causes deterioration in one of the other of those attributes. Often, though, the main issue is that the technique cannot meet the mission's requirement for the rate of false positives to be only one every few years. Beyond the current APDS mission, there are other related challenges ahead. One challenge is determination of threat agent viability, where the question is no longer whether the agent is present, but whether it is likely to cause infections. Another challenge is the detection of novel or emerging pathogens, which amounts to detecting an unknown and knowing that it is a threat. These lofty technical goals combined with the harsh reality of field operations will challenge workers in this area for some time to come. The importance of the mission makes it worthwhile.

ACKNOWLEDGEMENTS

The APDS program has been advancing for over a decade, and there have been many key contributors to the advances and successes. Most of them do not work on the project full time, but make critical contributions in their areas of expertise. The current contributors at LLNL include Tom Metz, Dean Hadley, Todd Corzett, Bruce Henderer, Vincent Riot, John Breneman, Elizabeth Wheeler, Chris Bailey, Todd Weisgraber, Shanavaz Nasarabadi, Christine Hara, Sally Hall, Staci Kane, Pejman Naraghi-Arani, and Jason Olivas, with additional support from Bill Benett, Dean Urone, Bob Paris, Kris Montgomery, and Julie Avila. Team members not mentioned above in the previous phase of the APDS150 and extensive field testing included Ben Hindson, Mary McBride, Julie Perkins, Dora Gutierrez, Dennis Imbro, Chris Spadaccini, Ujwal Sathyam, Candice Cook, Corey Chinn, Ramki Madabhushi, Sally Smith, Dave Cordes, Jack Regan, Stein Weissenberger, Wendy Wilson, Claudia Hertzog, and Erik Hofmann. Previous to that, other workers who took the project from proof-of-concept to APDS100 included Bill Colston, Rich Langlois, Steve Brown, Don Masquelier, Al Ramponi, Robert Johnson, Anne Marie Erler, Keith Burris, Pete Meyers, and Paul Sargis. The initial development was instigated by Don Prosnitz and conducted through its initial steps by Ray Mariella, Fred Milanovich, Robin Miles, Kodumudi Venkateswaran, and Les Jones.

There have also been many collaborators outside of LLNL who were critical to the development and successes: the CDC in the public sector and Northrop Grumman Corp. in the private sector are standouts. Other close industrial collaborations over the years have included Luminex Corp., Global FIA Inc., Research International Inc., Tetracore Inc., Radix BioSolutions Inc., and Biosearch Technologies Inc.

Critical phases of field testing were made possible by the New York City Police Department, Department of Public Health and Mental Hygiene, and Metropolitan Transit Authority; San Francisco Bay Area Rapid Transit and International Airport; Washington Metropolitan Area Transit Authority; Albuquerque International Sunport; and San Diego Air Pollution Control District.

Greatest thanks go to the local authorities including public health officials, facility and system operators, and law enforcement personnel who are involved in the current operational system. They cannot be named here, but their feedback enabled the transition from R&D prototype to operational, national-security asset, and their expertise keeps the system running.

The APDS was initiated with LLNL internal seed funding, then largely funded by the U.S. Department of Energy and more recently by the U.S. Department of Homeland Security. We thank the DHS for their continued support and vision. Related work that contributed to the APDS development path was funded by the U.S. Department of Defense. This work was performed under the auspices of the U.S. Department of Energy by Lawrence Livermore National Laboratory in part under Contract W-7405-Eng-48 and in part under Contract DE-AC52-07NA27344.

We apologize for any names that were omitted and contributions that were distorted.

This document was prepared as an account of work sponsored by an agency of the United States government. Neither the United States government nor Lawrence Livermore National Security, LLC, nor any of their employees makes any warranty, expressed or implied, or assumes any legal liability or responsibility for the accuracy, completeness, or usefulness of any information, apparatus, product, or process disclosed, or represents that its use would not infringe privately owned rights. Reference herein to any specific commercial product, process or service by trade name, trademark, manufacturer, or otherwise does not necessarily constitute or imply its endorsement, recommendation, or favoring by the United States government or Lawrence Livermore National Security, LLC. The views and opinions of authors expressed herein do not necessarily state or reflect those of the United States government or Lawrence Livermore National Security, LLC, and shall not be used for advertising or product endorsement purposes. The Department of Homeland Security's sponsorship of the production of this material does not constitute an endorsement of any products, services, policies, or activities of the authors.

References

[1] J. P. Fitch, E. Raber and D. R. Imbro, *Science* **302**, 1350 (2003).

[2] J. F. Regan, A. J. Makarewicz, B. J. Hindson, T. R. Metz, D. M. Gutierrez, T. H. Corzett, D. R. Hadley, R. C. Mahnke, B. D. Henderer, J. W. Breneman IV, T. H. Weisgraber and J. M. Dzenitis, *Anal. Chem.* **80**, 7422 (2008).

[3] D. A. Shea and S. A. Lister, *The BioWatch Program: Detection of Bioterrorism*, (Congressional Research Service Report No. RL 32152, 2003). Accessed online at http://www.fas.org/spg/crs/terror/RL32152.html on March, **31**, (2008).

[4] P. J. Meehan, N. E. Rosenstein, M. Gillen, R. F. Meyer, M. J. Kiefer, S. Deitchman, R. E. Besser, R. L. Ehrenberg, K. M. Edwards and K. R. Martinez, CDC MMWR *Recommendations Rep.* **53**(RR07), 1 (2004).

[5] T. P. Christie, Director, US Department of Defense Operational Test & Evaluation, Annual Report FY 2003. Accessed online at http://www.globalsecurity.org/military/ library /budget/ fy2003/fy03_DOTE_Annual_Report.pdf on April, **12**, 2008.

[6] J. Carrano, T. Jeys, D. Cousins, J. Eversole, J. Gillespie, D. Healy, N. Licata, B. Loerop, M. O'Keefe, A. Samuels, J. Schultz, M. Walter, N. Wong, B. Billotte, M. Munley, E. Reich and J. Roos, "Chemical and Biological Sensor Standards Study", *Defense Advanced Research Projects Agency Publication* (2004).

[7] J. A. Rice, *Mathematical Statistics and Data Analysis*, Wadsworth and Brooks/Cole: Belmont, 39 (1987).

[8] N. R. Beer, B. J. Hindson, E. K. Wheeler, S. B. Hall, K. A. Rose, I. M. Kennedy and B. W. Colston, *Anal. Chem.* **79**, 8471 (2007).

[9] M. T. McBride, S. Gammon, M. Pitesky, T. W. O'Brien, T. Smith, J. Aldrich, R. G. Langlois, B. Colston and K. S. Venkateswaran, *Anal. Chem.* **75**, 1924 (2003).

[10] W. J. Wilson, A. M. Erler, S. L. Nasarabadi, E. W. Skowronski and P. M. Imbro, *Mol. Cell. Probes* **19**, 137 (2005).

[11] J. W. Jacobson, K. G. Oliver, C. Weiss and J. Kettman, *Cytometry A* **69**, 384 (2006).

[12] K. B. Olson, *Emerg. Infect. Dis.* **5**, (16.4), 513 (1999).

[13] R. P. Mariella Jr., G. van den Engh, D. Masquelier and G. Eveleth, *Cytometry* **24**, 27 (1996).

[14] R. P. Mariella Jr., Z. Huang and R. G. Langlois, *Cytometry* **37**, 160 (1999).

[15] J. Ruzicka and G. D. Marshall, *Anal. Chim. Acta* **237**, 329 (1990).

[16] C. E. Lenehan, N. W. Barnett and S. W. Lewis, *Analyst* **127**, 997 (2002).

[17] B. J. Hindson, S. B. Brown, G. D. Marshall, M. T. McBride, A. J. Makarewicz, D. M. Gutierrez, D. K. Wolcott, T. R. Metz, R. S. Madabhushi, J. M. Dzenitis and B. W. Colston Jr., *Anal. Chem.* **76**, 3492 (2004).

[18] M. T. McBride, D. Masquelier, B. J. Hindson, A. J. Makarewicz, S. B. Brown, K. Burris, T. R. Metz, R. G. Langlois, K. W. Tsang, R. Bryan, D. A. Anderson, K. S. Venkateswaran, F. P. Milanovich and B. W. Colston Jr., *Anal. Chem.* **75** 5293, (2003).

[19] P. Belgrader, W. Benett, D. Hadley, J. Richards, P. Stratton, R. Mariella Jr. and F. Milanovich, *Science* **16**, 5413 (1999).

[20] P. Belgrader, C. J. Elkin, S. B. Brown, S. N. Nasarabadi, R. G. Langlois, F. P. Milanovich and B. W. Colston Jr., *Anal. Chem.* **75**, 3446 (2003).

[21] B. J. Hindson, M. T. McBride, A. J. Makarewicz, B. D. Henderer, U. S. Setlur, S. M. Smith, D. M. Gutierrez, T. R. Metz, S. L. Nasarabadi, K. S. Venkateswaran, S. W. Farrow, B. W. Colston Jr. and J. M. Dzenitis, *Anal. Chem.* **77**, 284 (2005).

[22] B. J. Hindson, A. J. Makarewicz, U. S. Setlur, B. D. Henderer, M. T. McBride and J. M. Dzenitis, *Biosens. Bioelectron.* **20**, 1925 (2005).

[23] B. J. Hindson, D. M. Gutierrez, K. D. Ness, A. J. Makarewicz, T. R. Metz, U. S. Setlur, W. J. Benett, J. M. Loge, B. W. Colston Jr., P. S. Francis, N. W. Barnett and J. M. Dzenitis, *Analyst* **133**, 248 (2007).

[24] U.S. Department of Homeland Security, Procure and Deploy an Autonomous Biodetection System Called Gen 3 BioWatch, draft solicitation number HSHQDC-09-R-00045D. Accessed online at https://www.fbo.gov/ on 2/20/2009.

Chapter Seventeen

Laser-Based Fabrication of Microflow Cytometers with Integrated Optical Waveguides

Mark Dugan,[1] Ali A. Said,[1] Tom Haddock,[1] Philippe Bado,[1,*] and Yves Bellouard[2]

[1]*Translume, 655 Phoenix Drive, Ann Arbor, MI 48108, USA*
[*]*pbado@translume.com*
[2]*Mechanical Engineering Department, Eindhoven University of Technology, Eindhoven, The Netherlands*

We present a novel microfabrication approach to create microflow cytometers. Femtosecond laser pulses are used in a direct-write embodiment to create, from fused silica monoliths, extremely robust flow cytometers with uncommon characteristics. One of the most interesting features associated with the femtosecond direct-write approach is the opportunity to fabricate optical and microfluidic elements in a single continuous step. This capability affects the design and performance of microflow cytometers in many ways. We expect that the direct-write process will create opportunities to expand the use of microflow cytometers outside traditional markets.

17.1 FLOW CYTOMETER MINIATURIZATION

Commercial flow cytometer manufacturers are slowly introducing smaller devices. The rationales for the use of miniaturized devices are numerous. The benefits sought through physical down-sizing include: reduction of consumables, smaller footprint, enhanced portability, integration with networking capabilities, and, most importantly, increased robustness and enhanced reliability. While the potential benefits of miniaturization seem vast, until now small flow cytometers or microflow cytometers are rare, their capabilities remain limited, and their ability to operate in demanding environments fall well short of the perceived avenues opened through miniaturization. Manufacturing issues are significant and have slowed advances in the field.

The Microflow Cytometer by J S Kim & F S Ligler

www.panstanford.com
978-981-4267-41-0

17.2 MICROFABRICATION APPROACHES AND THEIR RELEVANCE TO MICROFLOW CYTOMETERS

To date, the fabrication of microsystems relies mainly on two technology platforms that can be combined or used separately. The first technology platform is based on the surface micromachining of silicon. This platform relies on cleanroom-based photolithographic processes inherited from the microelectronics industry. Devices fabricated by surface micromachining are made through successive steps of thin layer deposition and selective material removal. Through a series of complex, but well-established fabrication steps, one can build intricate devices layer by layer. The second technology platform is microassembly fabrication. Parts produced by surface micromachining or other microfabrication techniques are assembled together to form a device. There, the main challenges are to precisely position shaped parts and assemble them in a reliable manner through various bonding or joining processes. The smaller the component sizes and the more diverse the materials used, the more challenging the assembly and fixturing.

While these two approaches are commonly used throughout the industry and are supported by a very large knowledge base, neither one is particularly applicable to the fabrication of microflow cytometers. The case against the photolithographic approach is rather obvious: silicon, being opaque in the visible, is clearly not the material of choice for a flow cytometer that relies on optical techniques to perform its function. Furthermore, with all surface technologies, the resulting devices are planar. While silicon-based MEMS are commonly perceived as three-dimensional devices, their out-of-plan dimensions rarely surpass ten microns. Additionally, material characteristics such as wetting, biocompatibility, etc. are potential issues for flow cytometry applications. In addition to the technological challenges mentioned above, there is also a very significant financial element arguing against the use of planar fabrication processes: it is well-known that developing novel devices using photolithographic processes is a hugely expensive undertaking that can be justified only through the economy of scales brought by mass production. At this point in time, the market for microflow cytometers is too small by many orders of magnitude to benefit from the economy of scale brought by photolithographic processes.

The argument against microassembly fabrication is that the approach requires more hands-on experience to be fully appreciated. It is generally difficult to fabricate devices through the assembly of many small components. This fabrication approach often requires a significant amount of skilled labor (traditionally addressed by moving manufacturing to lower cost locations). Microassembly challenges are significantly amplified when fabricating optical instruments. With most optical microsystems there is an inherent requirement to keep tight alignment between various elements during the initial assembly and during the final joining. This is very difficult to achieve using multiple elements made of different materials, a characteristic of most flow cytometers. The telecommunications industry has invested large sums of money and many years to solve this fabrication issue; yet to date the fabrication of many telecommunication devices requires extensive skilled labor and is marred by poor yields.

To summarize, fabricating a microflow cytometer is challenging, and traditional micro-device fabrication processes are ill matched to the task.

17.2.1 Direct-write Fabrication Approach

Recognizing this problem, we have been pursuing a very different manufacturing approach. Rather than fabricating a microflow cytometer through an assemblage of small elements and a multiplicity of materials, we have chosen to start with a solid block of an appropriate homogeneous material that is subsequently modified on a local scale to give it desirable functional properties.[1] The starting element (substrate) is no longer thought of simply as one part of an assembly forming a system, but rather as a system on its own. This objective is accomplished through local function integration, which is achieved through the controlled exposure of the substrate to an intense field from an ultrafast laser.

Fused silica is selected as the preferred base material due, in part, to it being a high optical quality glass. Several aspects motivate this choice: fused silica, also called fused quartz, is a highly homogeneous amorphous material, transparent from the deep UV to the shortwave infrared (from 190 nm to approximately 3.5 μm). Fused silica has the lowest UV-induced fluorescence coefficient of any commercial optical glass compositions. It is compatible with all bio- and industrial-fluids (except hydrofluoric acid). For all of these reasons, fused silica is the preferred material to fabricate conventional flow cytometers. In addition, fused silica has excellent elastic properties making it a suitable material for micromechanics. This last property is presently unused in the design of (micro) flow cytometers.

In sharp contrast to silicon, established processes to micromachine fused silica are quite limited. Glass microfabrication techniques that meet the miniaturization and systems integration requirements to fabricate microflow cytometers are not fully realized. In order to fabricate microflow cytometers out of fused silica, novel fabrication processes had to be developed. These processes depart in fundamental ways from accepted practices.

Unlike photolithographic processes, which form devices through a multilayer approach, we work with a thick (up to several mm) solid monolith within which fine spatial regions of the material are selectively modified through the controlled exposure to an ultrafast laser. Such material modification can be manifest through a change in various optical or mechanical properties as well as an enhanced chemical etching susceptibility. This approach eliminates all needs for primary positioning and joining. It makes possible the fabrication of three-dimensional volumetric devices of extreme robustness. Most importantly, the various constituents of a device fabricated this way remain aligned even when subjected to intense shocks, vibrations, and temperature extremes. On this account the direct-write fabrication approach is unrivaled.

17.3 DEVELOPMENT OF THE DIRECT-WRITE FABRICATION TECHNIQUE

At the heart of the direct-write fabrication process is the interaction of the silica glass with the intense electric field from a focused femtosecond laser. This interaction can modify the silica network resulting in local density and stress changes and on a fine scale a periodic chemical reorganization. Modifications of these material characteristics induce changes in the optical properties of index of refraction, scattering, and birefringence of the base material. When combined with wet chemical etching, these regions of laser-induced modifications can be removed at controlled rates, resulting in the creation of structural frames, as well as dynamic mechanical elements. Additional processes are used as needed for secondary operations. Individual or hybrid optical, mechanical, or fluidic functionalities are produced through local changes in the material physical characteristics and/or through selective material removal.

17.3.1 Prior Work — Ablation

The use of femtosecond lasers to machine glass and more specifically to micromachine fused silica was first investigated in the early 1990's.[2,3] These studies were concerned exclusively with laser-driven ablation. It turned out that ablation is not a very attractive process from a device fabrication point of view, as there are numerous shortcomings associated with it, especially when working with brittle materials such as glass. The ablation process generates undesirable cracks and creates high surface roughness. These laser ablation shortcomings are well documented. Several parties are pursuing various enhancements in the use of femtosecond pulses to reduce these shortcomings. However, so far, the results are wanting — while it is true that affected zones are much less prominent when using femtosecond pulses, they are never the less present. Similarly, microcracks are smaller but still create very significant scattering. In addition, inherently the ablation process creates a boundary surface that prevents further work directly underneath it. While there are ways to mitigate every one of these shortcomings, mutually they represent a considerable barrier to quality manufacturing and as a result, to date, we know of no device of significance manufactured from glass using femtosecond ablation alone.

17.3.2 FemtoWrite™ and FemtoEtch™

It had been known for some time that the index of refraction of doped glass could be changed by exposure to light. The discovery by Hill of the photosensitivity of silica fibers at UV wavelengths has had a significant impact on the manufacturing of some telecommunication devices.[4] However the real breakthrough came with the discovery from Hirao and his coworkers that femtosecond pulses can change the index of refraction of undoped fused silica.[5] Hirao immediately recognized that this process could be applied to the formation of subsurface waveguides. Numerous other groups followed on this early work. We extended this process to the manufacturing of high-quality, low-loss, single mode straight and curved waveguides.[6]

This index change process which we refer to as *femtoWrite*™ illustrated in Fig. 17.1.

An extremely short laser pulse is focused inside the fused silica substrate. At the focal point, the optical field is so intense that, through multiphoton or tunneling ionization and subsequent impact ionization, an electron-plasma is formed. Yet, since the pulse is extremely short ($\sim$100 fs) and has little energy content ($<$ 1 μJ, the glass re-solidifies on a short time scale compared to network vibrations and heat distribution. This process has been extensively studied. Raman spectroscopy data[7,8,9] indicate that the laser-illuminated zone contains a distribution of SiO_2 ring structures (3-8 member rings) weighted toward the smaller member rings when compared to the untreated material (average is a 6 member ring). The increase in smaller size siloxane ring structures is associated with a densified region and an elevated index of refraction.[8,10] Adjusting the femtosecond laser pulse parameters, one can maximize this phase and tailor the local refractive index. This local index change can be turned into an optical waveguide by carefully directing the laser at selected contiguous points within the glass substrate. While the change in the index of refraction obtained to date might appear weak ($\Delta n < 1\%$),[6] it corresponds to that of a commercial single-mode optical fiber.

Interaction with femtosecond pulses can also locally increase glass susceptibility to hydrofluoric acid (HF) etching. Initially demonstrated in photo-etchable

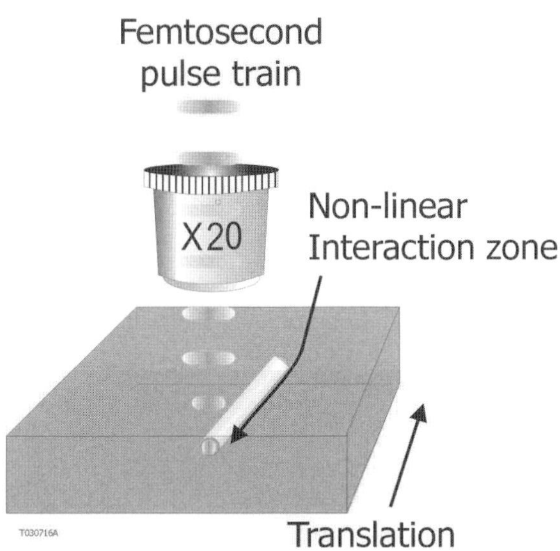

Figure 17.1. A schematic illustration of the *femtoWrite*™ process. Color reference – pg. 360.

glass,[11] the process was later shown to apply particularly well to fused silica.[12] This process, which is closely related to our *femtoWrite*™ process as illustrated in Fig. 17.2, can be use to selectively machine fused-silica substrates. Regions that have been exposed to the femtosecond laser pulses etch away much faster than the unexposed regions. The etching rate ratio approaches 100 to 1. This two-step process, which we referred to as *femtoEtch*™, is used to form three-dimensional structures whose geometrical complexity is limited only by a requirement for the HF to be able to easily reach the laser-illuminated zones.[13] This requirement does impact the manufacturing of tunnels. It is difficult to fabricate tunnels longer than a few millimeters. Consequently, when a long tunnel is called for, one generally fabricates a surface channel that is later capped by bonding a fused silica lid.

These two processes – *femtoWrite* and *femtoEtch* – form the basis of the direct-write approach as we have implemented it to fabricate various small fused-silica devices and instruments. They are of particular interest for optofluidics applications, including microflow cytometry.

The two processes are closely related, as the same femtosecond laser is used to change the index of refraction and to selectively pattern and control the HF etching. Consequently, both processes can be carried out jointly without any need for realignment or repositioning of the substrate. One can define in a single step both optical and mechanical features. This approach dramatically simplifies the overall fabrication of complex glass-based microdevices and eliminates significant alignment issues associated with sequential fabrication processes.

While this manufacturing approach greatly simplifies some key issues associated with the fabrication of glass-based small devices, it does have some limitations that impact the fabrication of microflow cytometers. The finite working distance and the spatial and temporal aberrations associated with the large numerical aperture (NA) focusing optics hampers working deeper than approximately 8 mm below the substrate top surface. To circumvent, at least partially, this limitation, one often uses a transverse geometry (*i.e.*, beam perpendicular to the device main features) to fabricate long channels, as illustrated in Fig. 17.2.

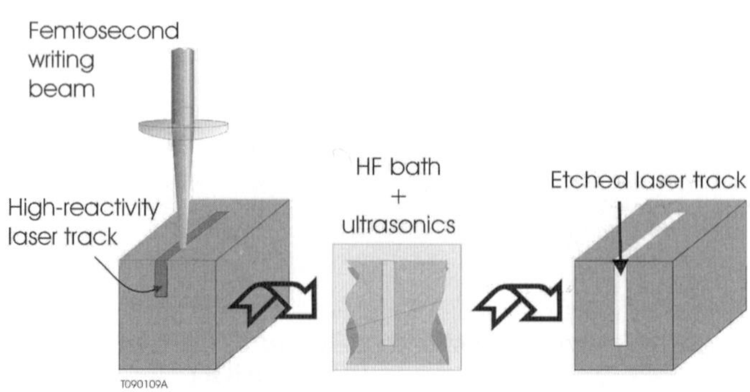

Figure 17.2. A schematic illustration of the *femtoEtch* process. Color reference – pg. 360.

17.4 APPLICATION OF THE DIRECT-WRITE APPROACH TO THE FABRICATION OF MICROFLOW CYTOMETERS

At its core a (micro) flow cytometer is composed of two subsystems: (i) a network of micro-fluidic channels whose function is to condition and bring the flow stream of interest to the analysis area and (ii) an optical system whose function is to detect and analyze any element of interest in the flow stream. These core subsystems, when connected to at least one source and one detector, form the basis of all (micro) flow cytometers.

Using only the *femtoWrite*™ and *femtoEtch*™ laser-based processes one can fabricate a complete flow cytometer core. We femtowrite waveguides that carry input and output optical signals; and femtoetch microfluidic channels that deliver the liquid stream to the interrogation point(s). Fabrication of a structural framework, interfacing to the local environment, and other functionalities are also created with these two laser-based processes.

17.4.1 Fabricating Flow Channels with the Direct-Write

The design and fabrication of flow channels are rather straightforward. The flow must be laminar, which is typically achieved by restricting the cross-section of the flow channels. Common dimensions are in the tens to hundreds of microns. These cross-sectional dimensions are easily reached with the direct-write laser-based fabrication process. The flow channel length is dictated not by fundamental flow requirements, but rather by other constraints; it can extend from a few millimeters to many centimeters. The cross-section geometry of flow cytometers is typically rectangular (This feature is often driven by today's manufacturing limitations). This is also the most common microfluidic channel geometry associated with the direct-write fabrication process, although one can easily fabricate more complex cross-section geometries as shown in Fig. 17.3. It should also be obvious to the reader that with the direct-write fabrication approach, the channel cross section does not have to be constant but can be locally adapted to meet specific functionality and flow requirements.[14,15] This extreme design flexibility cannot be achieved with conventional fabrication techniques.

In all flow cytometers there is a requirement to surround the fluid stream of interest with a sheath — which can be either a liquid or a gas. Conceptually simple, the formation of a three-dimensional dynamically stable and uniform sheath is difficult to implement. It is a significant cost element in the fabrication of flow cytometers. It is sometimes achieved by instilling the liquid of interest at the center of the sheathing flow using some type of "injection nozzle". Alternatively two (or more) secondary side channels bring the sheathing around the liquid of interest. The channels associated with the sheath generation are generally of dimensions similar to that of the main channel (gas-based sheath may be obtained with smaller channels).

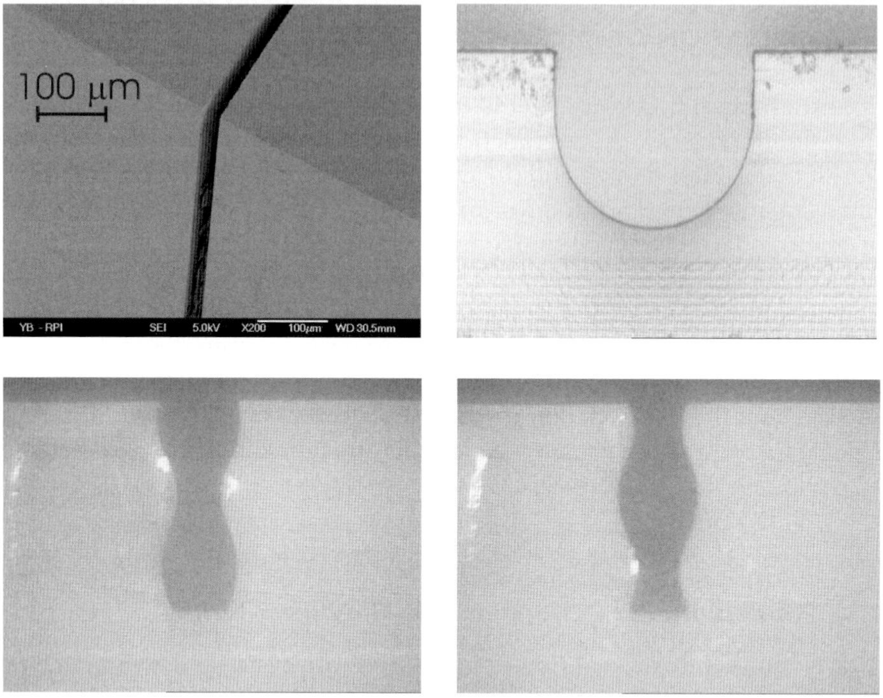

Figure 17.3. A collection of microfluidic channels created with the *femtoEtch*™ process. Top left, a deep straight channel. Top right, a channel with vertical walls and a rounded floor. Bottom left, a channel with a waist at mid-height. Bottom right, a channel with maximum width at mid-height. Color reference – pg. 361.

The secondary channels can either surround the main channel, an approach that requires the formation of partially- or fully-buried channels (i.e., tunnels), or the sheathing fluid must be made to twirl around the main stream. As flow cytometers operate in the laminar regime, inducing a swirling motion is a relatively complex undertaking that can be obtained by locally shaping the microfluidic channels.[16] This local shaping is challenging to achieve using traditional flow cytometer fabrication processes, but rather simple to implement with the direct-write approach, as shown in Fig. 17.4.

17.4.2 Fabricating Optics with the Direct-Write

With the direct write laser-based approach, optical elements can be created in two ways: (a) one can form optical waveguides by locally changing the index of refraction of the fused silica substrate, or (b) using the *femtoEtch*™ process one can selectively remove material to create refractive (or diffractive) optical elements.

In fused silica, the maximum change in the index of refraction obtained to date with the *femtoWrite*™ process is a fraction of one-percent. This low value entails some limitations on the minimum radius of curvature of our waveguides to

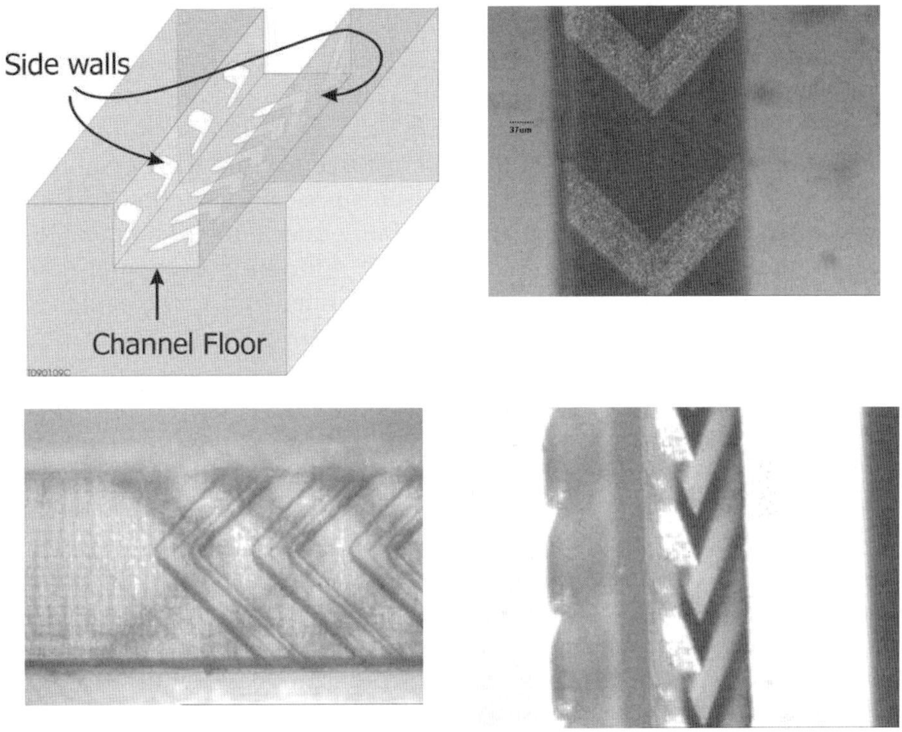

Figure 17.4. Microfluidic channel with shaped walls and floor. Top left, a schematic representation of the channel with location of various patterns. Top right, floor with chevrons. Bottom left, sidewalls with chevrons (the back wall is out of focus). Bottom right, a side view of chevrons in a sidewall. Color reference – pg. 361.

avoid excessive radiative loss (typically greater than 15 mm for single mode infrared wavelengths and 5 mm for multimode visible wavelengths), and on the maximum numerical aperture of the waveguides (NA < 0.14). In most flow cytometers, the collection optics are designed around large NA lenses. To date one cannot duplicate this optic relying solely on the *femtoWrite*™ process. However the index change induced by the *femtoWrite*™ process can be tailored to create collimators, soft lenses and other elements. These elements are used to condition optical signals prior to their insertion in the flow stream.

The *femtoEtch*™ process can be used to create single or compounded spherical, cylindrical, and aspheric lenses. Combining *femtoEtch*™ and *femtoWrite*™, one can create complex optical systems.

The optical waveguides created with the *femtoWrite*™ process are of high optical quality, with loss lower than 0.1 dB/cm (measured at 1550 nm).[6] The elements formed with the *femtoEtch*™ are more wanting, with a surface roughness that introduces significant unwanted scattering. This is a limitation that affects the performances of microflow cytometers fabricated with femtosecond direct-write processes.

17.4.3 Integrating Optical and Microfluidic Systems

One of the most interesting features associated with the femtosecond direct-write approach is the opportunity to fabricate optical and microfluidic elements in a single continuous step. As a result, one can position a multitude of these various elements with relative accuracy of the order of one micron. Furthermore, these elements, being imbedded in a block of fused silica, will never move and thus will never need to be re-aligned. This capability affects the design and performances of microflow cytometers in many ways. Efficient collection of transmitted and scattered signals is obtained through integrated optics (waveguide or free space) placed very closely to the flow stream. Current flow cytometers often use a cylindrical lens to illuminate the fluid, creating a line focusing across the flow stream. A second cylindrical lens system is used on the output side to collect the small angle scattered signal. A beamstop, placed just after the flow stream, is used to block the unwanted intense unscattered transmitted signal from entering the forward-scatter detectors. This element is critical to the good performance of the flow cytometer. When designing this complex optical system, one must take into account manufacturing alignment tolerancing, as well as short- and long-term alignment drift. Trade-offs must be reached. The beamstop must be oversized or alternatively the instrument must include means to reposition it periodically. The incoming beam is typically oversized in order to insure that the full stream core is illuminated; yet it must not be too large to reduce accidental coincidence detection. These are only a few of many tradeoffs the optical designer must consider. Ultimately, these considerations add significantly to the complexity of modern flow cytometers, drive up manufacturing costs, and prevent deployment in all but the most stable environment.

To be able to position very accurately and in an immutable way the various elements that form the optical system surrounding the flow stream drastically simplifies the flow cytometer design and fabrication. For example the incoming beam size can be decreased, which will lower the required input optical power. In current instruments, the illumination spot is roughly cylindrical with dimensions of about 20 microns along the flow axis and about 100 microns perpendicular to the flow axis. A more desirable illumination region would have dimensions of about 2 by 50 microns.[17] In standard instruments, using such a small illumination region could lead to severe problems when the system drifts. However, with the femtosecond direct-write approach, system drift is not of concern and therefore one can design flow cytometers around a very small illumination region. The beam stop, being perfectly aligned with what approaches an ideal point source, can also be downsized. In fact one can conceive of eliminating the beamstop all together and placing one, or many, collecting optics (waveguided or otherwise) slightly off the main optical axis. Scattered signals can be collected using an array of tightly placed optical elements. While conceptually simple, this is in fact very difficult to implement using conventional bulk optical elements that will eventually drift. Accurate angular scattering data can yield information on cell morphology through

Mie scattering signatures. The wealth of data associated with multiple-angle scattering is generally not accessed in current flow cytometers.

Current instruments use large NA optics to maximize the collection of the scatter signal. Angular resolution is sacrificed in order to improve the signal-to-noise ratio. Improving the optical signal strength through better permanent alignment can achieve the same result with smaller NA optics. Thus, with the femtosecond direct-write approach one may be able to obtain a good signal-to-noise ratio and good angular resolution.

The use of waveguides in place of conventional bulk optics in flow cytometers was first proposed in 1986.[18] However, to date this has generally been achieved by attaching or bonding optical fibers to the flow cytometer body. This hybrid approach sacrifices many of the advantages associated with the direct-write approach. While there has been some work with integrated optics,[19] it has generally been implemented in ways that are not conductive to commercial products.

The tight integration of microfluidic and optical systems discussed above is further complemented by the direct writing of the flow cytometer mechanical frame (body) as well as anchoring points for optoelectronics elements and ancillary fiber attachment ports and the like. This combined capability is unprecedented, and its full implication to the field of microflow cytometry has not yet been fully realized. Figure 5 shows a concept microflow cytometer with associated elements fabricated with the femtosecond laser direct-write processes. This exemplary monolithic device shows at the top two angled side channels bringing the sheath around the central channel. The side channel cross-sections are shaped (Fig. 17.5(A)) to wrap the sheath around the central channel and to hydrodynamically focus the flow stream. Often injection chambers are fabricated by conventional techniques that share a propensity to cause micro turbulence at sharp edges. While the dividing wall of the branch shown in Fig. 17.5(B) is sharp by design, it could have been shaped in any way. Also, vent chambers (not shown) can be placed in the injection chamber as needed to bleed off priming bubbles, either by natural buoyancy or by vacuum extraction.

The device mid-section shows two interrogation points (Fig. 17.5(C)), each equipped with its own optical analytical capability. The optical systems are formed of waveguides and lenses. This capability is seen clearly in Fig. 17.5(D) where a complex lens shape was created using the *femtoEtch*™ process. They are interfaced to the outside world via optical fiber ports. The close placement of signal-reception waveguides will permit accurate detection of the angular scattering distribution from cells. Two angled waveguide signal inputs are shown exiting a flow channel in Fig. 17.5(E). Accurate angular scattering data will yield information on cell morphology through Mie scattering signatures. Due to their small footprint, numerous waveguide-based interrogation points can be written in a single glass monolith both at the same flow point for parallel analysis or along the flow line for serial analysis. This sort of parallel or serial processing cannot be done with conventional systems, even those using optical fiber, due to problems of space and the reliability of alignment of the systems. With system reliability going as the inverse

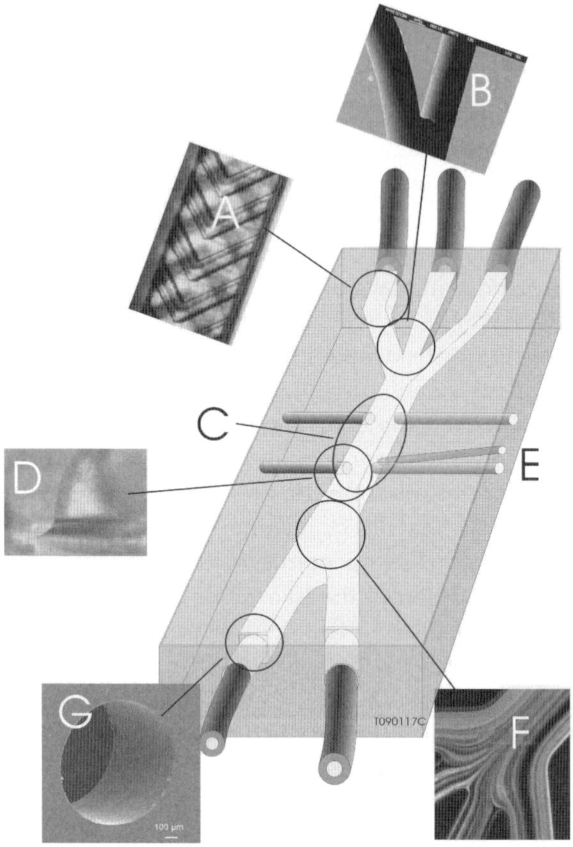

Figure 17.5. A schematic representation of a microflow cytometer with an overlay of various relevant elements fabricated by femtosecond direct write. Color reference – pg. 362.

power of the number of elements, the requirements for alignment and reliability of multiple-component systems are stringent. This limitation does not apply to devices fabricated with the femtosecond direct-write approach as it produces (optical) components that cannot go out of alignment.

To further improve the performance of microflow cytometers fabricated with the direct-write approach, one needs to reduce stray light noise within monolithic fused silica. Several techniques have been tested, including curving the waveguides away from the collection region toward the detectors, so unwanted stray light emanating from the scattering region will be orthogonal to the orientation of the detectors; absorbing coatings have also been used to further reduce noise from stray light. These techniques are found in various fluid-sensing systems now operating in the field.

The device final section shows an optically driven sorter (Fig. 17.5(F)). The ability to sort cells in the flow stream using optical techniques has been demonstrated.[20] Finally the flow exits through interfacing ports (Fig. 17.5(G))

The device shown in Fig. 17.5 is of course only one variant of a very large set of designs that can be achieved with the femtosecond laser direct-write manufacturing approach.

17.5 ADDRESSING THE PRESENT LIMITATIONS OF THE DIRECT-WRITE

The direct-write integrated monolithic fabrication capability has the potential to drastically change the design, manufacturing, and deployment of microflow cytometers. However this approach has some limitations impacting commercialization. These limitations will have to be addressed before the femtosecond direct-write approach is more generally accepted in flow cytometer fabrication.

We have briefly mentioned two issues that are relevant to the performance of the microflow cytometer optical system; namely the limited index change that can be reached with the *femtoWrite*™ and the optical surface quality associated with our *femtoEtch*™ process. We now present some means to address these issues.

17.5.1 Limited Index of Refraction

The main optical parameters characterizing a waveguide are cross-section size, delta n, optical loss, and birefringence. The size and the delta n of a waveguide are related to its ability to support guided propagation of light of a given wavelength. With the direct-write process one has significant latitude as to the cross-section size and shape but one can generate only small change in the index of refraction (Δn). We have done extensive work to increase the index change.[6] For fused silica the change in the index of refraction is a function of the femtosecond laser pulses intensity, as illustrated in Fig. 17.6. However above some threshold intensity value, a form of laser damage that is incompatible with optical waveguiding replaces the uniform index change process. We have demonstrated that the index change is cumulative; one can progressively raise the delta n through multiple laser exposures (at a given laser intensity) up to some saturation value. Using that knowledge we are able to generate an index change (6×10^{-3}) that corresponds to that found in standard single mode telecommunication optical fibers.

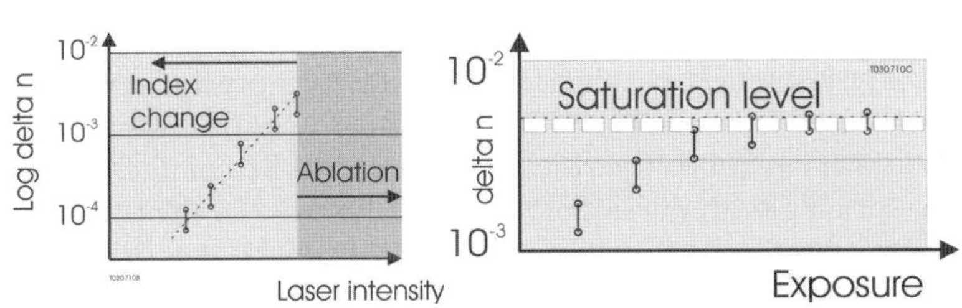

Figure 17.6. Change in the index of refraction of fused silica as a function of the femtosecond laser pulses intensity (left), and the cumulative exposure (right).

The underlying physical mechanisms behind the increased index of refraction are yet to be fully understood. Using a novel scanning thermal microscope and a Raman spectrometer, we recently observed in laser-affected zones, a localized sharp decrease of the thermal conductivity correlated with an increased presence of low-number SiO_2 rings.[9] This study suggests that the femtosecond laser pulses densified the fused silica by approximately 8%, a value that is based on the shift of relevant Raman band.[21] Assuming this level of densification and using previous data from a study on shock-densified silica,[10] one would extrapolate a local increase of refractive index of about 3×10^{-2}. This result is five-fold higher than our measured increase of the refractive index in fused silica exposed to femtosecond laser pulses. To explain this discrepancy, we formulate two hypotheses. One possibility is that the refractive index measurements reported so far are averaged values of the laser-affected zone and its immediate surroundings. This explanation is plausible considering the small size of the zone affected by the focused femtosecond laser and the limited spatial resolution of local refractive index measurement technique.[22] A second possibility is that the structure of "femtosecond laser-densified" fused silica is different from the structure of shock-densified silica, therefore preventing the use of shock-densified silica as a direct comparison. The former laser-induced densification may have a more abrupt transition between the exposed and unexposed material resulting in a larger density gradient. This transition region, in turn, would have a compressive-to-tensile stressed transition relative to the bulk material. As a consequence, the local index of refraction would have a stress-induced contribution. In any case, there is substantial evidence that the fused silica densification obtained so far with femtosecond pulses is appreciably less than the densification obtained with other processes, such as mechanical densification. Obtaining higher densification and higher index of refraction with femtosecond lasers would have significant commercial implications.

The refractive index change also governs the loss associated with propagation of the optical signal along a curved waveguide. An optical signal propagating down a waveguide with a small Δn will undergo substantial radiative loss. For the index change induced in fused silica with the femtosecond direct write process, losses become substantial when the waveguide radius of curvature falls below 15 mm for single mode propagation at 1550 nm. The minimum bend radius can be further reduced for multimode and/or shorter wavelength propagation. This limitation can be bypassed using an assembly formed of waveguides and a reflecting surface. The reflecting surface can be fabricated by etching out an air trench and either filling it with a reflective material or leaving it empty and operating in total internal reflection as shown in Fig. 17.7. Conceptually simple, the implementation of this embodiment requires that the waveguide tips be enlarged to reduce the diffraction associated with the unguided pathway between the two waveguides. Further the sequence of operations — writing and etching — has to be optimized to allow for the etching of the reflecting trench without engendering any degradation of the optical waveguides. It has been shown that a very thin region of untreated material will protect waveguides during the etching process.

In Fig. 17.7 the waveguides shown are terminated 25 micron from the wall of the microfluidic trench.

17.5.2 Optical Surface Quality

There is a noticeable surface roughness associated with the *femtoEtch*™ process. This roughness, measured with an atomic force microscope, is of the order of 100 nm to 500 nm (RMS value) as illustrated in Fig. 17.8. This surface roughness is

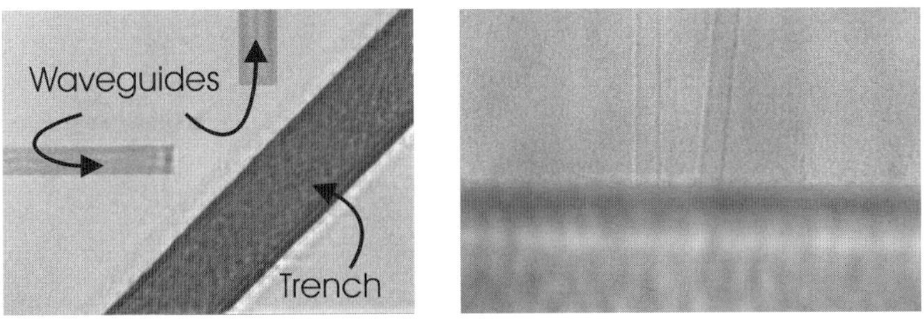

Figure 17.7. A total internal reflection mirror assembly formed of waveguides and a reflecting surface (left). Two waveguides terminated near a deep microfluidic trench (right). The waveguides can be located 20 to 1000 microns below the top surface. Color reference – pg. 362.

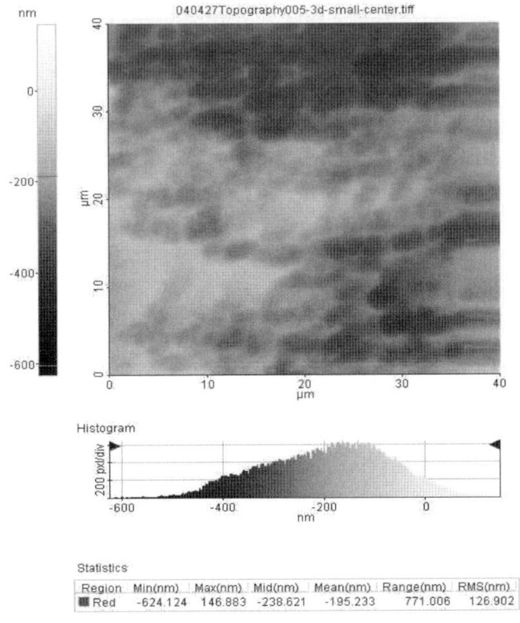

Figure 17.8. Surface roughness data collected with an atomic force microscope. The RMS roughness, measured on a vertical wall (surface parallel to the laser beam) is of the order of 100 to 500 nm. Color reference – pg. 363.

further characterized by some long-range modulation (correlation spacing of the order of 1 to 15 microns) related to the laser-writing pattern. In general, surfaces that are perpendicular to the writing laser beam (bottom of microfluidic trench for example) are rougher than surfaces that are parallel to the beam.

The typical roughness associated with the microchannels fabricated by *femtoEtch*™ is sufficiently weak that it does not appreciably affect the flow of the fluid stream even if the sheathing fluid was not present to prevent direct contact. However, this roughness significantly affects the optical characteristic of microflow cytometers fabricated by direct-write. The surface roughness introduces optical loss and scattering. While the optical losses are not noteworthy for many applications, the scattering introduces optical noise that drastically reduces the sensitivity of microflow cytometers. The surface roughness needs to be reduced by a factor of three, at least, in order to manufacture optical surfaces of the quality needed for many microflow cytometer applications.

Several approaches are being pursued to reduce the surface roughness associated with the *femtoEtch*™ process. Over-etching does reduce surface roughness, but only slowly. The device needs to be etched for extended periods before any significant improvement can be detected. Thermal smoothing as also been suggested. Global smoothing is probably not a very practical solution, in part due to the extremely high melting temperature of fused silica. There is however a significant body of work regarding the use of low-intensity CO_2 lasers to locally polish optical glasses.[23,24] This process has been shown to be quite successful for materials, such as fused silica, which have a small coefficient of thermal expansion. Because of the strong absorbance of fused silica in the infrared, CO_2 laser radiation is extremely well suited for this task. The laser beam melts a micron-thin surface layer that flows under the action of surface tension. Surface imperfections (i.e. roughness) that are initially present are smoothed over during the melting process. Others have reported on laser polishing protocols and have demonstrated techniques capable of smoothing silica surfaces with a 1 micron scale roughness down to 1 nm levels with no unwanted effect on the as-machined net surface shape.[25] One has yet to successfully adapt this process to the geometry associated with microflow cytometers.

17.6 BONDING

As mentioned previously the roofs of the microfluidic channels are formed by bonding a fused-silica lid to the main glass body. While there are numerous techniques available to bond fused silica to fused silica, many are not directly applicable to the fabrication of microflow cytometers. The requirement that the microfluidic channels not be permanently plugged by the bonding agent disqualifies many of the most common bonding approaches. Thermal bonding and optical contact bonding are two techniques that are compatible with microflow cytometer applications.

17.6.1 Optical Contact Bonding

This low temperature process involves bringing in contact two highly polished surfaces in such a way that van der Waals attraction between the two bonding surfaces forms a solid bond. The resulting parts are sometimes annealed at a relatively low temperature in a subsequent step to further strengthen the bond through the formation of ionic and/or covalent bonds. There are demanding requirements associated with this process. The surfaces to be bonded must be brought in perfect physical contact or incomplete or no bonding will form. The surfaces intended for optical contacting must be extremely flat and highly polished. They must also be carefully cleaned. The clean, polished surfaces are aligned and carefully mated in a clean room. Under the proper condition a bond forms immediately, which in many cases can be problematic if correct alignment was not achieved initially. Optical contact bonding is quite expensive due to the extreme surface preparation requirement.

17.6.2 Thermal Bonding

Thermal bonding is an old process that is experiencing a revival, as it is used to fabricate lightweight hollowed mirror blanks for large space-based telescopes. Conceptually the process is quite simple: the fused silica parts to be bonded are placed in physical contact and placed into a high-temperature furnace. The furnace temperature is raised above the softening point (1665°C) then taken back down near the annealing temperature (1140°C) and cooled in a controlled manner down to room temperature. The time necessary for this step varies depending on the glass thickness. Since the bonding process takes place at an elevated temperature some deformation of the fused silica plates can occur. It is important to control accurately the temperature sequence to prevent non-uniform bonding, thickness variation and substrate sagging. Possible misalignment may occur during the thermal cycle. In any case the bonded assembly must be slowly cooled down to relieve internal stresses after it is formed.

To fabricate microflow cytometers we have used a low temperature variant where fusion bonding is done at temperatures only slightly above the annealing point of the fused silica.[26]

Recently direct femtosecond laser bonding of fused silica has also been demonstrated. This approach is still in its infancy and presently is too slow to be commercially viable for the fabrication of microflow cytometers.[27]

17.6.3 Manufacturing Cost

The femtosecond laser direct-write fabrication is less expensive than traditional flow cytometer fabrication techniques. However it ought to be pointed out that because this manufacturing approach is based on direct-write processes that generate parts in a serial fashion, it is unlikely to reach the very low manufacturing costs generally associated with mass-production based on photolithographic processes.

17.7 FUTURE DEVELOPMENT RELATED TO THE DIRECT-WRITE APPROACH AND THEIR IMPACT ON THE FABRICATION OF MICROFLOW CYTOMETERS

While the direct-write approach cannot yet create all the features found on high-end laboratory flow cytometers, it provides a means to fabricate important features and capabilities that are not found on today's instruments. With this approach one can cleanly integrate micromechanical elements and electrodes, or one can multiplex functionalities through serial or parallel addition and even duplication of a desirable feature. For example, one can fabricate, within one flow cytometer body, multiple successive analytical points with associated optics. Alternatively, one can introduce the means to simultaneously monitor a point of interest from multiple angles. One can even integrate extremely sensitive interferometric optical measurement capabilities or combine our devices with optical tweezers. All of these features can be provided in a package that is insensitive to the environment. Using the direct-write approach, we have already demonstrated many of the features, although not necessarily in a microflow cytometry context.

17.7.1 Micromechanical Elements

It is common knowledge that glass is brittle. The presence of surface flaws such as microcracks leads to catastrophic failure. However, with the femtosecond laser direct-write manufacturing processes, these defects can be significantly suppressed, leading to outstanding mechanical behavior.[28] As a result, fused silica glass can be turned into an attractive material for the fabrication of micromechanical components. We have demonstrated that ultimate tensile strength in excess of 2.5 GPa can be obtained, which compares favorably to silicon for the manufacturing of MEMS type devices. Fused silica does not show any plastic deformation but fails while the deformation is elastic. This is a very useful characteristic in applications where a calibrated force needs to be applied or sensed. We have fabricated various instruments (force sensor, linear translation stage, etc.) that incorporate fused silica flexures. These flexures have been subjected to lengthy tests, as shown in Fig. 17.9.

Today's flow cytometers do not include (micro) mechanical elements with the exception of add-on mechanical sorters. Standard fabrication techniques do not provide an easy means to integrate mechanical elements with flow cytometers. In contrast, with the direct-write approach, one can rather simply integrate fused-silica MEMS with our monolithic fused-silica microflow cytometers. One can fabricate, as an integral part of the microflow cytometer body, a means to divert cells of interest from the main flow stream (a sorter) or a related mechanical element capable of testing mechanical and physical characteristics of a cell, such as its membrane elasticity, on the fly. These are just two exemplary micromechanical elements out of a vast array that can be fabricated with our direct-write processes within a flow cytometer.

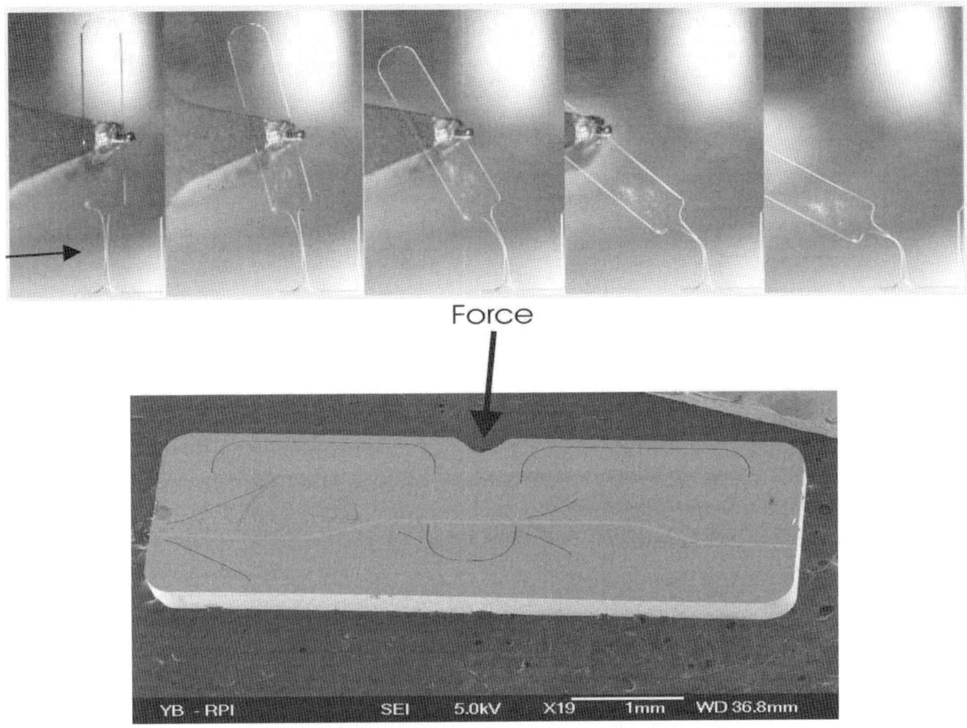

Figure 17.9. Thick MEMS fabricated out of fused silica with the femtosecond direct write processes. On top a glass flexure (thickness at neck is approximately 100 micron) photographed at various points during a bending test. On bottom a force sensor fabricated within a fused-silica monolith. The force, applied at the notch, is measured optically using an integrated optical waveguide (overlaid in orange). Color reference – pg. 363.

17.7.2 Novel Integrated Optical Capabilities

Today flow cytometers rely on small-angle forward scattering and fluorescence to measure the physical and chemical characteristic of biological cells. While these analytical procedures are well established, they do not always provide unequivocal information, and they are not the only techniques of interest. In contrast, with the direct-write manufacturing approach, it is quite simple to integrate an optical interferometer within the flow cytometer body. This interferometer can be used to detect and characterize cells through change in the relative optical path between the two arms of the interferometers, as illustrated in Fig. 17.10.

Due to the extreme environmental sensitivity associated with interferometric measurements, it would be impractical to have a similar capability in a non-integrated device. This technique can of course be combined with more traditional analytical techniques such as small-angle forward scattering and fluorescence.

The combination of scattering measurements and optical trapping is another potential worthwhile combination that is difficult to implement with standard fabrication processes. With the direct-write approach, optical trapping and optical

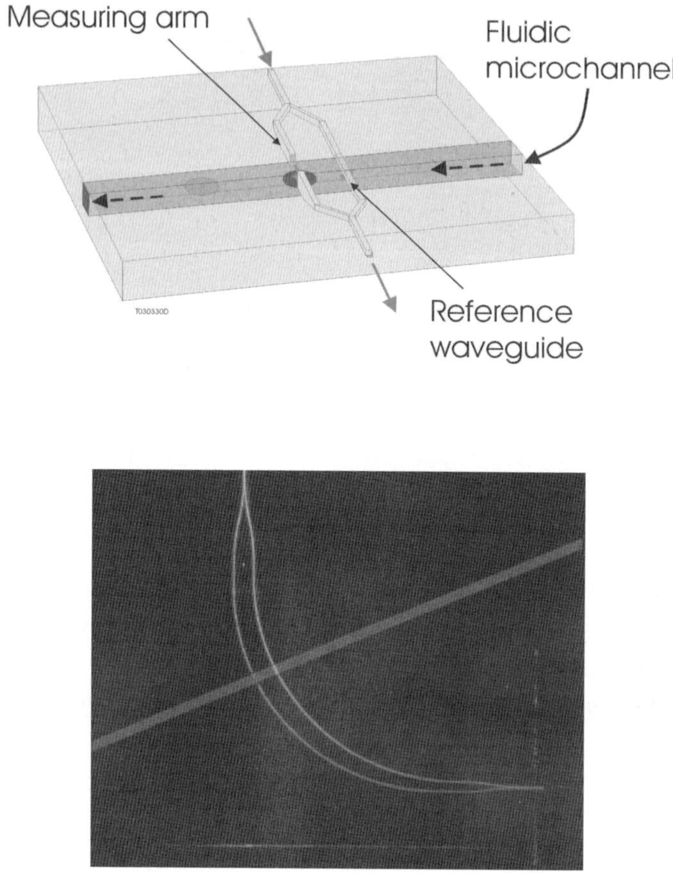

Figure 17.10. Schematic representation of a device incorporating an integrated optical interferometer to detect and characterize particles moving down a microfluidic channel (top). Curved unbalanced Mach-Zehnder interferometer fabricated with the *femtoEtch*™ process (bottom). This image has been digitally enhanced and the flow channel overlaid. Color reference – pg. 364.

sorting capabilities can be included in a microflow cytometer. The use of waveguide optical output arrays to steer the incoming cells within the flow stream by optical focusing will maintain a positional accuracy significantly better than can be achieved by allowing the cells to travel freely in the stream. This cell-steering capability has already been demonstrated.[29] This work was accomplished in an open-loop mode without the need of any feedback signal.

17.7.3 Additional Capabilities

Microflow cytometers fabricated by femtosecond direct-write can be fitted with electrodes to detect, characterize and deflect cells by electrical means. One can introduce detection and volumetric characterization capability based on impedance

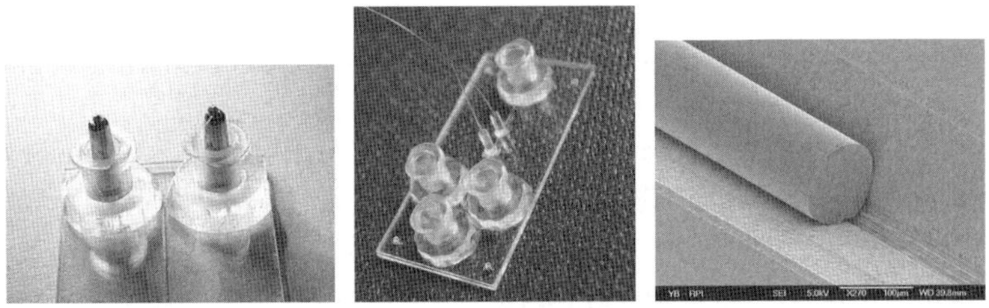

Figure 17.11. Various types of interface for microflow cytometers. On the left, two electrical contact points formed by filling a channel with metal. Center- two fibers, with collimator, bonded perpendicular to the main channel of a microflow cytometer. On the right, close-up view of a V-groove formed in the body of a microflow cytometer using the *femtoEtch*™ process (fiber diameter: 125 micron).

change similar to that of a Coulter counter. Similarly one can produce the equivalent of an electrostatic sorter. The addition of a cell sorting capability to a flow cytometer makes it possible to isolate purified population of cells. One can integrate within the body of our microflow cytometers an element that generates microdroplets that are charged and subsequently deflected using integrated electrodes. The microdroplets can be formed using a vibrating micromechanical element or ultrasonic waves.[30] Note that our demonstrated ability to shape the cross-section of our microfluidic channels can be used to create ultrasonic resonators, thus increasing the efficiency of any ultrasonic process.

Electrodes are often formed through patterned metal deposition on the surface of the microflow cytometers. While this process is compatible with the direct-write fabrication approach, it requires a mask. To eliminate the costs and delay associated with mask fabrication, an alternate microelectrode fabrication process has been developed- one can create subsurface electrodes by filling microchannels of the appropriate shape with molten metal as shown in Fig. 17.11. With this approach the electrodes can be brought in very close contact with microfluidic channels. Interfacing with the external world is done via luer connectors or equivalent. Moreover these electrodes are robust, while surface electrodes must be handled with great care.

Using the femtosecond direct-write method, one can also attach optical fibers to the body of microflow cytometers. The fibers can be aligned with waveguides using V-grooves or insertion port that is an integral part of the microflow cytometer body, as illustrated in Fig. 17.11.

17.7.4 Related Manufacturing Processes

We are presently using only two processes to modify the fused silica monolith locally as the basis of our microflow cytometers. Additional functionalities can be added with the help of several processes demonstrated by others and us. Fused silica is an amorphous material with a uniform structure that forbids second-order nonlinear processes. This lack of internal order can be lifted using thermal poling to create a preferential material orientation, thus allowing non-linear effects.[31] We previously reported on the use of femtosecond laser direct writing together with thermal poling, to fabricate a waveguide electro-optic (EO) modulator in bulk fused silica.[32] Others have demonstrated the use of femtosecond laser pulses to add metal to fused silica substrates.[33]

17.8 CONCLUSION AND OUTLOOK

In summary, while not all aspects of traditional flow cytometers have been duplicated using the femtosecond direct-write, this unconventional manufacturing approach provides a means to create microflow cytometers with unique features and capabilities. Most importantly, monolithic fused silica devices fabricated with the femtosecond direct-write are characterized by a robustness that is unprecedented and can operate in the most extreme environments without ever having to be realigned or recalibrated. We expect that microflow cytometers fabricated using this monolithic approach will not replace traditional flow cytometers, but rather they will find their own niche and will generally expand the use of microflow cytometers outside traditional markets.

References

[1] Y. Bellouard, A. Said, M. Dugan and Ph. Bado, Fabrication of high-aspect ratio, microfluidic channels and tunnels using femtosecond laser pulses and chemical etching. *Opt. Express* **12**, 2120–2129 (2004).

[2] J. Ihlemann, Excimer laser ablation of fused silica. *Appl. Surf. Sci.* **54**, 193–200 (1992).

[3] D. Du, X. Liu, G. Korn, J. Squier and G. Mourou, Laser-induced breakdown by impact ionization in Si02 with pulse widths from 7 ns to 150 fs. *Appl. Phys. Lett.* **64**, 3071–3073 (1994).

[4] K. O. Hill, Y. Fujii, D. C. Johnson and B. S. Kawasaki, Photosensitivity in optical fiber waveguides: application to reflection filter fabrication. *Appl. Phys. Lett.* **32**, 647–649 (1978).

[5] K. M. Davis, K. Miura, N. Sugimoto and K. Hirao, Writing waveguides in glass with a femtosecond laser. *Opt. Lett.* **21**, 1729–1731 (1996).

[6] Ph. Bado, A. A. Said, M. Dugan, T. Sosnowski and S. Wright, Dramatic improvements in waveguide manufacturing with femtosecond lasers. *Proceedings of the 18th annual National Fiber Optic Engineers Conference (NFOEC 2002)*, Dallas (TX), 1153–1158 (2002).

[7] J. W. Chan, T. Huser, S. Risbudand and D. M. Krol, Structural changes in fused silica after exposure to focused femtosecond laser pulses. *Opt. Lett.* **26**, 1726–1728 (2001).

[8] D. M. Krol, Femtosecond laser modification of glass *J. Non-Cryst. Solids* **354**, 416–424 (2008).

[9] Y. Bellouard, E. Barthel, A. A. Said, M. Dugan and Ph. Bado, Scanning thermal microscopy and Raman analysis of bulk fused silica exposed to low energy femtosecond laser pulses. *Opt. Express* **16**, 19520–19534 (2008).

[10] N. Kitamura, Y. Toguchi, S. Funo, H. Yamashita and M. J. Kinoshita, Refractive-index of densified silica glass. *J. Non-Cryst. Solids* **159**, 241–245 (1993).

[11] Y. Kondo, T. Suzuki, H. Inouye, K. Miura, T. Mitsuyu and K. Hirao, Three-dimensional microscopic crystallization in photosensitive glass by femtosecond laser pulses at non-resonant wavelength. *Jpn. J. Appl. Phys.* **37**, 94–96 (1998).

[12] A. Marcinkevičius, S. Juodkazis, M. Watanabe, M. Miwa, S. Matsuo, H. Misawa and J. Nishii, Femtosecond Laser-assisted three-dimensional microfabrication in silica. *Opt. Lett.* **26**, 277–279 (2001).

[13] Y. Bellouard, A. Said, M. Dugan and Ph. Bado, Monolithic three-dimensional integration of micro-fluidic channels and optical waveguides in fused silica. *MRS Proceedings* **782**, Boston (2003).

[14] P. B. Howell, D. R. Mott, S. Fertig, C. R. Kaplan, J. P. Golden, E. S. Oran and F. S. Ligler, A microfluidic mixer with grooves placed on top and bottom of the channel. *Lab Chip* **5**, 524–530 (2005).

[15] A. D. Stroock, S. K. W. Dertinger, A. Ajdari, I. Mezic, H. A. Stone and G. M. Whitesides, Chaotic mixer for microchannels. *Science* **295**, 647–651 (2002).

[16] P. B. Howell, J. P. Golden, L. R. Hilliard, J. S. Erickson, D. R. Mott and F. S. Ligler, Two simple and rugged designs for creating microfluidic sheath flow. *Lab Chip* **8**, 1097–1103 (2008).

[17] H. M. Shapiro, *Practical Flow Cytometry.* John Wiley (2003).

[18] H. M. Shapiro and M. Hercher, Flow cytometers using optical wave-guides in place of lenses for specimen illumination and light collection. *Cytometry* **7**, 221–223 (1986).

[19] G. B. Lee, C. H. Lin and G. L. Chang, Micro flow cytometers with buried su-8/sog optical waveguides. *Sensors And Actuators A-Physical* **103**, 165–170 (2003).

[20] R. W. Applegate, J. Squier, T. Vestad, J. Oakey and D. W. M. Marr, Optical trapping, manipulation, and sorting of cells and colloids in microfluidic systems with diode laser bars. *Optics Express* **12**, 4390 (2004).

[21] H. Sugiura and T. Yamadaya, Raman-scattering in silica glass in the permanent densification region. *J. Non-Cryst. Solids* **144**, 151–158 (1992).

[22] P. Oberson, B. Gisin, B. Huttner and N. Gisin, Refracted near-field measurements of refractive index and geometry of silica-on-silicon integrated optical waveguides. *Appl. Opt.* **37**, 7268–7272 (1998).

[23] P. A. Temple, W. H. Lowdermilk and D. Milam, Carbon dioxide laser polishing of fused silica surfaces for increased laser-damage resistance at 1064 nm. *Appl. Optics* **21**, 3249–3255 (1982).

[24] Y. M. Xiao and M. Bass, Thermal stress limitations to laser fire polishing of glasses. *Appl. Optics* **22**, 2933–2936 (1983).

[25] K. M. Nowak, H. J. Baker and D. R. Hall, Efficient laser polishing of silica micro-optic components. *Appl. Optics* **45**, 162–171 (2006).

[26] T. W. Hobbs, M. Edwards and R. VanBrocklin, Current fabrication techniques for ULE and fused silica lightweight mirrors. *Proceedings of SPIE Vol. 5179 Optical Materials and Structures Technologies*, Ed. William A. Goodman SPIE, Bellingham, WA, 1–11 (2003).

[27] I. Miyamoto, A. Horn, J. Gottmann, D. Wortmann and F. Yoshino, Fusion welding of glass using femtosecond laser pulses with high-repetition rates. *J. of Laser Micro/Nanoengineering* **2**, 57–63 (2007).

[28] Y. Bellouard, A. Said and Ph. Bado, Integrating optics and micro-mechanics in a single substrate: A step toward monolithic integration in fused silica. *Opt. Express* **13**, 6635–6644 (2005).

[29] R. W. Applegate, J. Squier, T. Vestad, J. Oakey, D. W. M. Marr, Ph. Bado, M. A. Dugan and A. Said, Microfluidic sorting system based on optical waveguide integration and diode laser bar trapping. *Lab Chip* **6**, 422–426 (2006).

[30] C. Grenvall, P. Augustsson, H. Matsuoka and T. Laurell, Multiple node ultrasonic standing wave separation in microchannels improves lipid discrimination from complex bio-suspensions. *The Proceedings of the MicroTAS Conference*, San Diego, paper M41A, (2008).

[31] R. A. Myers, N. Mukherjee and S. R. J. Brueck, Large second-order nonlinearity in poled fused silica. *Opt. Lett.* **16**, 1732–1734 (1991).

[32] G. Li, K. A. Winick, A. A. Said, M. Dugan and Ph. Bado, Waveguide electro-optic modulator in fused silica fabricated by femtosecond laser direct writing and thermal poling. *Optics Lett.* **31**, 739–741 (2006).

[33] R. Haight, P. Longo and A. Wagner, Metal deposition with femtosecond light pulses at atmospheric pressure. *J. Vacuum Science and Technology A* **21**, 649–652 (2003).

Chapter Eighteen

Systems Integration

Jason S. Kim, Joel P. Golden and Frances S. Ligler*

Center for Bio/Molecular Science and Engineering, Naval Research Laboratory, 4555 Overlook Ave. SW, Washington DC, USA
**frances.ligler@nrl.navy.mil*

Robust, portable microflow cytometers to be used at the point-of-care and in the field must be designed and developed with the integration of many miniaturized components. Optical and fluid control components necessary for the function of microflow cytometers are decreasing in size, but are also becoming microfabricated into the particle focusing chip. Technologies such as polymer lenses, optofluidic waveguides, organic light emitting diodes, and monolithic microvalves are some of the innovations being explored for microflow systems. Automation of sample preparation and processing is another challenge to system integration for portability and expansion into diagnostics. This chapter describes some of the system integration solutions for optics and detection as well as the possibilities for integration of sample preparation and processing components.

18.1 THE IMPORTANCE OF SYSTEMS INTEGRATION TO MICROFLOW CYTOMETRY

Flow cytometry systems have come a long way since the first flow cytometer was patented in 1968.[1] Large commercial laboratory systems have as many as 4 lasers and 18 fluorescence detectors. They also include sorting capability, based on the principles developed by Len Herzenberg, and can sort into microtiter plates as well as into individual tubes. The kinds of analyses that can be performed with such devices continually expand our knowledge of basic molecular and cell biology.[2–4] However, the newest generation of commercial flow cytometers is much less complex and much more reliable. They fit on the benchtop and do not require a dedicated operator. They are not only less expensive, but also more robust than larger traditional systems. They utilize diode lasers so well integrated with the rest of the optics that there is little, if any, requirement for alignment by the operator. The number of lasers is usually limited to two and the capability for sorting has been

The Microflow Cytometer by J S Kim & F S Ligler

www.panstanford.com
978-981-4267-41-0

abandoned, but these user-friendly cytometers can be used for many of the assays run routinely on more sophisticated systems.

The next generation of flow cytometers is coming on line for more specialized applications including point-of-care and point-of-use analyses. The sample stream is aligned in the laser beams by microfluidic focusing techniques rather than the nozzle-based sample injection of the first two generations (See chapters in particle focusing). The lasers and detectors occupy a smaller footprint and are more amenable to insertion into a common housing or substrate. The fluidics and optics under development for microflow cytometers are being selected with a new vision for systems integration. Ultimately, microflow cytometers will be available that automatically perform all the sample processing and assay steps, as well as the analysis, in a portable device at the point of use.

Clearly, the microflow cytometers will not have all the capabilities of the complex laboratory systems. Initially, devices will be manufactured with specific applications in mind, and the functions necessary for each application will be included in the system. Examples include (i) automated on-chip PCR synergized with point-of-care diagnostics, (ii) side light scatter as well as near-angle light scatter for the classification of unprocessed marine algae, and (iii) immunostaining for homeland security systems.

The challenges for system design of microflow cytometers escalate in proportion to the demands of the customer. If the cytometer must be easily portable, then integration of the optics and fluidics onto a single support becomes more important. Use outside the laboratory necessitates the use of casings that keep the electronics (see Chapter 13) and optics clean and dry, and reliance on battery or solar power significantly affects the selection of pumps and valves (see Chapter 9) and light sources. Electronic interfaces must suit the customer, even if that customer is miles from the point-of-use, as would be the case if the cytometer was included in a continuous monitor for water purity on the space shuttle or on a buoy at the mouth of a river. The customer also determines the cost constraints, which are a function not only of the market value of the analysis performed, but also of the number of devices that can be sold. Microflow cytometers will open new markets as the technology becomes accessible to a much broader customer base. However, the systems must provide information that is actionable and must be designed to meet customer requirements.

18.2 OPTICAL COMPONENTS FOR INTEGRATED MICROFLOW CYTOMETERS

Two relatively recent areas of research provide new paradigms for the design of optical microdevices.[5,6] Microphotonics strives to integrate the required optical devices on a single chip, and focuses on device size, functionality, and fabrication technology. Optofluidics combines optics and microfluidics on the same chip and utilizes microfluidic structures to impart novel functionalities to the optical components. The following subsections will provide examples of optical components, based on microphotonics or optofluidics, which could be included in microflow cytometers to make them smaller, cheaper, or more robust, or to impart functionalities not described in currently reported devices.

18.2.1 Waveguides

The problems with directing the excitation light through the air and using a lens to focus the light onto the interrogation region of a microflow cytometer are that there is light scattering at the air-solid and solid-liquid interfaces and requirements for alignment are not conducive to portability or use by non-technically trained operators. The use of waveguides offers tighter control of the transport of both the excitation and emitted light.

Optical fibers have been used in standard flow cytometers to carry light from lasers and deliver it to the interrogation region. However, only a few investigators have used optical fibers as waveguides for light delivery and collection in microflow cytometers, presumably due to the increased complexity of fabricating cytometer substrates with integrated optical fibers and alignment of the fibers after insertion. The channels for fiber emplacement must be fabricated to exactly the appropriate depth to align the fiber core with the sample stream. Chang *et al.* described such an alignment for an unsheathed sample stream and measured particle size based on the length of time for a particle to pass the fiber.[7] Golden *et al.* fabricated a much more complex device using two individual fibers to deliver light from two lasers and two more fibers to collect emission monitored at all four different wavelengths.[8] In both reports, the ends of the fibers were placed as close to the wall of the fluid channel as possible and in direct contact with the fluid passing through the channel. The refractive index of the water reduced the numerical aperture of the fibers compared to the numerical aperture in air, focusing the beam somewhat even without a lens.

A variety of techniques for fabricating waveguides in solid substrates has been reported in the last ten years. The first of these used high energy sources to modify the substrate, changing the refractive index in the exposed region to create a waveguide. Wong and Pun used electron beam writing to create waveguides in epoxy novolak resin and polymethyl methacrylate (PMMA).[9] Kim and Kim also created waveguides in PMMA, but they used deep X-ray lithography and a synchrotron source to produce ridge waveguides with 45° micromirrors on the end.[10] Said *et al.* reported waveguides and other optical structures created in sol gel glasses using femtosecond laser pulses.[11] That microfluidic channels can also be created using this technique makes it particularly attractive. Betthiol *et al.* used proton beam writing to make waveguides in PMMA, SU-8, and Foturan.[12] Sullivan *et al.* used low-power, CW lasers to write buried waveguides in photopolymers.[13]

An alternative to the direct writing techniques is to mold or deposit the waveguides on the substrate. Zhao *et al.* molded ridge waveguides on a solid substrate (SiO_2) using an elastomeric mold which was removed after crosslinking the polyurethane waveguides.[14] Lien *et al.* filled PDMS channels with a PDMS of higher refractive index in such a way as to align both fluidic channels and waveguides on a single PDMS substrate.[15] Lee *et al.* filled silver-lined PDMS channels with polyurethane containing a dye.[16] The reflective cladding reduced loss generated by the roughness of the milled molds to 0.1 dB/cm at 632 nm. Chang-Yen *et al.* demonstrated ridge waveguides molded into the PDMS substrate in a manner completely compatible with integration into microfluidic systems.[17] Hajj-Hassan *et al.* used laser direct-write deposition of an ink that was thermally removed to leave a polymer waveguide behind.[18] The waveguide was clad with a ruthenium-containing xerogel and used for oxygen sensing.

Direct-write and molding techniques produce waveguides that are integrated into the substrates, reducing alignment to an initial device design problem. However, the smoothness of the waveguide walls and the refractive index difference between the waveguide and the surrounding material significantly affect the light loss from the waveguide. When reported, the propagation losses in the waveguides mentioned above vary from 0.2 to 0.6 dB/cm at visible wavelengths, compared to less than 0.05 for appropriate optical fibers.[19]

Optofluidic waveguides provide a possible alternative that has not yet been explored for microflow cytometers. Liquid-core waveguides can be fabricated in the same types of substrates as have already been used to fabricate microflow cytometers. Schmidt and Hawkins review the concepts[20] and fabrication methods[21] for making liquid-core waveguides. Approaches they describe using both index-guided channels and anti-resonant reflecting claddings could provide integrated waveguides for use with microflow cytometers. In a second approach, Yin *et al.* use a liquid-core waveguide to confine both the sample and the light in the same volume.[22] Excitation light is delivered by a solid-core waveguide perpendicular to the liquid-core waveguide and fluorescence is collected from the end of the liquid-core waveguide. The third approach for optofluidic waveguides is described as a liquid-liquid (L2) waveguide and is probably not particularly appropriate for use with microflow cytometers.[23] If the sample "core" stream was also to serve as the waveguide core, the sheath fluid would be required to have a refractive index less than 1.33.

18.2.2 Lenses

Optical fiber waveguides can be integrated with lenses to improve excitation and light collection efficiency. Lenses are used in traditional systems to focus the light on the interrogation region of flow cytometers and to collect scatter and fluorescence signals. They can also serve the same purpose in integrated optical systems. We suggest two approaches for including lenses in integrated systems. The first possibility is to form the lenses at the end of optical fiber waveguides, while the second, more elegant approach is to build the lenses directly into the substrate with the fluidic channels.

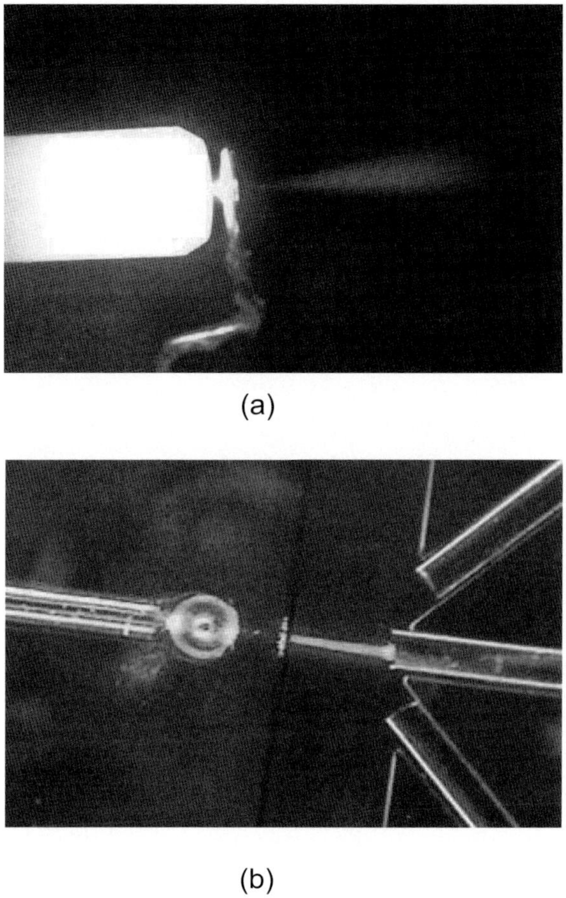

Figure 18.1. Optical fiber with output focused by ball lens. (a) ball lens aligned with optical fiber in air, visualized with vapor from dry ice in water. (b) ball lens aligned with optical fiber (on left) in a PDMS microflow cytometer design. The channel was filled with dye to visualize the output beam. Color reference – pg. 365.

An example of the output of an optical fiber focused by a ball lens is shown in Fig. 18.1. Figure 18.1(a) shows the effective focusing region of a 300 μm ball lens at the end of a 65 μm core multimode fiber in a standard ST connector in air. Figure 18.1(b) shows a channel design made in PDMS that includes a setting for a 200 um ball lens. The channel was filled with Cy5 dye to show the beam location. The design placed the ball lens about 30 μm away from the fiber end, yielding an output beam that was roughly collimated. The dump fiber on the far side of the channel guides the excess excitation light away from the channel, preventing it from scattering inside the channel and contributing to unwanted background signal.

An example of on-chip polymer lenses is provided by Seo and Lee.[24] Figure 18.2 shows SEM images of single and compound microlenses focusing light

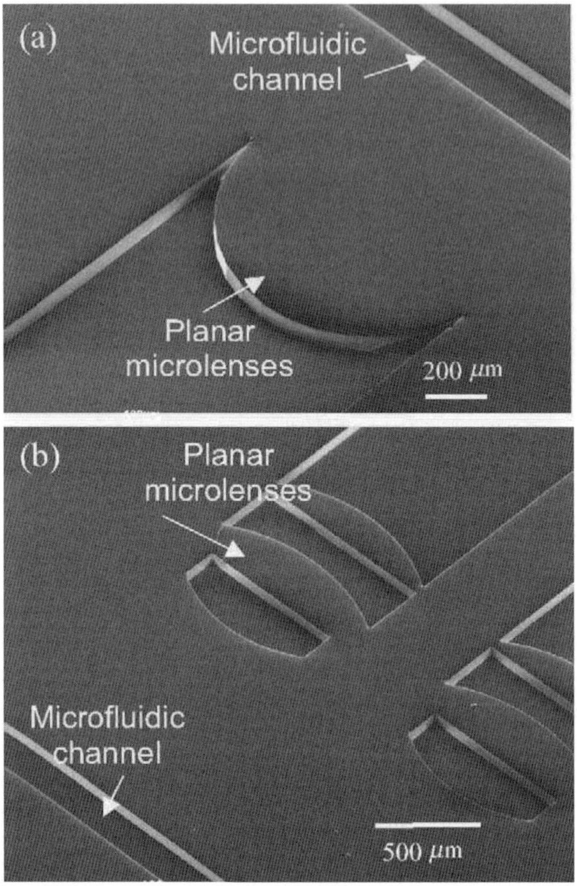

Figure 18.2. Scanning electron microscopy images of (A) single microlens and (B) compound microlenses prepared by soft lithographic techniques. [Reprinted from J. Seo and L. P. Lee *Sensors and Actuators B: Chemical* **99**(2-3), 615–622 (2004) with permission from Elsevier.]

on a microfluidic channel in a PDMS substrate. The use of the lenses increases the fluorescence measured in the channel by more than a factor of 6. Levy and Samai review work on other forms of integrated lenses that can be tuned using electrowetting or pressure.[25]

18.2.3 Filters

Current systems rely on combinations of dichroic mirrors and filters to distinguish the various wavelengths of fluorescence emitted from particles in the interrogation region of a flow cytometer. Depending on the number of fluorescent signals to be discriminated, various combinations of short-pass, band-pass, and long-pass filters are employed. Most commercial cytometer manufacturers incorporate these filters into cassettes or small, light-proof containers, along with the detectors, in order to make the systems more robust. In future systems, we envision the potential for integrating the filters with the waveguides used to carry the light to the detectors.

The basic concept for integrating filters into waveguides is not new. Dyes have been included in waveguides as filters for visible wavelengths for decades. Several investigators have extrapolated this approach to integrated optical systems. Hofmann *et al.* used dye-doped PDMS films to provide long-pass filters for organic photodiodes in disposable fluorescence detection systems.[26] Lee *et al.* doped molded polyurethane waveguides with a pinacyanol iodide dye, causing selective propagation losses of 71.8 dB/cm at 590 nm and 0.3 dB/cm at 760 nm.[16] Bliss used dye leached from the surrounding PDMS into channels filled with high refractive index fluids to make wavelength-selective waveguides for fluorescence detection.[27] The possibility of incorporating dye solutions into photonic-crystal waveguides for reconfigurable optical waveguide/filters is also intriguing.[28]

The concept of using gratings to reflect light out of a waveguide and achieve spectral separation of reflected wavelengths has also been utilized extensively. However, we are not aware of the use of this approach in flow cytometry. The potential for using tunable optical filters with waveguides (e.g. d'Allesandro *et al.*).[29] is appealing if the filters are adapted to visible wavelengths. In this design, a holographic grating is placed between a waveguide and a cover. The grating is composed of alternating stripes of polymer and liquid crystal with coplanar electrodes to tune the grating properties. Schueller *et al.* has reported a microfluidic diffraction grating fabricated in PDMS, which uses fluids to tune the diffraction and functions at optical wavelengths with switching times of ~50 milliseconds.[30]

18.2.4 Light Sources

Light sources for microflow cytometers can be placed in a docking station for the fluidic chip and the excitation light channeled into the interrogation region of the flow channel by lenses (current practice) or waveguides. Alternatively, the light sources can be integrated into the chip. In the first case, diode and frequency-doubled lasers offer excellent power and a variety of visible wavelengths, but size, weight and cost are limitations. For integration of light sources into the microfluidic substrate, the light source must be very small due to limitations on practical chip size, inexpensive so that chip replacement is feasible, and capable of producing coherent emission at reasonable power levels.

Organic Light Emitting Diodes (OLEDs) have been developed primarily for large-area displays and lighting but they have also been integrated into fluorescence-based sensing systems.[31–33] However, if OLEDs or even LEDs are to be used with microflow cytometers, the lack of coherent emission must be overcome with excellent focusing mechanisms that can also be incorporated on chip. One attractive measure to address this problem is to embed the OLED in the waveguide, as proposed by Gather *et al.*[33] They used photolithographic methods to build an architecture in which the active layers of the OLED also form the waveguide, avoiding any losses from coupling between the waveguide and the light source. Chapter 11 on polymer photonics can be referenced for further details.

Photonic crystal lasers offer a very attractive alternative. Such lasers are made using the same technologies as microfluidic systems, and thus the fabrication methods are highly compatible for on-chip integration with the microfluidics and waveguides.[28,34] The capacity for bending light around corners and tuning the wavelengths with fluids further increases their potential utility in small integrated systems. It should be possible to focus the emitted light into a very small numerical aperture and align the excitation light in the interrogation region of a microflow cytometer. The power output is a function of the pump power; whether or not sufficient power can be generated for microflow cytometry is still a question that remains to be answered.

Micron-scale dye lasers have been reported for over a decade, but the first integration of a dye laser into a microfluidic chip, along with other sensing elements, relied on the development of microfluidic technologies. This work and more recent progress are reviewed by Li and Psaltis.[35] An example of the integration of a dye laser with a microfluidic sensors system is shown in Fig. 18.3. While dye lasers are certainly small and amenable to systems integration, they must still be pumped with lasers of relatively high power, and it is not clear whether or not sufficient output power can be generated for operating microflow cytometers.

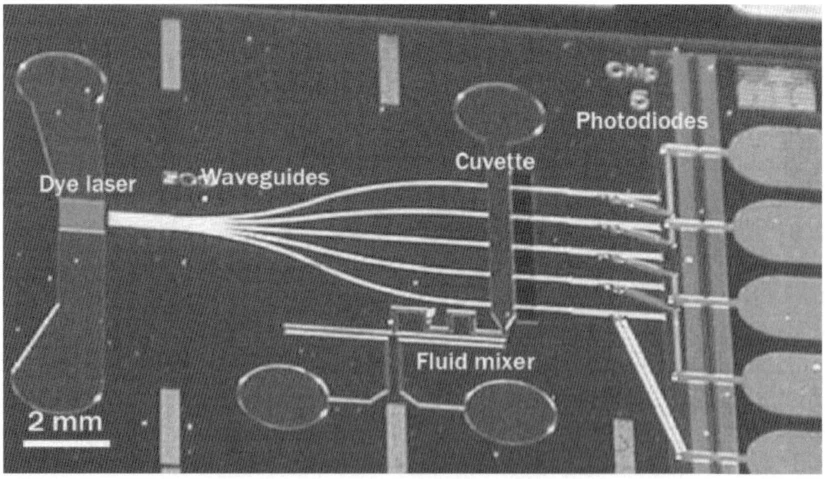

Figure 18.3. Photograph of the lab-on-chip device with integrated microfluidic dye laser, optical waveguides, microfluidic network and photodiodes. The metallic contact pads for the photodiodes are seen on the far right. The chip footprint is 15 mm by 20 mm. [S. Balslev, A. M. Jorgensen, B. Bilenberg, K. B. Mogensen, D. Snakenborg, O. Geschke, J. P Kutter and A. Kristensen, Lab-on-a-chip with integrated optical transducers.*Lab Chip* **6**, 213 (2006). Reproduced by permission of The Royal Society of Chemistry.]

18.2.5 Detectors

Most flow cytometers require the sensitivity of photomultipliers for the fluorescence measurements. However, several of the systems described in this book have employed photodiodes or avalanche photodiodes for signal measurement. These detectors can be inserted into silicon, glass or other hard substrates using standard microfabrication technology. They can also be used on these and other substrates as bare dies using laser-direct write methods. The primary challenges are then focusing the signal on the detector and achieving the required sensitivity.

Organic photodiodes offer the potential for printing low cost detectors onto the substrate with the microfluidics.[37,38] de Mello and colleagues have shown that organic photodiodes can measure chemiluminescence in microfluidic channels[26,32] Bürgi *et al.* and Kraker *et al.* demonstrated the integration of both OLEDs and organic photodiodes into a single sensor chip.[33,39] Wojciechowski *et al.* measured fluorescence from immunoassays performed on the opposite side of a glass slide from the organic photodiodes with a sensitivity comparable to that of a CCD camera.[40] The organic photodiodes can be easily placed in the immediate proximity of the interrogation region of a microflow cytometer, possibly simplifying the focusing issues. However, they are not as sensitive as photomultiplier tubes, so may not be appropriate for all applications.

18.3 PUMPS AND VALVES

Fluid control is undoubtedly a necessity for flow cytometry and sorting systems to generate sheath flow and transport particles for detection. Pumps and valves, like optical and detection components, have been decreasing in size as part of the effort to reduce overall size of the system. Yet, even the smallest commercially available pumps and valves require incorporation of additional transport features (such as tubing), increasing the sample volume for detection. Reduction of the

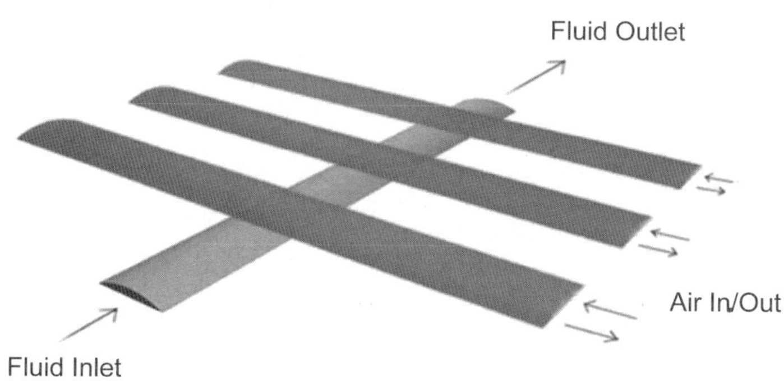

Figure 18.4. Diagram of a peristaltic micropump composed of multiple microvalves controlled by air pressure to deform the elastomeric structures created via soft lithography. [Reprinted from Unger *et al.*, *Science* **288**(5463), 113–116 (2000) with permission from AAAS.] Color reference – pg. 365.

size of the fluid control system is particularly important because of its potential to dilute samples and affect assay performance.

Chapter 9 can be referenced for greater detail on pumps and valves for building microflow cytometers. Here, we will discuss the introduction of micropumps and microvalves and their integration into systems for flow cytometry and sorting. Unger *et al.* published a paper in *Science* introducing microfabricated pumps and valves.[41] Multilayer soft lithography was used to create pneumatic valves, controlling fluid flow in a channel by deforming soft elastomers with air pressure (Fig. 18.4). This led to a logical progression toward the creation of a peristaltic micropump through use of multiple pneumatic microvalves.

In 2002, Fu *et al.* integrated these developments into a microfabricated cell sorter capable of differentiating and collecting enhanced GFP expressing *E. coli* from a mixed population.[42] A chip was fabricated with soft lithographic techniques and mounted onto a microscope equipped with a PMT to detect fluorescence. The incorporated micropumps and microvalves were used to control flow velocity and trap, sort, and recover cells.

18.4 SAMPLE PROCESSING

Scientists often forget the many seemingly simple tasks that are involved in preparing and processing samples before and after analysis. At the point-of-care and in resource limited environments, equipment and instruments ubiquitous to a laboratory environment are not always convenient or available. Additionally, these sample processing steps require training and equipment familiarity that most developers (likely achieving advanced degree training) overlook. The steps may include tasks that require common glassware or plasticware for the preparation of reagents or culture of cells. Procedures to enrich samples for greater analytic sensitivity, using microcentrifuges, column chromatography, or PCR are basic for teaching and research environments in universities, but not easily accessible a local doctor's office or in an African clinic. Yet, steps can be simplified with integration of sample processing for analysis to occur rapidly in any country. Thus, the integration of these types of processes will make microfluidic devices valuable to a broader audience of users.

18.4.1 *Sample Pre-Processing*

In order to simplify assays for end users, sample processing features can be integrated with a microflow cytometer. Assays are significantly improved by sample pre-concentration or removal of impurities because of the increase in sample signal or reduction of background. Yet, common benchtop tasks, such as centrifugation or dialysis are often challenging on a microfluidic chip. Maximizing assay performance is important as microflow cytometers continue to try to perform at the levels of larger systems using miniaturized, developmental components.

18.4.1.1 Target Preconcentration

With bead-based flow cytometry being used for detection and analysis, magnetic beads have found utility as an ideal platform to perform assays for targets preconcentrated using magnetic affinity purification. Yang *et al.* has performed a magnetic bead-based immunoassay for virus detection in a microflow cytometer (Fig. 18.5).[43] In this device, antibody-labeled magnetic beads were mixed with a virus containing sample. Application of a permanent magnetic allowed impurities to be washed away by buffer. A fluorescent tracer antibody was introduced and mixed with the magnetic beads bound with virus. Another application of the permanent magnet allowed immobilization of the magnetic beads and removal of excess tracer antibody. The magnetic bead sample was introduced into the integrated microflow cytometer for analysis and sorting. This technology has also been reported with electromagnetic microcoils for detection of Dengue viruses in concentrations as low as 10^{-2} pfu/mL.[44]

18.4.1.2 Partial Purification

Jandik *et al.* has developed a microfluidic method, known as the H-Filter™, to remove impurities based on molecular weights.[45] Differences in molecular weights affect the diffusion speed of molecules within lamellar flow. Particles or cells

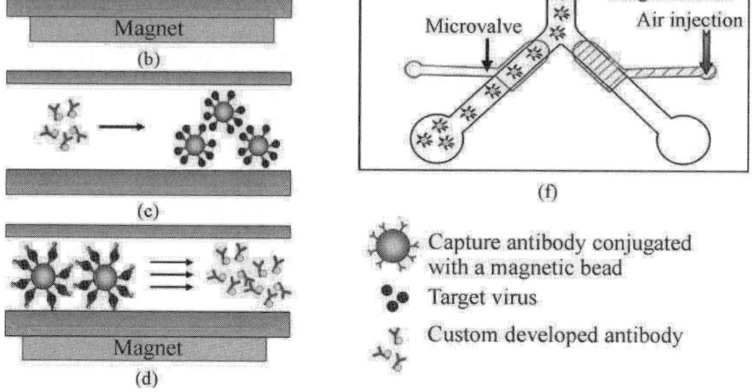

Figure 18.5. Schematic representation of magnetic bead-based immunoassay in a microflow cytometer: (a) interaction of target with antibody-conjugated magnetic bead, (b) magnet-assisted wash of impurities, (c) addition of tracer antibody, (d) and magnet-assisted wash of excess tracer antibody, (e) release of beads into microflow cytometer, and (f) use of pneumatic microvalves. [Reprinted from Yang *et al.*, *Biosens. Bioelectron.* **24**(4), 861–868 (2008). with permission from Elsevier.]

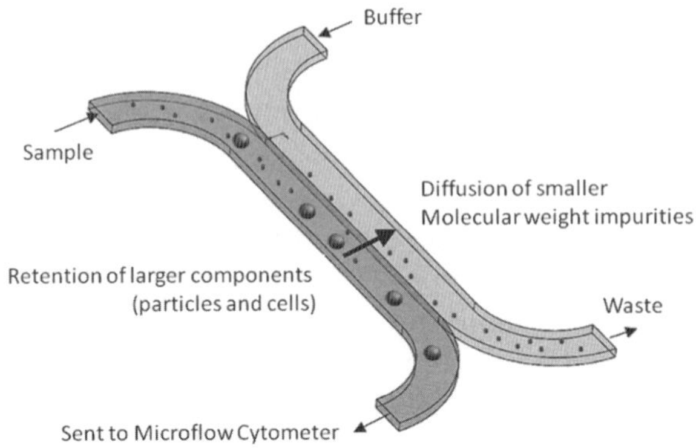

Figure 18.6. H-Filter™ structure for purification based on rapid diffusion of smaller molecular weight impurities from larger samples (i.e. cells or particles) in a dual stream laminar channel. [Reprinted from Yager *et al. Nature* **442**(7101), 412–418 (2006) with modifications; permission from Nature Publishing Group.] Color reference – pg. 366.

incubated with small molecular weight reagents (such as fluorescent tags) are introduced into the H-Filter™ structure, where the smaller, unbound molecules will diffuse into the wash stream running adjacent to the sample stream (Fig. 18.6).[46] The dimensions of the microfluidic structure can control the rate of small molecule diffusion and ultimately the purification of the sample. The now purified sample can be introduced into the device, such as a microflow cytometer, with significantly reduced background signal.

18.4.1.3 Reagent Addition and Mixing

The manual preparation and addition of reagents to assays are additional steps that require training and equipment such as pipetters and microwell plates. Such requirements are trivial for most laboratories, but are meaningful at the point-of-care and in resource-scarce environments. On-chip storage of reagents can further reduce need for bulky equipment like refrigerators. Downstream of the storage area, mixing is necessary during reagent reconstitution and preparation for the assay.

Garcia *et al.* demonstrated the preservation of a functional protein in a dry-reagent storage cavity within the wall of a microfabricated channel.[47] The enzyme was stabilized using a non-reducing disaccharide, trehalose, and dextran. The dry protein was reconstituted in the channel by addition of buffer and shown to cleave a fluorescence-producing peptide substrate. Later, a porous polyester pad was incorporated onto a chip to store dry reagents for an immunoassay.[48] Linder *et al.* used air spacers within fluidic tubing to load reagents within a cartridge. The application of pressure to one side of a series of plugs dispenses the reagents into the microfabricated channel for mixing and analysis.[49]

Continuous flow mixing within microfluidic channels is highly dependent upon diffusion and often requires external turbulence to enable greater mixing

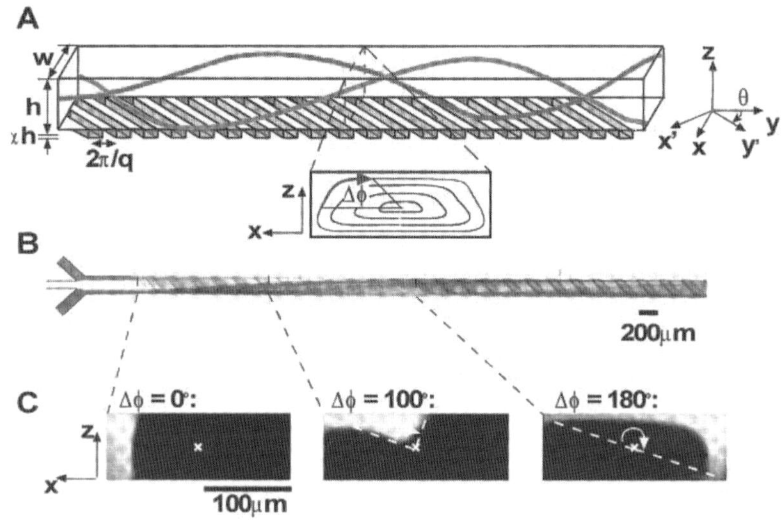

Figure 18.7. (a) Schematic of a mixing by twisting flow in a microchannel as a result of groove structures. (b) Optical micrograph showing the mixing this structure organization. (c) Fluorescent confocal microscope images of the cross sections within the micrograph. [Reprinted from Stroock *et al.*, *Science* **295**(5555), 647–651 (2002) with permission from AAAS.] Color reference – pg. 366.

efficiency. Stroock and Whitesides recognized the challenge of mixing in laminar flow and demonstrated the use of passive mixers ("groove" and "herringbone" structures) within microchannels (Fig. 18.7).[50] Later, this work was further advanced by Howell *et al.* using a computer model to optimize the organization of grooves and herringbones within the channels.[51] Recently, Williams *et al.* examined the herringbone mixer in detail.[52] These types of passive mixers could play an important role in the preparation of samples.

The same need for turbulence is true for mixing in droplet-based devices, but differs in the type of turbulence needed to generate mixing within the plug-flow. For example, Bringer *et al.* showed that microdroplets formed in a winding channel will mix their contents by stretching, folding, and reorientation.[53] Fundamental studies are underway with application of such ideas to high-throughput screening devices occurring in parallel.[54]

Alternatively, active mixers may be used to prepare samples more rapidly than passive mixing schemes. Often, this may require the engineering and miniaturization of larger, traditional mixers into a device. But, there have been recent advances that are more innovative from a microfabrication standpoint. One such example is Yaralioglu's integration of tranducers into a microchannel to create ultrasonic mixing.[55] Another example is the fabrication of a microscale magnetic stir bar into the microchannel.[56] Mixing remains a subtle, but important issue in sample preparation, as it often affects assay reproducibility and speed.

18.4.1.4 Microfluidic Analysis in Droplet-based Systems

Microfluidic droplet devices have the ability to generate microreactors with unique throughput and scaling benefits;[57,58] merging microdroplet manipulation with microflow cytometry would be useful for analysis of the processes occurring within the individual microdroplets. Rapid drug discovery and screening are applications being actively pursued, as high throughput droplet assays on bacteria, mammalian cells, and small organisms have been demonstrated.[59,60]

Early on, droplet devices performed colorimetric analysis, utilizing a pH indicator for acid-base reactions[61] or the orange-color formation during bromination of alkenes.[62] More recently, Ismagilov and coworkers have reported the development of the "chemistrode", described as a droplet-based microfluidic device capable chemical stimulus and recording of multiple spectroscopic modalities (Fig. 18.8).[63] This device can measure secretions of a tissue using droplet-based microfluidics to introduce stimuli and perform fluorescence correlation spectra, MALDI-mass spectra, and dual fluorescence microscopy profiles of the individual droplets.

Droplets have been used for a variety of biomolecular reactions, such as PCR or protein synthesis.[57] These reactions utilized the confined environment of the emulsion droplets. Griffiths and coworkers developed a process known as *in vitro* compartmentalization (IVC) to use microdroplets to select clones from a gene library.[64] In these experiments, droplets contained individual genes and an *in vitro* transcription/translation mixture, producing an enzyme that converted a substrate into a fluorescent product. The droplets with fluorescence, indicating the desired protein expression, were sorted and collected by fluorescence activated cell sorting (FACS).

Droplets have been used to encapsulate a variety of organelles,[65] cells,[66,67] and small organisms.[68] Huebner *et al.* encapsulated individual *E. coli* cells within droplets using a T-junction microfluidic structure and detected those that contained cells and/or expressed fluorescent protein (Fig. 18.9).[69] Kline *et al.* used a droplet-forming device to perform blood typing by contrast analysis of the droplets containing agglutinating blood cells.[70] Recently, a microdroplet-forming device combined bacteria and a viability indicator to show the ability to detect and screen for drug susceptibility of methicillin-resistant *Staphylococcus aureus* (MRSA).[71] Nanoliter volumes increase local concentration and reduce bacterial detection time, while spacer plugs separate multiple antibiotic containing droplets within the same device.

18.4.2 Sample Post-Processing

Beyond integration of system components for miniaturization and sample preparation, processing after analysis will lead to new and exciting opportunities to create systems of advanced function. Previous chapters have reviewed the cutting edge with Raman-activated and optical tweezer technologies for sorting. Yet, the more mature pressure-driven sorters have demonstrated advances in miniaturization and simple, monolithic construction. Later, commercial chips may begin to incorporate functions like polymerase chain reaction (PCR) for nucleic acid enrichment or culture of analyzed cells. Already, literature shows the demonstration of integration of these advanced functions in microflow cytometers.

18.4.2.1 Sorting

In section 3 micropumps and microvalves were discussed and pressure-driven cell sorters were introduced. In recent years, pressure-driven sorters have even been used to sort and screen whole-animals, *Caenorhabditis elegans*. Pressure-driven sorters are more developed than the Raman-activated or optical tweezer sorters described in previous chapters, thus advancements toward systems integration and commercialization are progressing rapidly.

With the previously described pneumatic micropumps, Fu *et al.* initially presented an integrated microflow cytometer with sorting capabilities. The sorting was driven by the closing and opening of the microvalves in conjunction with the action of the pumps. One of the drawbacks of this system, often seen in developmental devices, was the mounting of the chip onto a large microscope with 100X magnification objective to detect fluorescence to drive the sorting algorithm. Later, Yang *et al.* expanded this technology and created a microflow cytometer/sorter that utilized flow switching at the sample outlet for cell sorting (Fig. 18.10).[72,73] In this work, a CCD camera was used in place of the a microscope. In 2007, Rohde *et al.* presented a microflow device that could sort and analyze *C. elegans* using

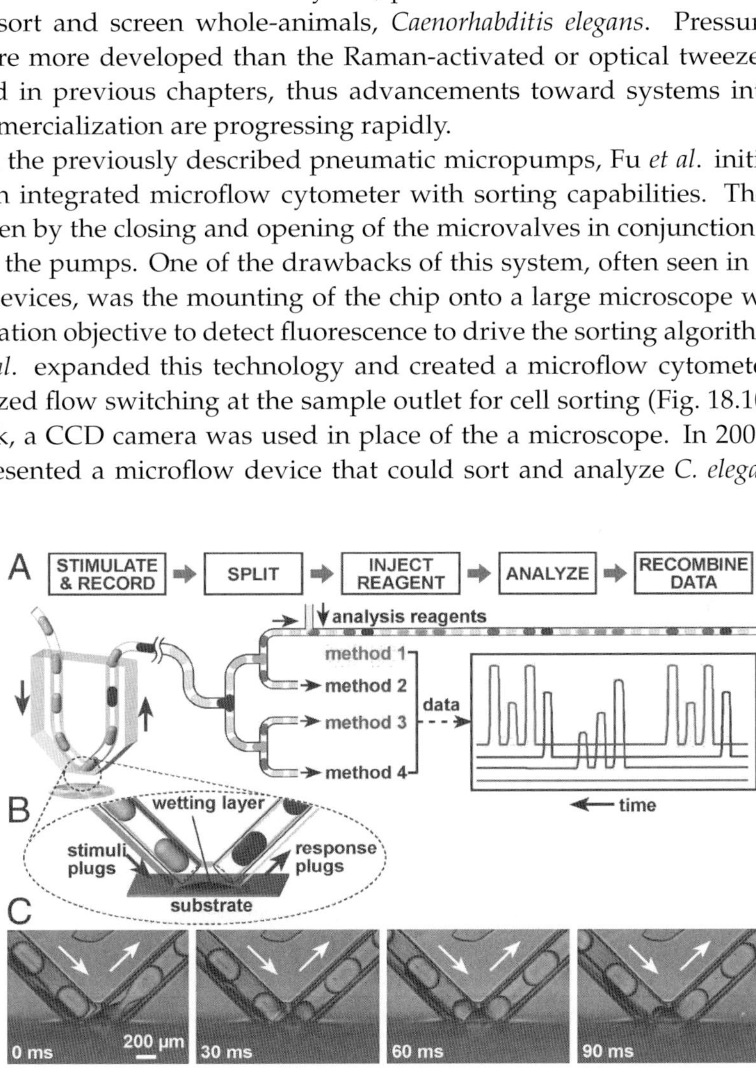

Figure 18.8. (a) Schematic representation of the "chemistrode" microdroplet-based device capable of stimulus of a tissue, separation of droplets for multimodal analysis, and recombination of data. (b) Zoomed in region of detection in which the droplet plugs interact with the target tissue. (c) Optical micrographs showing time course of droplets at the interaction region. [Reprinted from Chen *et al.*, *Proc. Natl. Acad. Sci. USA* **105**(44), 16843–16848 (2008). with permission from the National Academy of Sciences.]

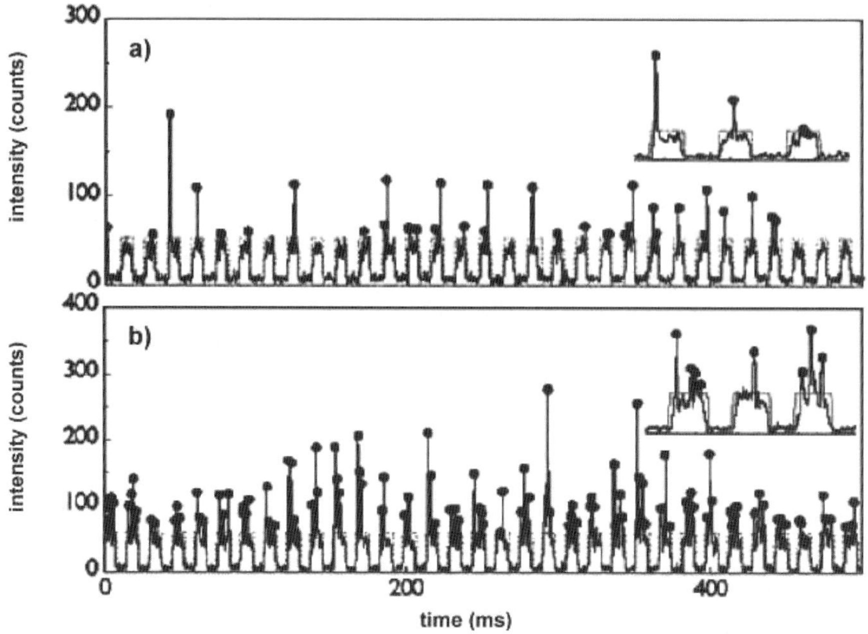

Figure 18.9. Laser-induced fluorescence analysis of droplets containing individual fluorescent protein expressing *E. coli* at (a) low and (b) high concentrations of cells. [Reprinted from Huebner *et al.*, *Chem. Commun. (Camb.)* **12** 1218–1220 (2007) with permission from the Royal Society of Chemistry.]

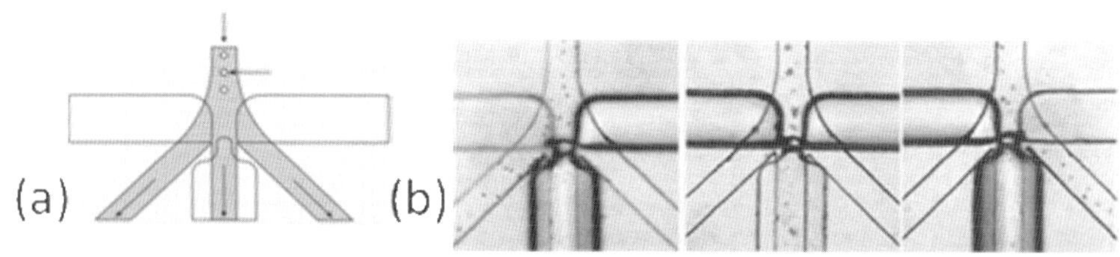

Figure 18.10. (a) Schematic of the three microvalve cell sorter used by Yang *et al.* (b) Optical micrographs of the flow switching seen by the action of the pneumatic microvalves at the outlet. [Reprinted from Yang *et al.*, *Meas. Sci. Technol.*, **17**, 2001–2009 (2006) with permission from Institute of Physics Publishing.] Color reference – pg. 367.

a series of pneumatic valves.[74] The worms were immobilized for analysis using suction from a micrograting, before sending to collection or waste.

Currently, NanoEnTek, Inc. is marketing a single-use, cartridge-based microflow cytometer know as the C-Box. This device is based on two-dimensional hydrodynamic focusing prevalent in most microfluidic chips.[75] The group has demonstrated the incorporation of a pressure-driven sorting function in recent years.[76] To generate particle focusing in this chip design, the sheath inlets sandwich the sample inlet. The fluid flows toward two outlets for sorting. Yet, what is unique about the design is the presence of an additional channel before the sorting outlets. This channel controls the pressure of one of the sheath streams and can divert the sample stream to one outlet or the other, ultimately driving the sorting of particles and cells.

18.4.2.2 On-chip Polymerase Chain Reaction (PCR)

PCR has generated significant interest for on-chip incorporation because of its profound impact use in molecular biology, clinical diagnostics, and forensic science. The repetitive process of DNA amplification, relying on temperature cycles and thermally stable enzymes, is highly amenable to microfluidic technology. As a microflow cytometer deals with many biological samples, PCR would be a helpful feature upstream for sample amplification or downstream to process sorted cells and particles rapidly.

The first on-chip PCR devices were performed in microwells on silicon substrates with thermocycling to achieve the desired DNA amplification.[77,78] Continuous-flow systems were developed for improved system integration.[79] In these systems, a sample moves through a channel with regions of varying temperatures to accomplish annealing, extension, and denaturation of the DNA. Later, a device containing a serpentine channel that elegantly guided samples unidirectionally through the three temperature regions showed sample amplification in 90 seconds via 20 cycles (Fig. 18.11).[80] This design has been popular and often the basis for further advances, as reviewed by Auroux *et al.*[81]

System integration of additional features is an ever-present goal in lab-on-a-chip research, PCR being no exception. Many groups have integrated useful

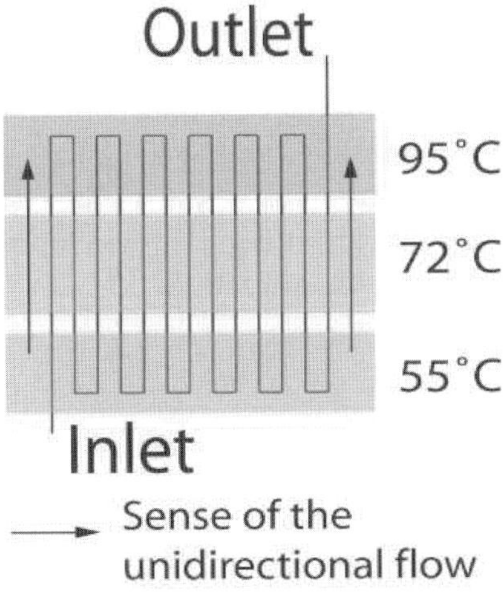

Figure 18.11. A schematic representation of continuous flow PCR, where the sample passes through three thermocycling regions in unidirectional flow. [P. A. Auroux, Y. Koc, A. deMello, A. Manz and P. J. Day, Miniaturised nucleic acid analysis. *Lab Chip* **4**(6), 534 (2004). Reproduced by permission of The Royal Society of Chemistry.]

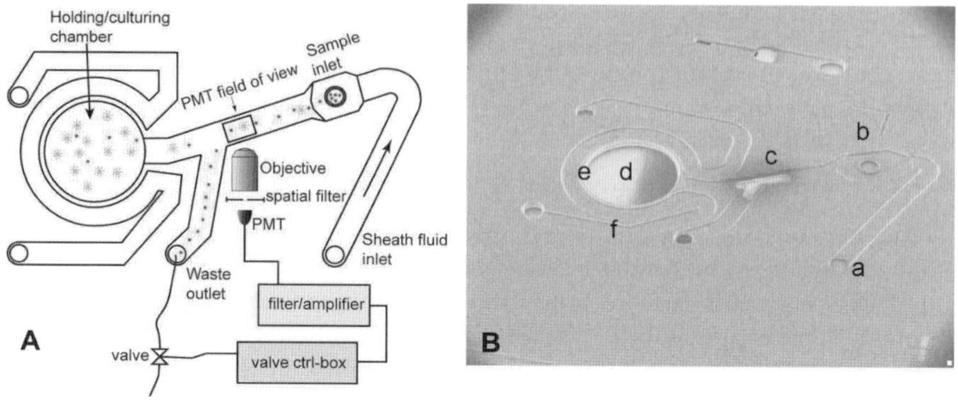

Figure 18.12. (a) Schematic of a microfluidic flow cytometer/sorter with integrated cell culture chamber. (b) A scanning electron micrograph of the device. [A. Wolff, I. R. Perch-Nielsen, U. D. Larsen, P. Friis, G. Goranovic, C. R. Poulsen, J. P. Kutter and P. Telleman, Integrating advanced functionality in a microfabricated high-throughput fluorescent-activated cell sorter. *Lab Chip* **3**(1), 22 (2003). Reproduced by permission of The Royal Society of Chemistry.] Color reference – pg. 367.

functions. Waters *et al.* prepared a microchip PCR device capable of cell lysis, addition of intercalating fluorescent dye, and electrophoretic analysis.[82,83] Liu, Grodzinksi, and coworkers developed a PCR device integrated with magnetic bead-based cell concentration and purification, cell lysis, DNA hybridization, and electrochemical detection.[84] Li and Yeung have reported an on-chip PCR system that could count cells using gravity flow and laser-induced fluorescence, as well as perform cell lysis and electrochemical detection of PCR products.[85] This system has demonstrated correlation of PCR-amplified product to individual cells. Integration with a true microflow cytometer with multiple optical detection channels would yield more significant data as the PCR data could be correlated to the optical data from each cell. Sorting capabilities would enhance such an integrated system to allow selection of desired populations based on the optical information. Merging microfluidic technologies that are still under development may be challenging, but the end product would be highly attractive.

18.4.2.3 Cell Culture

After sorting cells in a microflow cytometer, it may be desirable to culture the cells for further analysis. Integration of the cell culture within the device can reduce cell death due to harsh handling or the rigors of traveling through the cell sorter. Increase in cell survival after sorting can allow smaller samples to be run through the device, a factor that is particularly important when dealing with limited quantities. Wolff *et al.* prepared an integrated microflow cytometer capable of sorting GFP-positive yeast cells, then cultured those cells for several days and observed division and growth (Fig. 18.12).[86]

18.5 CONCLUSIONS

Developers of microflow cytometers constantly address the challenges of shrinking feature sizes while adding new functions. Most begin with the goal of microfabricating the fluidics, with success measured by the degree of particle focusing achieved and the coefficients of variation in the measurements. Although most microflow cytometers still use off-chip optics and fluidics, the integration of optical and fluidic control components onto the fluidic substrate is beginning. However, systems integration is challenging, and issues of material compatibility, reliability of the miniaturized components, feasibility of manufacturing, and cost must be addressed.

Before all the optical, fluidic, and electronic components are integrated, it is more likely that we will see the microfluidic components of the cytometer integrated with sample processing components. Pre-processing and post-processing components have already been miniaturized for a number of analyses, and combining these with a microflow cytometer is a natural way ahead. Such combinations can be fabricated while keeping the easily replaceable microfluidics separate from the more expensive optics and fluidic control components.

It will be interesting to see which systems find a market. Integration of a microflow cytometer with other functions in automated systems will provide additional information. However, simpler devices that generate more limited types of information may be easier to fully automate and cheaper to produce. The simplest of such cytometers, ones that just count cells, are already being marketed in Korea[75] and the US[87,88] as point-of-care devices. Our prediction is that a wide range of microflow cytometers will evolve specifically designed for a variety of on-site applications.

ACKNOWLEDGMENTS

The work presented here was performed under NIH grant UO1 A1075489 and ONR/NRL 6.2 work unit 6336. The views presented here are those of the authors and do not represent the opinion of the US Navy, the Department of Defense, the National Institutes of Health, or the Department of Health and Human Services.

References

[1] W. Gohde and W. Dittrich, Flow-through chamber for photometers to measure and count particles in a dispersion medium, US3761187, (1968).

[2] L. A. Sklar, *Flow Cytometry for Biotechnology*, Oxford University Press: New York, (2005).

[3] J. P. Robinson, *Handbook of Flow Cytometry Methods*, Wiley-Liss: New York, (1993).

[4] D. F. Keren, J. P. McCoy and J. L. Carey, *Flow Cytometry in Cinical Diagnosis. 3rd ed.* ASCP Press: Chicago, (2001).

[5] D. Psaltis, S. R. Quake and C. Yang, Developing optofluidic technology through the fusion of microfluidics and optics. *Nature* **442**(7101) 381–386 (2006).

[6] C. Monat, P. Domachuk and B. J. Eggleton, Integrated optofluidics: A new river of light. *Nat. Photon* **1**(2), 106–114 (2007).

[7] C. C. Chang, Z. X. Huang and R. J. Yang, Three-dimensional hydrodynamic focusing in two-layer polydimethylsiloxane (PDMS) microchannels. *Journal of Micromechanics and Microengineering* **17**(8), 1479–1486 (2007).

[8] J. P. Golden, J. S. Kim, J. S. Erickson, L. R. Hilliard, P. B. Howell, G. P. Anderson, M. Nasir and F. S. Ligler, Multi-wavelength microflow cytometer using groove-generated sheath flow. *Lab Chip* 10.1039/B822442K (2009).

[9] W. H. Wong, E. Y. B. Pun and K. S. Chan, Electron beam direct-write tunable polymeric waveguide grating filter. *Photonics Technology Letters, IEEE* **15**(12), 1731–1733 (2003).

[10] K. Joon-Sung and K. Jang-Joo, Fabrication of multimode polymeric waveguides and micromirrors using deep X-ray lithography. *Photonics Technology Letters, IEEE* **16**(3), 798–800 (2004).

[11] A. A. Said, M. Dugan, P. Bado, Y. Bellouard, A. Scott and J. R. Mabesa, Manufacturing by laser direct-write of three-dimensional devices containing optical and microfluidic networks. *Photon Processing in Microelectronics and Photonics III* **5339**, 194–204.12 (2004).

[12] A. A. Bettiol, S. V. Rao, E. J. Teo, J. A. van Kan and F. Watt, Fabrication of buried channel waveguides in photosensitive glass using proton beam writing. *Applied Physics Letters* **88**(17), 171106–171108 (2006).

[13] A. C. Sullivan, M. W. Grabowski and R. R. McLeod, Three-dimensional direct-write lithography into photopolymer. *Appl .Opt.* **46**(3), 295–301 (2007).

[14] X.-M. Zhao, S. P. Smith, S. J. Waldman, S. J. Whitesides and M. Prentiss, Demonstration of waveguide couplers fabricated using microtransfer molding. *Applied Physics Letters* **71**(8), 1017–1019 (1997).

[15] V. Lien, Y. Berdichevsky and L. Yu-Hwa, A prealigned process of integrating optical waveguides with microfluidic devices. *Photonics Technology Letters, IEEE* **16**(6), 1525–1527 (2004).

[16] K. S. Lee, H. L. Lee and R. J. Ram, Polymer waveguide backplanes for optical sensor interfaces in microfluidics. *Lab Chip* **7**(11), 1539–1545 (2007).

[17] D. A. Chang-Yen, R. K. Eich and B. K. Gale, A monolithic PDMS waveguide system fabricated using soft-lithography techniques. *Journal of Lightwave Technology* **23**(6), 2088–2093 (2005).

[18] M. Hajj-Hassan, T. Gonzalez, E. Ghafar-Zadeh, H. Djeghelian, V. Chodavarapu, M. Andrews and D. Therriault, Direct-dispense polymeric waveguides platform for optical chemical sensors. *Sensors* **8**(12), 7636–7648 (2008).

[19] http://www.oceanoptics.com/products/fiberspecs.asp (2007).

[20] H. Schmidt and A. Hawkins, Optofluidic waveguides: I. Concepts and implementations. *Microfluidics and Nanofluidics* **4**(1), 3–16 (2008).

[21] A. Hawkins and H. Schmidt, Optofluidic waveguides: II. Fabrication and structures. *Microfluidics and Nanofluidics* **4**(1), 17–32 (2008).

[22] D. Yin, E. J. Lunt, M. I. Rudenko, D. W. Deamer and A. R. Hawkins, H. Schmidt, Planar optofluidic chip for single particle detection, manipulation, and analysis. *Lab Chip* **7**(9), 1171–1175 (2007).

[23] D. B. Wolfe, R. S. Conroy, P. Garstecki, B. T. Mayers, M. A. Fischbach, K. E. Paul, M. Prentiss and G. M. Whitesides, Dynamic control of liquid-core/liquid-cladding optical waveguides. *Proceedings of the National Academy of Sciences of the United States of America* **101**(34), 12434–12438 (2004).

[24] J. Seo and L. P. Lee, Disposable integrated microfluidics with self-aligned planar microlenses. *Sensors and Actuators B: Chemical* **99**(2-3), 615–622 (2004).

[25] U. Levy and R. Shamai, Tunable optofluidic devices. *Microfluidics and Nanofluidics* **4**(1), 97–105 (2008).

[26] O. Hofmann, X. Wang, A. Cornwell, S. Beecher, A. Raja, D. D. Bradley, A. J. deMello and J. C. deMello, Monolithically integrated dye-doped PDMS long-pass filters for disposable on-chip fluorescence detection. *Lab Chip* **6**(8), 981–987 (2006).

[27] C. L. Bliss, J. N. McMullin and C. J. Backhouse, Integrated wavelength-selective optical waveguides for microfluidic-based laser-induced fluorescence detection. *Lab Chip* **8**(1), 143–51 (2008).

[28] H. Kurt and D. S. Citrin, Reconfigurable multimode photonic-crystal waveguides. *Opt Express* **16**(16), 11995–2001 (2008).

[29] A. d'Alessandro, D. Donisi, L. De Sio, R. Beccherelli, R. Asquini, R. Caputo, C. Umeton, Tunable integrated optical filter made of a glass ion-exchanged waveguide and an electro-optic composite holographic grating. *Opt. Express* **16**(13), 9254–9260 (2008).

[30] O. J. A. Schueller, D. C. Duffy, J. A. Rogers, S. T. Brittain and G. M. Whitesides, Reconfigurable diffraction gratings based on elastomeric microfluidic devices. *Sensors and Actuators A: Physical* **78**(2–3), 149–159 (1999).

[31] J. B. Edel, N. P. Beard, O. Hofmann, J. C. deMello, D. D. C. Bradley, and A. J. deMello, Thin-film polymer light emitting diodes as integrated excitation sources for microscale capillary electrophoresis. *Lab Chip,* **4**(2), 136–140 (2004).

[32] X. Wang, O. Hofmann, J. Huang, E. M. Barrett, R. Das, A. J. de Mello, J. C. de Mello and D. C. Bradley, Organic light emitting diodes and photodetectors: Towards applications in lab-on-a-chip portable devices. *Proceedings of SPIE: BioMEMS and Nanotechnology II* **6036**, 60361O–1–9 (2006).

[33] E. Kraker, A. Haase, B. Lamprecht, G. Jakopic, C. Konrad and S.Kostler, Integrated organic electronic based optochemical sensors using polarization filters. *Applied Physics Letters* **92**(3) (2008).

[34] O. J. Schueller, G. M. Whitesides, J. A. Rogers, M. Meier and A. Dodabalapur, Fabrication of photonic crystal lasers by nanomolding of solgel glasses. *Appl. Opt.* **38**(27), 5799–802 (1999).

[35] Z. Li, Z. Zhang, A. Scherer and D. Psaltis, Optofluidic microring dye laser. *2007 Digest of the Leos Summer Topical Meetings,* 70–71 (2007).

[36] S. Balslev, A. M. Jorgensen, B. Bilenberg, K. B. Mogensen, D. Snakenborg, O. Geschke, J. P. Kutter and A. Kristensen, Lab-on-a-chip with integrated optical transducers. *Lab Chip* **6**(2), 213–217 (2006).

[37] O. Inganas, L. S. Roman, F. L. Zhang, D. M. Johansson, M. R. Andersson and J. C. Hummelen, Recent progress in thin film organic photodiodes. *Synthetic Metals* **121**(1-3), 1525–1528 (2001).

[38] R. Pieler, E. Fureder, M. Sonnleitner, Printed photonics for lab-on-chip applications - art. no. 673919. *Electro-Optical Remote Sensing, Detection, and Photonic Technologies and Their Applications* **6739**, 73919–73919 (2007).

[39] L. Burgi, R. Pfeiffer, M. Mucklich, P. Metzler, M. Kiy and C. Winnewisser, Optical proximity and touch sensors based on monolithically integrated polymer photodiodes and polymer LEDs. *Organic Electronics* **7**(2), 114–120 (2006).

[40] J. R. Wojciechowski, L. C. Shriver-Lake, M. Y. Yamaguchi, M. Sonnleitner, E. Füreder, R. Pieler and F. S.Ligler, Organic photodiodes for biosensor miniaturization. *Anal. Chem.* In Press (2009).

[41] M. A. Unger, H. P. Chou, T. Thorsen, A. Scherer and S. R. Quake, Monolithic microfabricated valves and pumps by multilayer soft lithography. *Science* **288** (5463), 113–116 (2000).

[42] A. Y. Fu, H. P. Chou, C. Spence, F. H. Arnold, S. R. Quake, An integrated microfabricated cell sorter. *Anal. Chem.* **2002,** 74 (11), 2451–7.

[43] S. Y. Yang, K. Y. Lien, K. J. Huang, H. Y. Lei and G. B. Lee, Micro flow cytometry utilizing a magnetic bead-based immunoassay for rapid virus detection. *Biosens Bioelectron* **24**(4), 861–868 (2008).

[44] K. Y. Lien, J. L. Lin, C. Y. Liu, H. Y. Lei and G. B. Lee, Purification and enrichment of virus samples utilizing magnetic beads on a microfluidic system. *Lab Chip* **7**(7), 868–875 (2007).

[45] P. Jandik, B. H. Weigl, N. Kessler, J. Cheng, C. J. Morris, T. Schulte and N. Avdalovic, Initial study of using a laminar fluid diffusion interface for sample preparation in high-performance liquid chromatography. *J. Chromatogr A.* **954**(1,2), 33–40 (2002).

[46] P. Yager, T. Edwards, E. Fu, K. Helton, K. Nelson, M. R. Tam, B. H. Weigl, Microfluidic diagnostic technologies for global public health. *Nature* **442**(7101), 412–418 (2006).

[47] E. Garcia, J. R. Kirkham, A. V. Hatch, K. R. Hawkins and P. Yager, Controlled microfluidic reconstitution of functional protein from an anhydrous storage depot. *Lab Chip* **4**(1), 78–82 (2004).

[48] D. Y. Stevens, C. R. Petri and J. L. Osborn, Spicar-Mihalic, P. McKenzie, K. G. Yager, P., Enabling a microfluidic immunoassay for the developing world by integration of on-card dry reagent storage. *Lab Chip* **8**(12), 2038–2045 (2008).

[49] V. Linder, S. K. Sia and G. M. Whitesides, Reagent-loaded cartridges for valveless and automated fluid delivery in microfluidic devices. *Anal Chem.* **77**(1), 64–71 (2005).

[50] A. D. Stroock, S. K. Dertinger, A. Ajdari, I. Mezic, H. A. Stone and G. M. Whitesides, Chaotic mixer for microchannels. *Science* **295**(5555), 647–51 (2002).

[51] P. B. Howell, Jr. D. R. Mott, S. Fertig, C. R. Kaplan, J. P. Golden, E. S. Oran and F. S. Ligler, A microfluidic mixer with grooves placed on the top and bottom of the channel. *Lab Chip* **5**(5), 524–30 (2005).

[52] M. S. Williams, K. J. Longmuir and P. Yager, A practical guide to the staggered herringbone mixer. *Lab Chip* **8**(7), 1121–1129 (2008).

[53] M. R. Bringer, C. J. Gerdts, H. Song, J. D. Tice and R. F. Ismagilov, Microfluidic systems for chemical kinetics that rely on chaotic mixing in droplets. *Philos. Transact. A Math. Phys. Eng. Sci.* **362**(1818), 1087–104 (2004).

[54] Y. Kikutani, T. Horiuchi, K. Uchiyama, H. Hisamoto, M. Tokeshi and T. Kitamori, Glass microchip with three-dimensional microchannel network for 2 x 2 parallel synthesis. *Lab Chip* **2**(4), 188–192 (2002).

[55] G. G. Yaralioglu, I. O. Wygant, T. C. Marentis and B. T. Khuri-Yakub, Ultrasonic mixing in microfluidic channels using integrated transducers. *Anal. Chem.* **76**(13), 3694–3698 (2004).

[56] K. S. Ryu, K. Shaikh, E. Goluch, Z. Fan and C. Liu, Micro magnetic stir-bar mixer integrated with parylene microfluidic channels. *Lab Chip* **4**(6), 608–13 (2004).

[57] S. Y. Teh, R.L. H. Lin, Hung and A. P. Lee, Droplet microfluidics. *Lab Chip* **8**(2), 198–220 (2008).

[58] A. J. deMello, Control and detection of chemical reactions in microfluidic systems. *Nature* **442**(7101), 394–402 (2006).

[59] J. El-Ali, P. K. Sorger and K. F. Jensen, Cells on chips. *Nature* **442**(7101), 403–411 (2006).

[60] S. Haeberle and R. Zengerle, Microfluidic platforms for lab-on-a-chip applications. *Lab Chip* **7**(9), 1094–110 (2007).

[61] J. R. Burns and C. Ramshaw, The intensification of rapid reactions in multiphase systems using slug flow in capillaries. *Lab Chip* **1**(1), 10–5 (2001).

[62] Z. T. Cygan, J. T. Cabral, K. L. Beers and E. J. Amis, Microfluidic platform for the generation of organic-phase microreactors. *Langmuir* **21**(8), 3629–3634 (2005).

[63] D. Chen, W. Du, Y. Liu, W. Liu, A. Kuznetsov, F. E. Mendez, L. H. Philipson and R. F. Ismagilov, The chemistrode: a droplet-based microfluidic device for stimulation and recording with high temporal, spatial, and chemical resolution. *Proc. Natl. Acad. Sci. USA* **105**(44), 16843–16848 (2008).

[64] V. Taly, B. T. Kelly and A. D. Griffiths, Droplets as microreactors for high-throughput biology. *Chembiochem.* **8**(3), 263–272 (2007).

[65] M. He, J. S. Edgar, G. D. Jeffries, R. M. Lorenz, J. P. Shelby and D. T. Chiu, Selective encapsulation of single cells and subcellular organelles into picoliter- and femtoliter-volume droplets. *Anal. Chem.* **77**(6), 1539–1544 (2005).

[66] Y. C. Tan, K. Hettiarachchi, M. Siu, Y. R. Pan and A. P. Lee, Controlled microfluidic encapsulation of cells, proteins, and microbeads in lipid vesicles. *J. Am. Chem. Soc.* **128**(17), 5656–5658 (2006).

[67] M. Chabert and J. L. Viovy, Microfluidic high-throughput encapsulation and hydrodynamic self-sorting of single cells. *Proc. Natl. Acad. Sci. USA* **105**(9), 3191–3196 (2008).

[68] W. Shi, J. Qin, N. Ye, B. Lin, Droplet-based microfluidic system for individual Caenorhabditis elegans assay. *Lab Chip* **8**(9), 1432–1435 (2008).

[69] A. Huebner, M. Srisa-Art, D. Holt, C. Abell, F. Hollfelder, A. J. deMello and J. B. Edel, Quantitative detection of protein expression in single cells using droplet microfluidics. *Chem. Commun. (Camb.)* **12** 1218–1220 (2007).

[70] Kline, T. R. Runyon, M. K. Pothiawala, M. Ismagilov, R. F., ABO, D blood typing and subtyping using plug-based microfluidics. *Anal. Chem.* **80**(16), 6190–6197 (2008).

[71] J. Q. Boedicker, L. Li, T. R. Kline and R. F. Ismagilov, Detecting bacteria and determining their susceptibility to antibiotics by stochastic confinement in nanoliter droplets using plug-based microfluidics. *Lab Chip* **8**(8), 1265–1272 (2008).

[72] S. -Y. Yang, S. -K. Hsiung, Y.- C. Hung, C. -M. Chang, T. -L. Liao and G. -B.Lee, A cell counting/sorting system incorporated with a microfabricated flow cytometer chip. *Meas. Sci. Technol.*, **17**, 2001–2009 (2006).

[73] C. H. Wang and G. B. Lee, Automatic bio-sampling chips integrated with micro-pumps and micro-valves for disease detection. *Biosens. Bioelectron.* **21**(3), 419–25 (2005).

[74] C. B. Rohde, F. Zeng, R. Gonzalez-Rubio, M. Angel and M. F. Yanik, Microfluidic system for on-chip high-throughput whole-animal sorting and screening at subcellular resolution. *Proc. Natl. Acad. Sci. USA* **104**(35), 13891–13895 (2007).

[75] S. Chung, S. J. Park, J. K. Kim, C. Chung, D. C. Han and J. K. Chang, Plastic microchip flow cytometer based on 2-and 3-dimensional hydrodynamic flow focusing. *Microsystem Technologies-Micro-and Nanosystems-Information Storage and Processing Systems* **9**(8), 525–533 (2003).

[76] H. W. Bang, C. N. Chung, J. K. Kim, S. H. Kim, S. Chung, J. Park, W. G. Lee, H. Yun, J. Lee, K. C. Cho, D. C. Han and J. K. Chang, Microfabricated fluorescence-activated cell sorter through hydrodynamic flow manipulation. *Microsystem Technologies-Micro-and Nanosystems-Information Storage and Processing Systems* **12** (8), 746–753 (2006).

[77] M. A. Northrup, M. T. Ching, R. M. White and R. T. Watson, DNA amplification with a microfabricated reaction chamber. *Transducers '93* (1993).

[78] P. Wilding, M. A. Shoffner and L. J. Kricka, PCR in a silicon microstructure. *Clin. Chem.* **40**(9), 1815–1818 (1994).

[79] H. Nakano, K. Matsuda, M. Yohda, T. Nagamune, I. Endo and T. Yamane, High speed polymerase chain reaction in constant flow. *Biosci. Biotechnol. Biochem.* **58** (2), 349–352 (1994).

[80] M. U. Kopp, A. J. Mello and A. Manz, Chemical amplification: continuous-flow PCR on a chip. *Science* **280**(5366), 1046–1048 (1998).

[81] P. A. Auroux, Y. Koc, A. deMello, A. Manz and P. J. Day, Miniaturised nucleic acid analysis. *Lab Chip* **4**(6), 534–546 (2004).

[82] L. C. Waters, S. C. Jacobson, N. Kroutchinina, J. Khandurina, R. S. Foote, and J. M. Ramsey, Multiple sample PCR amplification and electrophoretic analysis on a microchip. *Anal. Chem.* **70**(24), 5172–5176 (1998).

[83] L. C. Waters, S. C. Jacobson, N. Kroutchinina, J. Khandurina, R. S. Foote, and J. M. Ramsey, Microchip device for cell lysis, multiplex PCR amplification, and electrophoretic sizing. *Anal. Chem.* **70**(1), 158–162 (1998).

[84] R. H. Liu, J. Yang, R. Lenigk, J. Bonanno and P. Grodzinski, Self-contained, fully integrated biochip for sample preparation, polymerase chain reaction amplification, and DNA microarray detection. *Anal. Chem.* **76**(7), 1824–1831 (2004).

[85] H. Li and E. S. Yeung, Selective genotyping of individual cells by capillary polymerase chain reaction. *Electrophoresis* **23**(19), 3372–3380 (2002).

[86] A. Wolff, I. R. Perch-Nielsen, U. D. Larsen, P. Friis, G. Goranovic, C. R. Poulsen, J. P. Kutter and P. Telleman, Integrating advanced functionality in a microfabricated high-throughput fluorescent-activated cell sorter. *Lab Chip* **3**(1), 22–27 (2003).

[87] http://www.invitrogen.com/site/us/en/home/brands/Product-Brand/Countess-Automated-Cell-Counter.html. 2008.

[88] www.micronics.net/products/microcyt.php. 2009.

Color Index

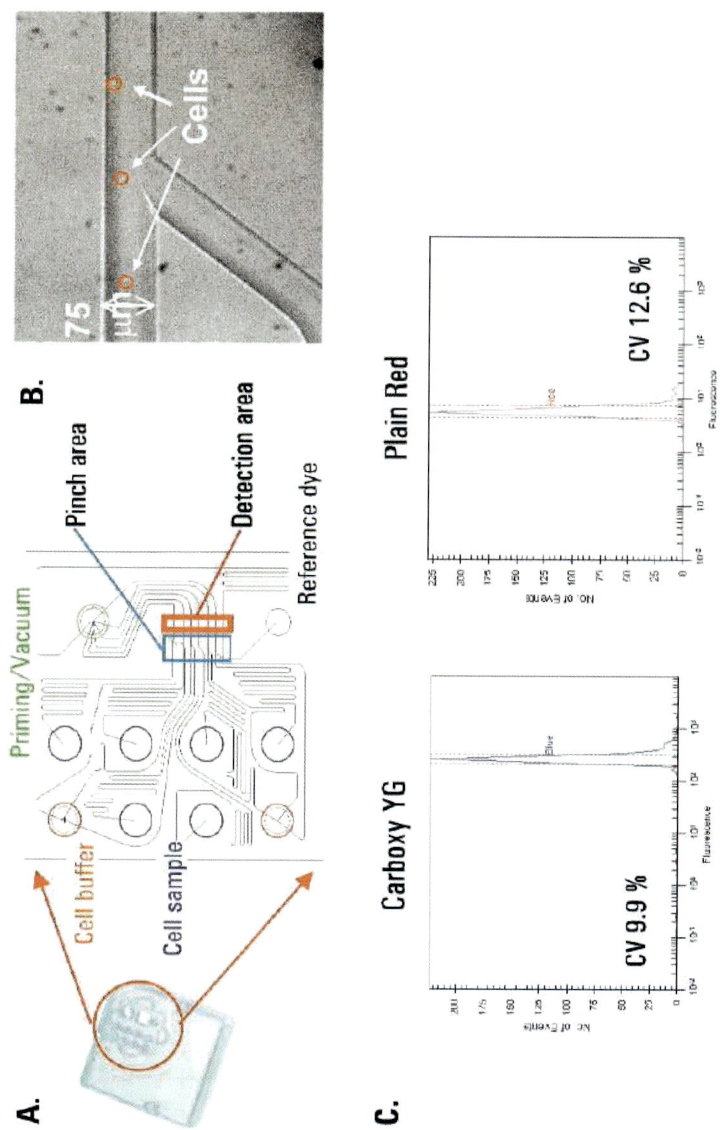
A.
Priming/Vacuum
Cell buffer
Cell sample
Pinch area
Detection area
Reference dye
B.
75 µm
Cells
C.
Carboxy YG
CV 9.9 %
Plain Red
CV 12.6 %

Figure 2.1

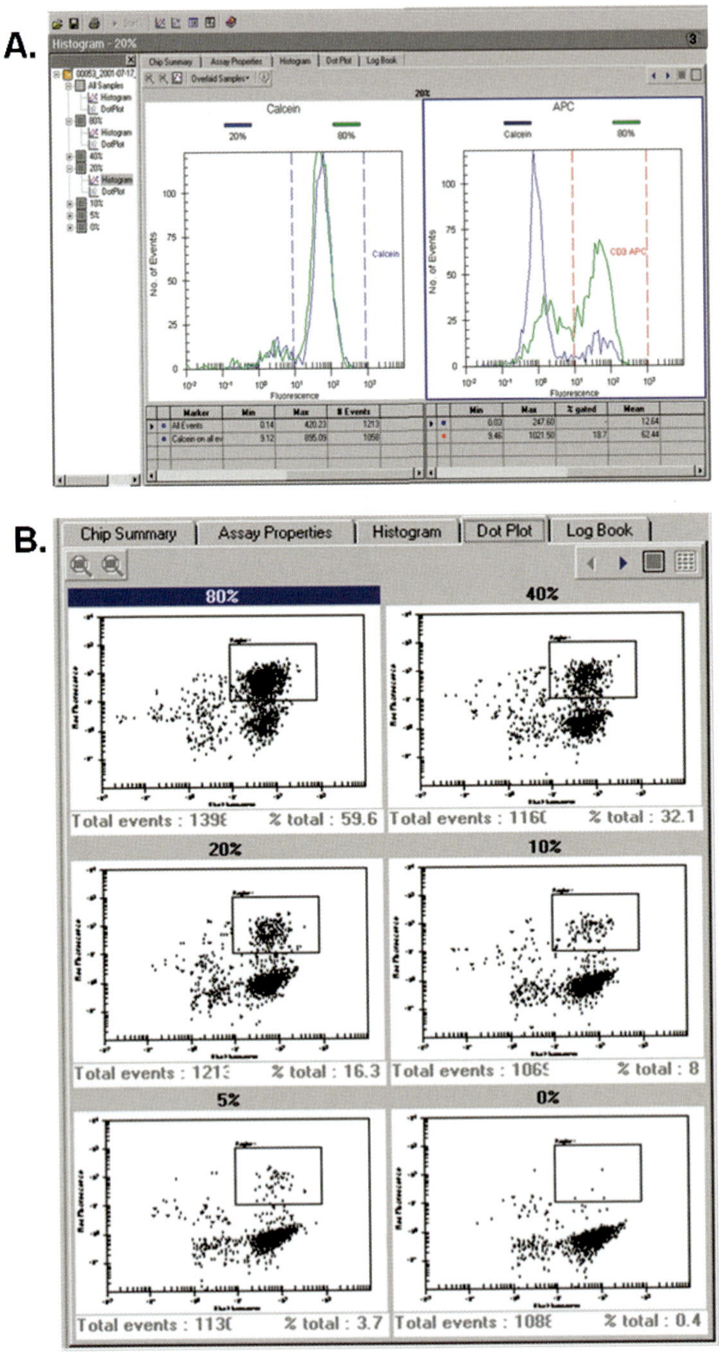
A.
Histogram - 20%
Chip Summary
Assay Properties
Histogram
Dot Plot
Log Book
Overlaid Samples
All Samples
Histogram
DotPlot
80%
40%
20%
10%
5%
0%
Calcein
20%
80%
APC
Calcein
80%
No. of Events
Fluorescence
CD3 APC
Marker
Min
Max
Events
All Events
0.14
420.23
1213
9.12
895.09
1058
Min
Max
% gated
Mean
0.03
247.60
12.64
9.46
1021.50
18.7
62.44
B.
Chip Summary
Assay Properties
Histogram
Dot Plot
Log Book
80%
Total events : 139
% total : 59.6
40%
Total events : 116
% total : 32.1
20%
Total events : 121
% total : 16.3
10%
Total events : 106
% total : 8
5%
Total events : 113
% total : 3.7
0%
Total events : 108
% total : 0.4

Figure 2.2

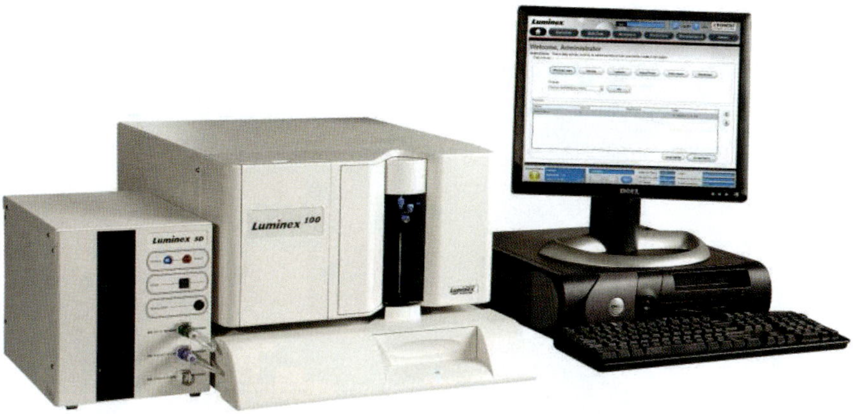

Figure 3.2

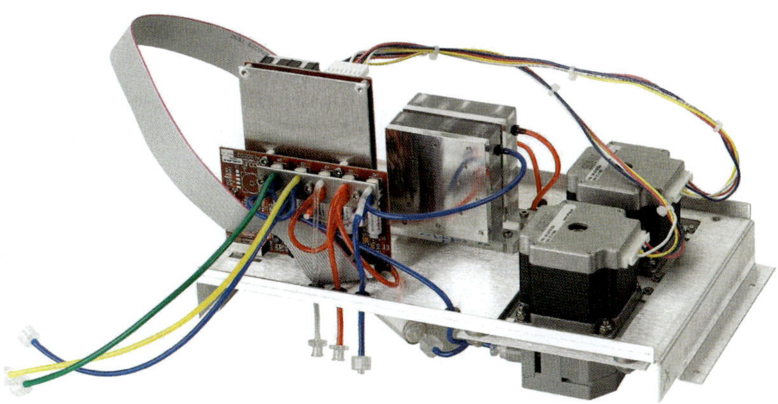

Figure 4.3

Figure 4.4

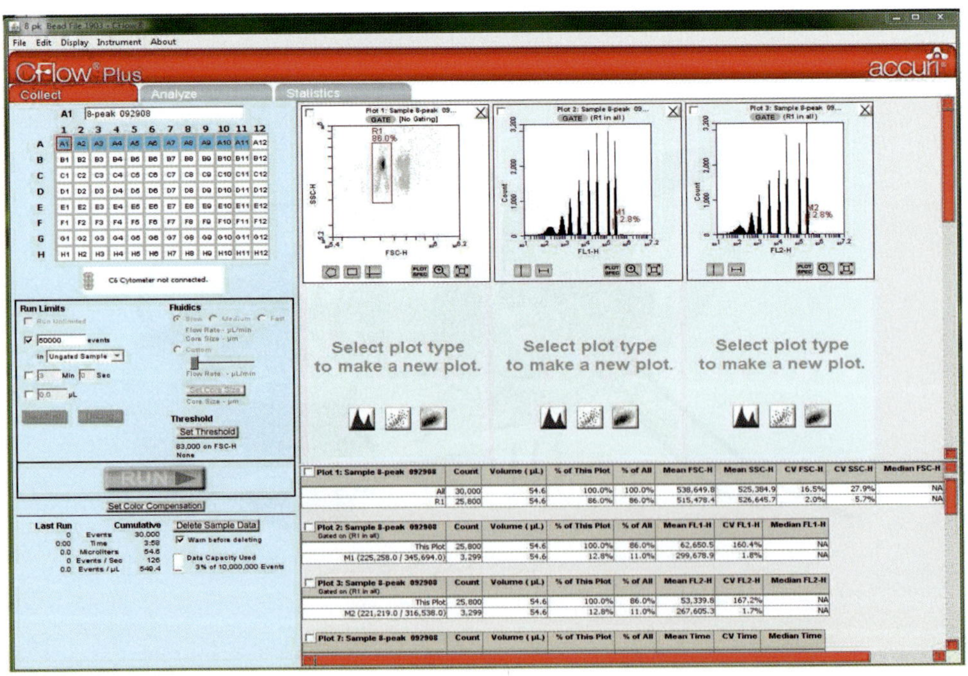

Figure 4.5

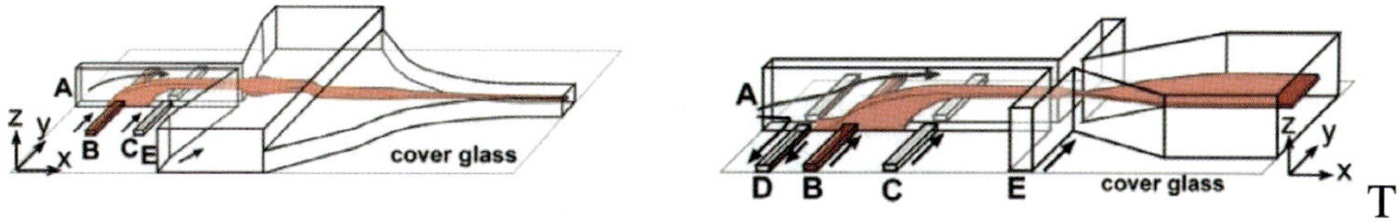

Figure 8.1

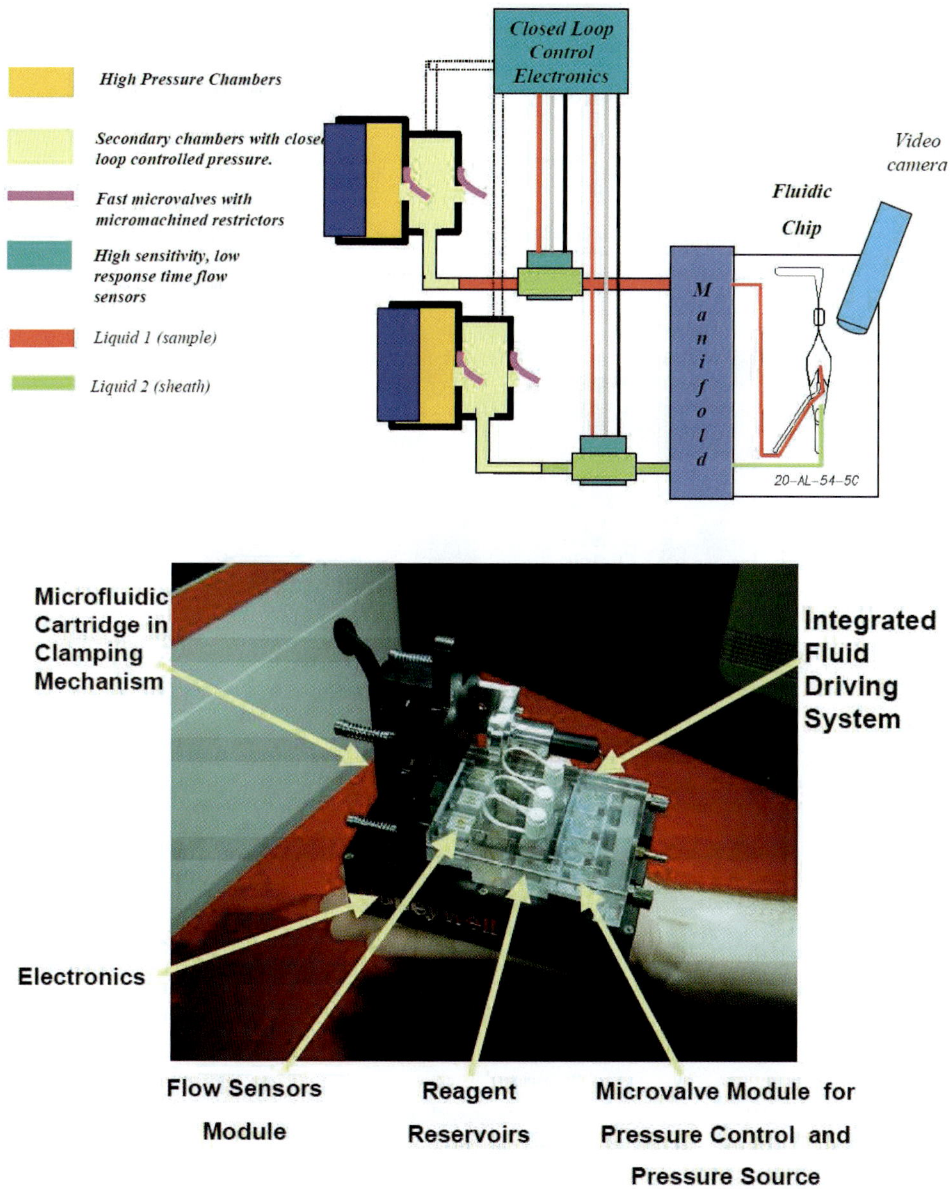

Figure 9.1

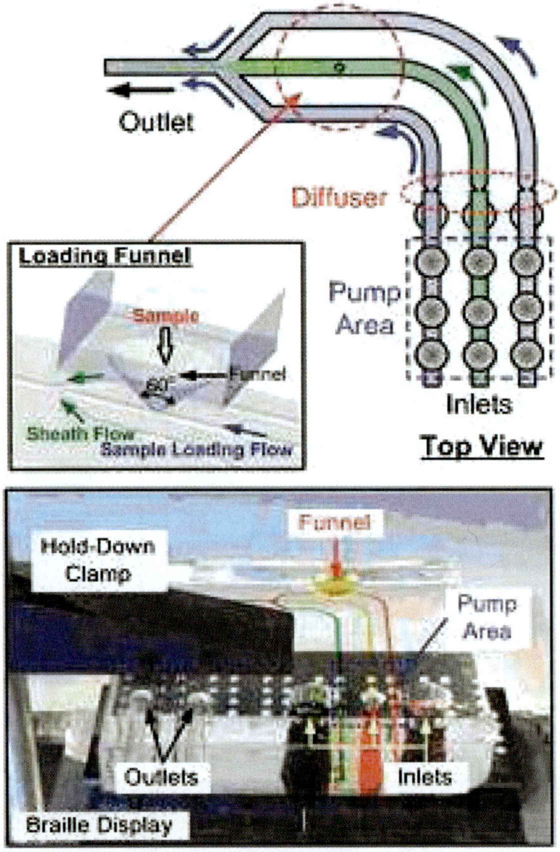

Figure 9.3

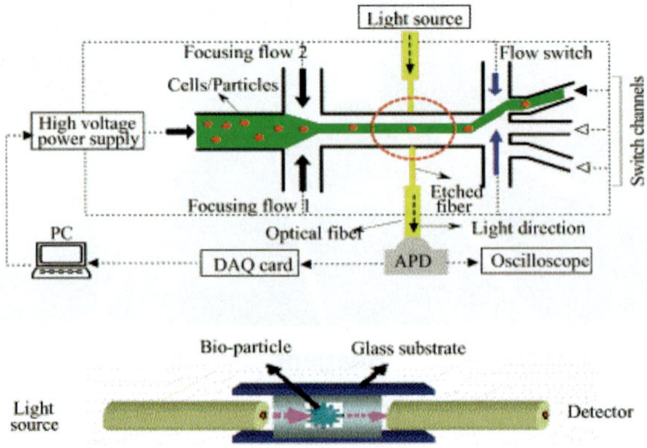

Figure 10.2

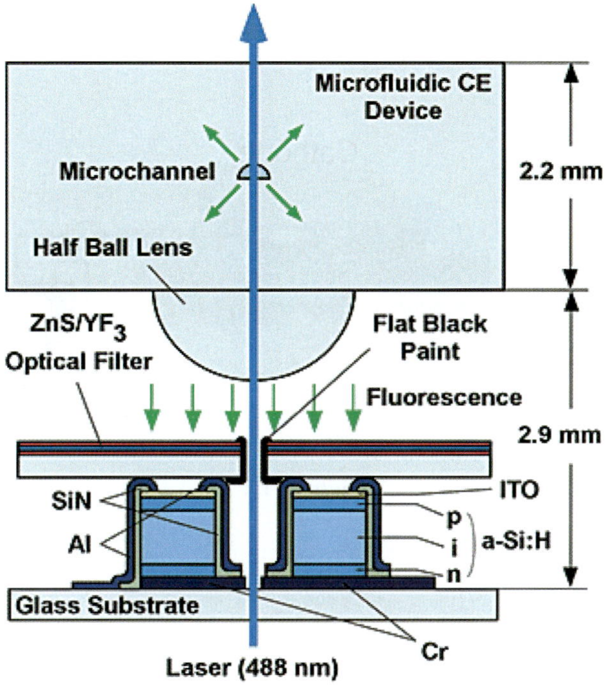

Figure 10.3

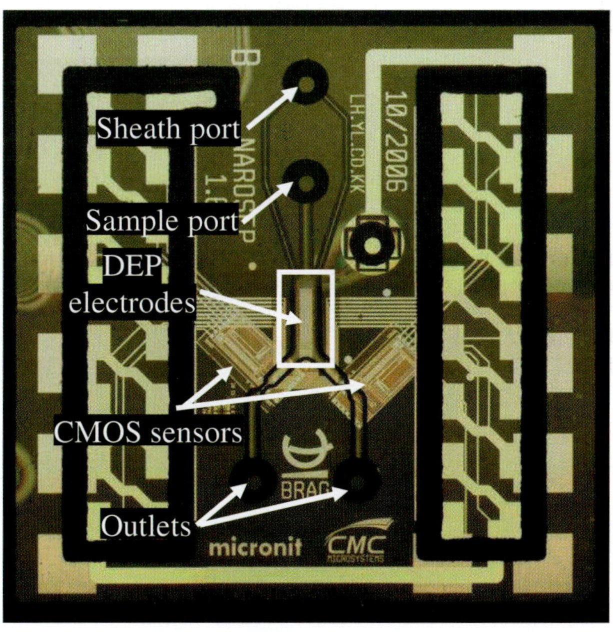

Figure 10.4

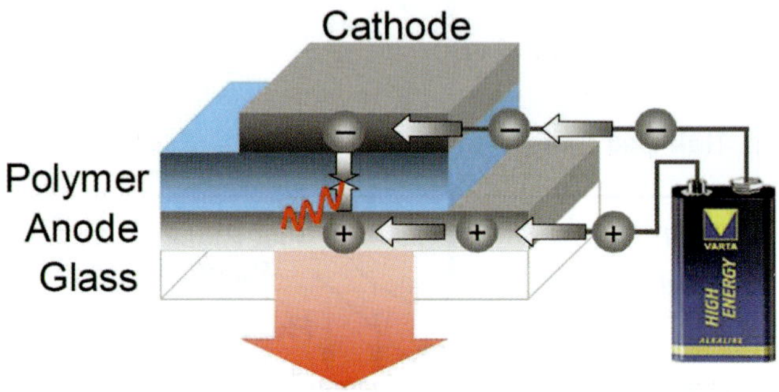

Figure 11.2

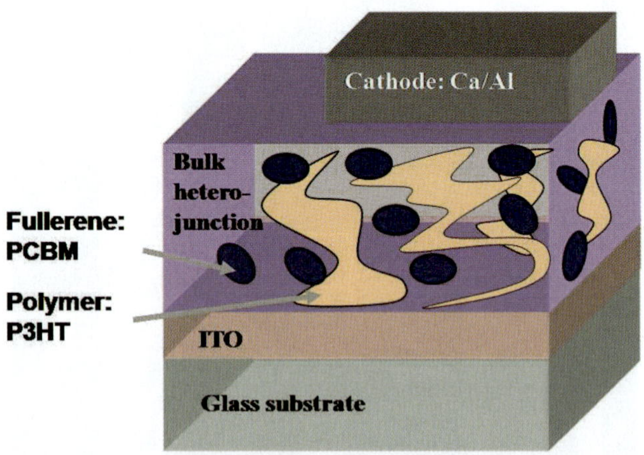

Figure 11.4

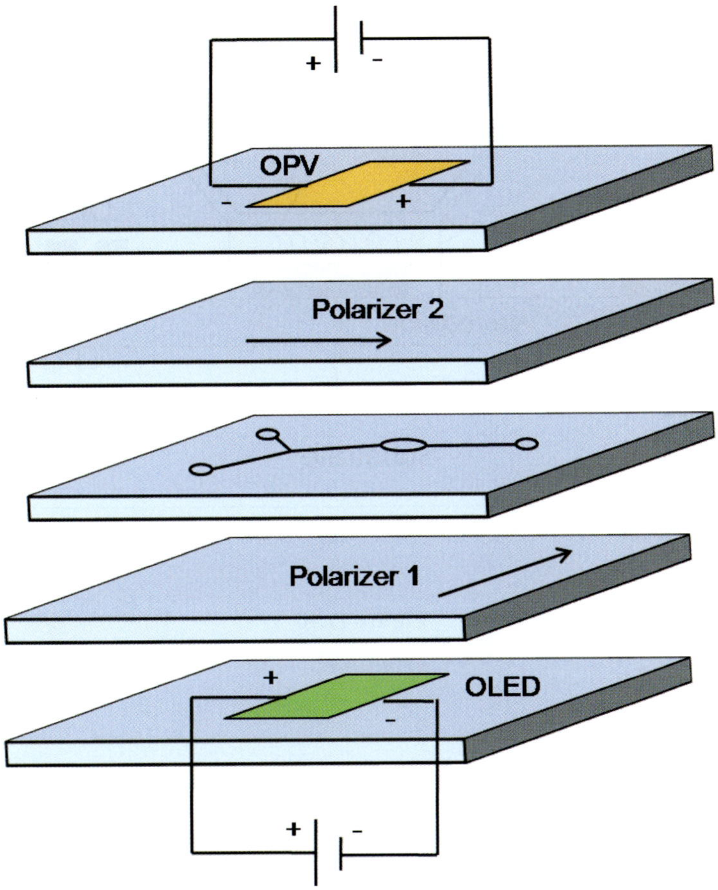

Figure 11.8

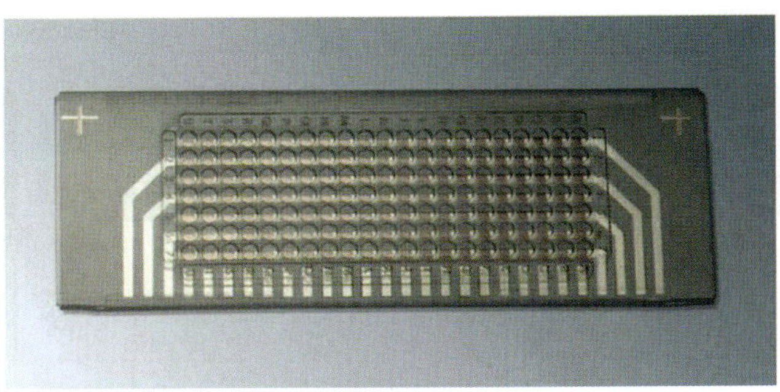

Figure 11.9

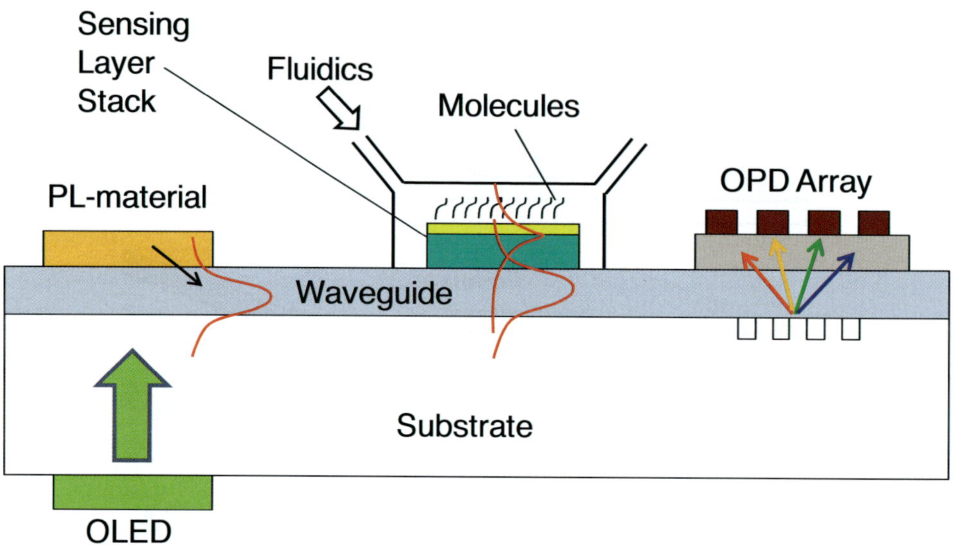

Figure 11.10

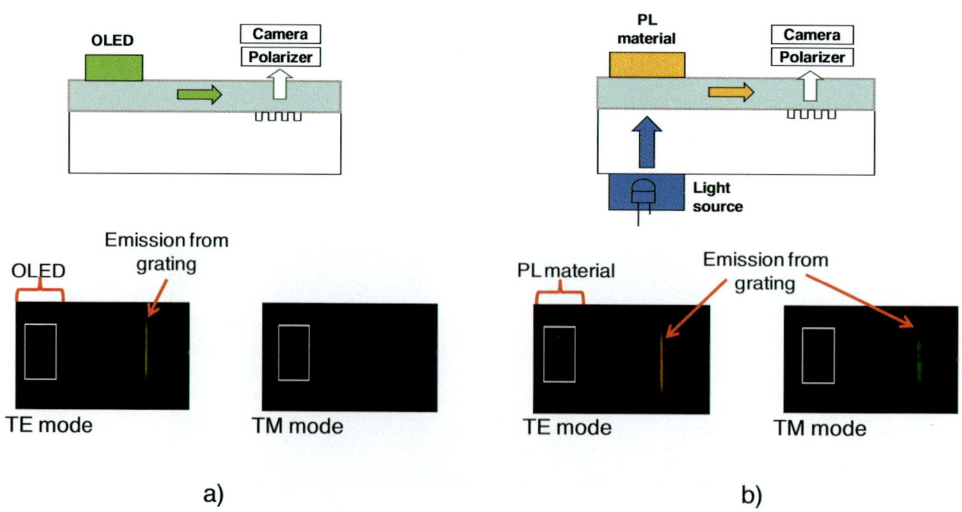

Figure 11.11

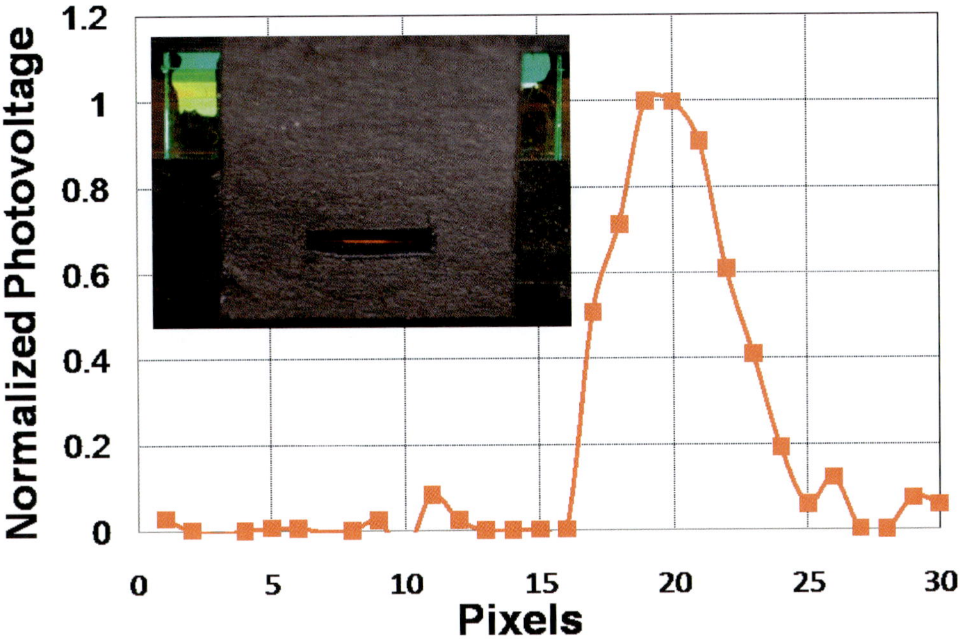

Figure 11.13

Figure 11.14

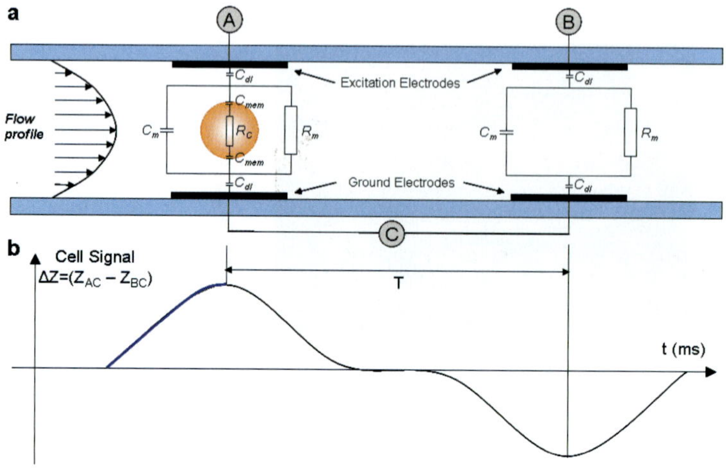

Figure 12.1

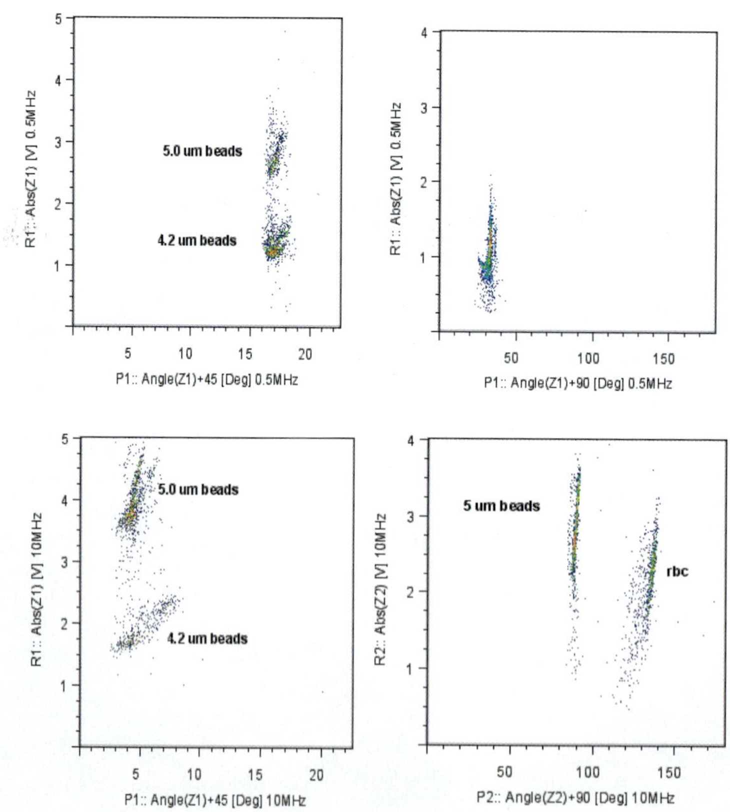

Figure 12.2

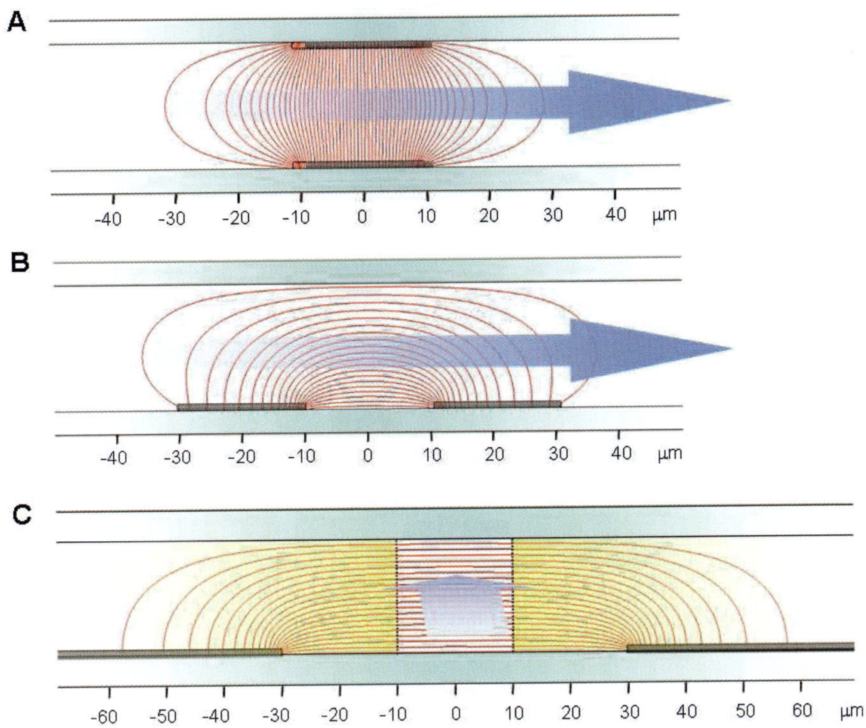

Figure 12.3

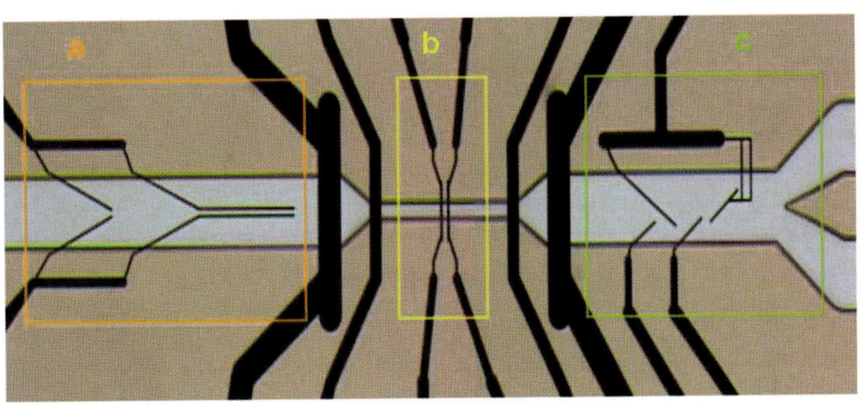

Figure 12.4

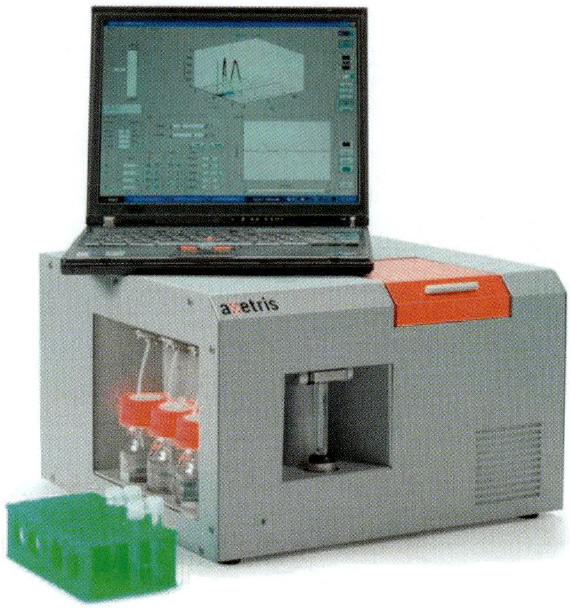

Figure 12.6

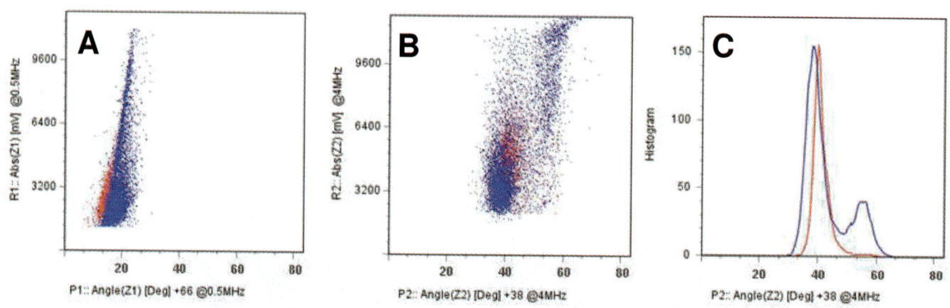

Figure 12.7

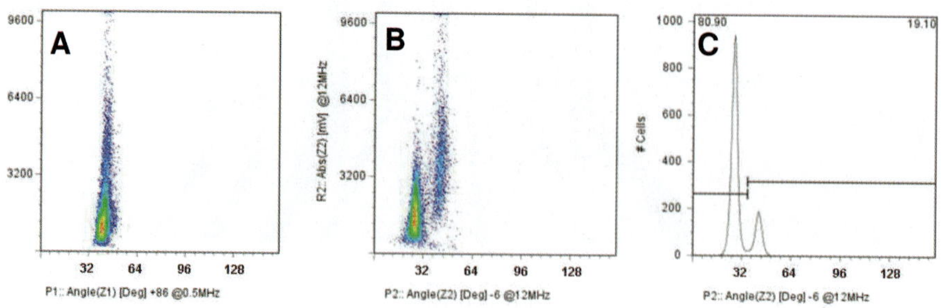

Figure 12.8

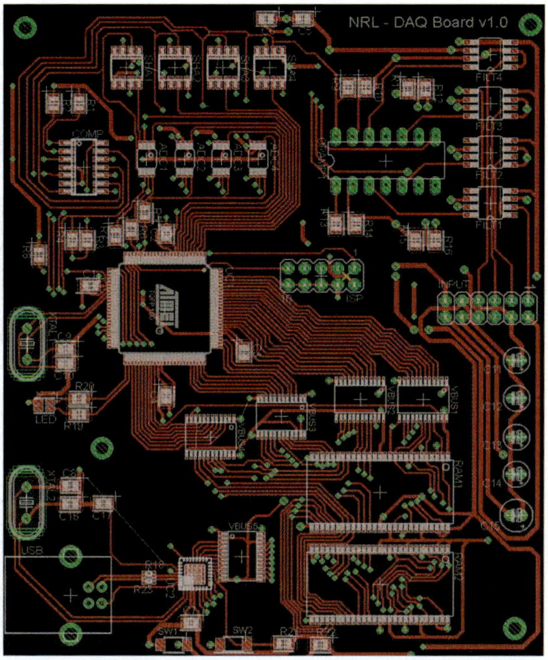

(a)

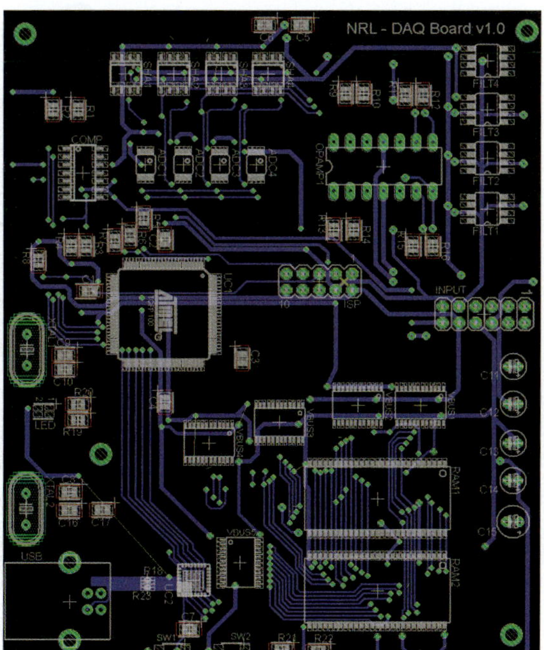

(b)

Figure 13.3

(a)

(b)

Figure 13.4

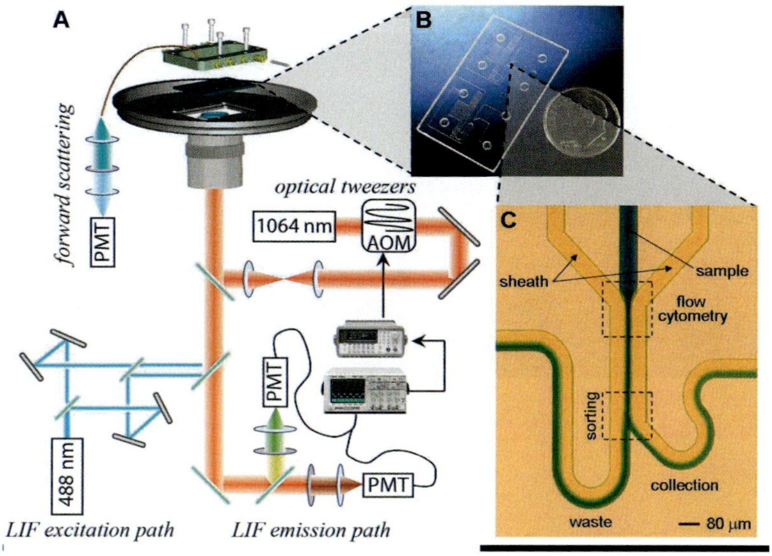

Figure 14.1

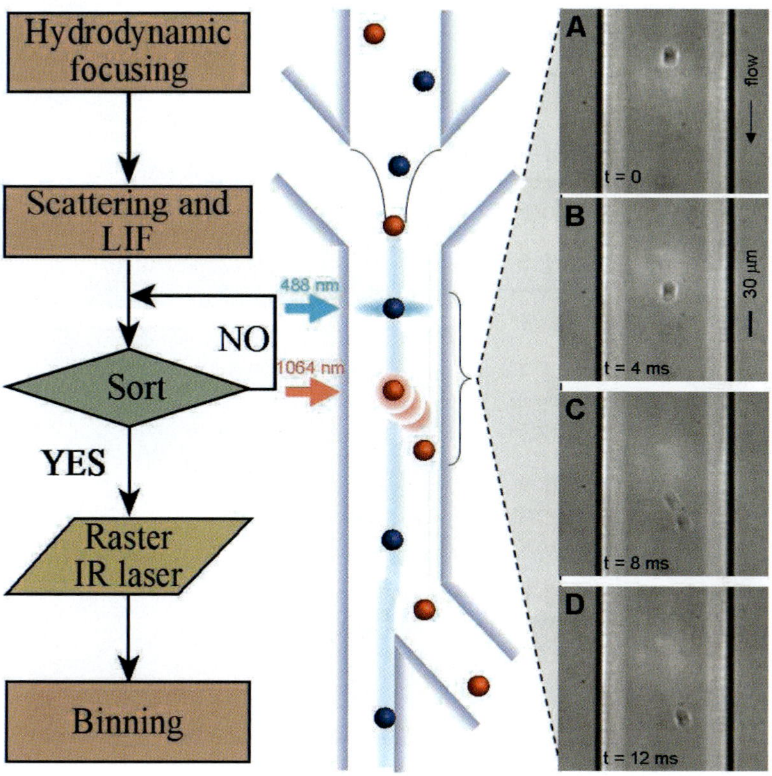

Figure 14.2

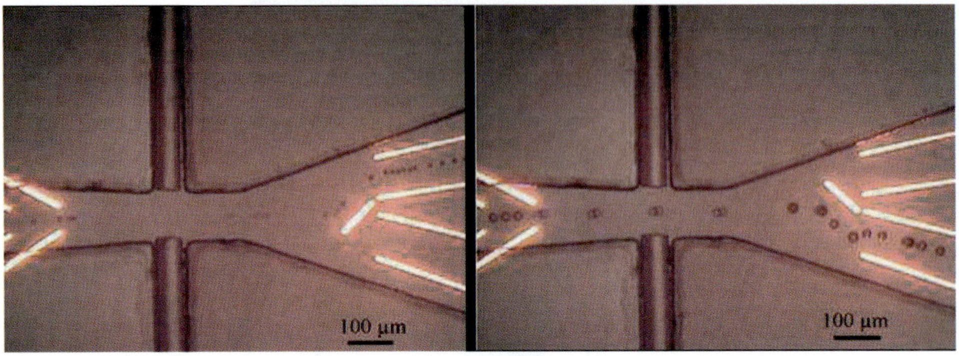

Figure 14.4

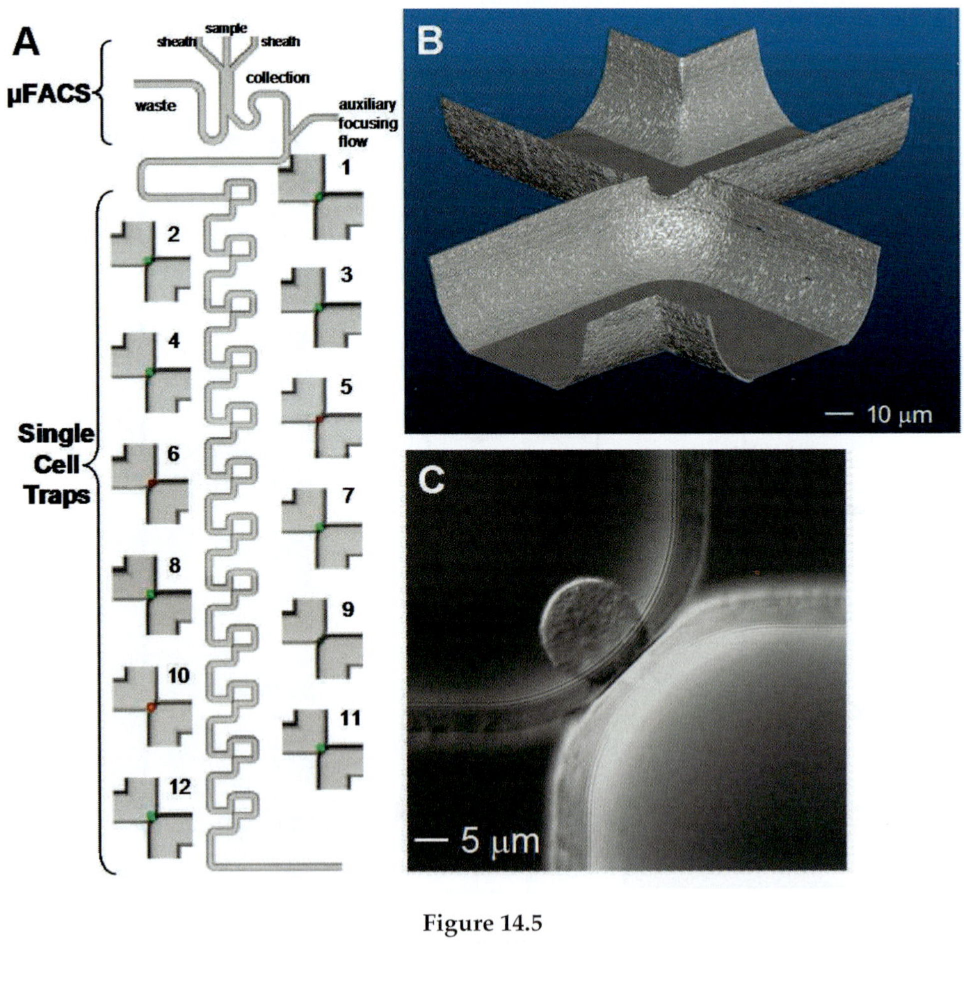

Figure 14.5

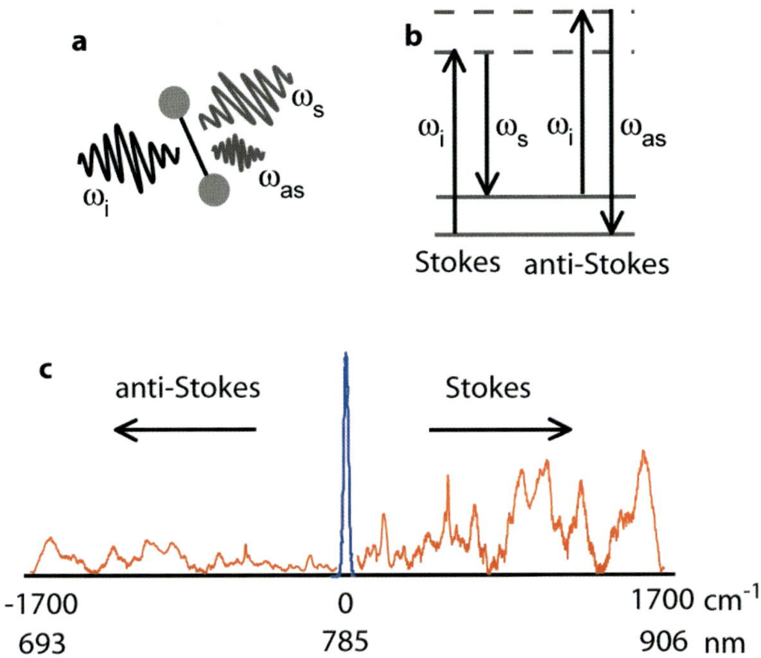

Figure 15.1

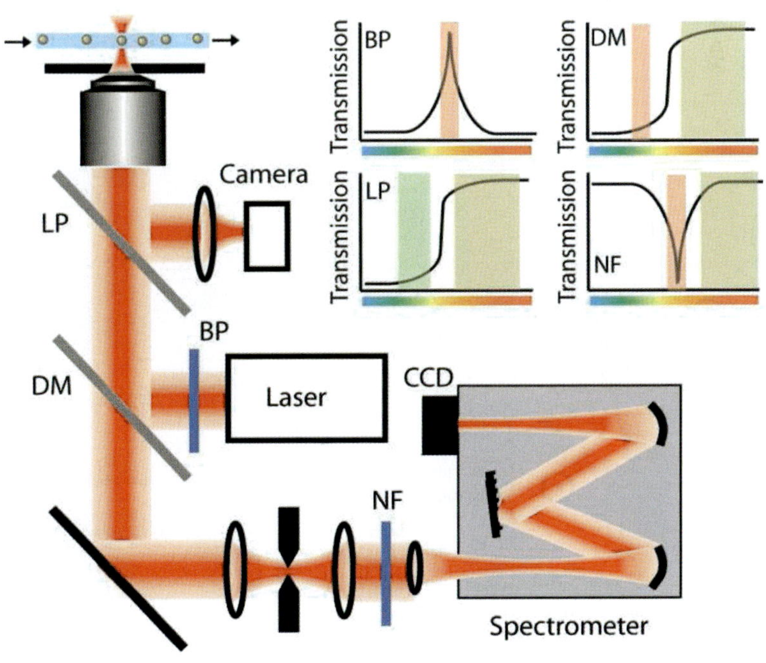

Figure 15.2

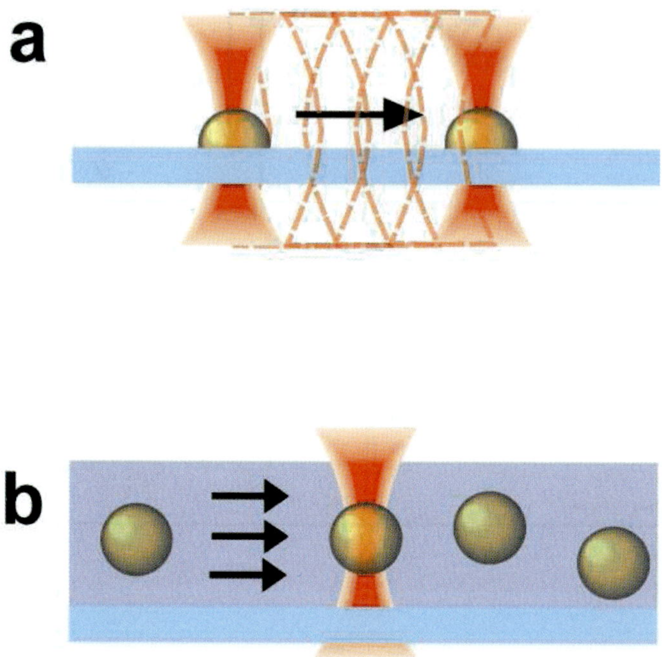

Figure 15.3

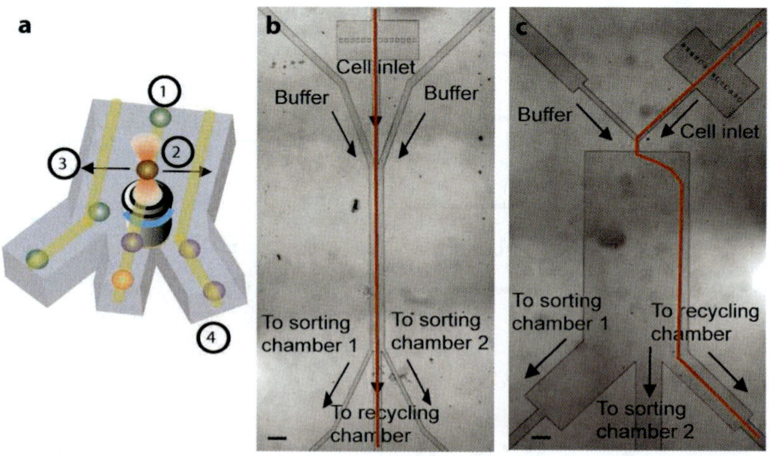

Figure 15.4

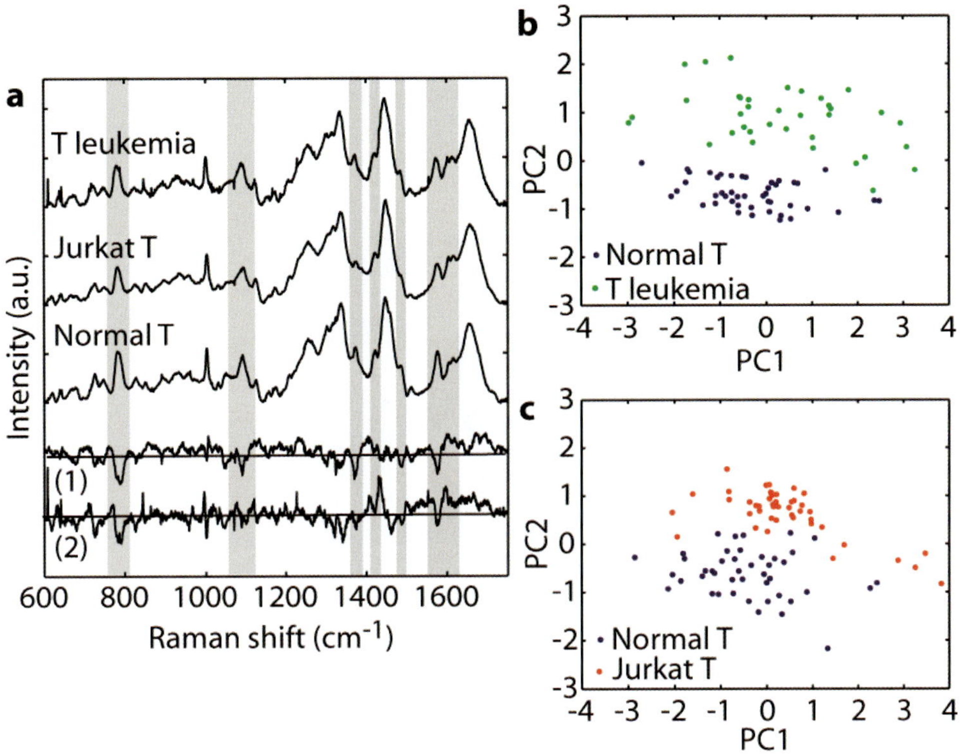

Figure 15.5

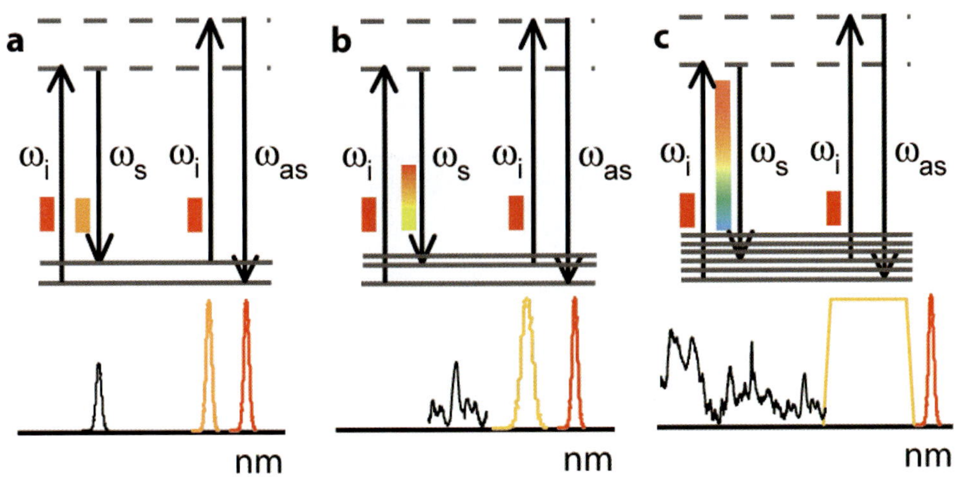

Figure 15.6

Figure 16.4

Figure 16.10

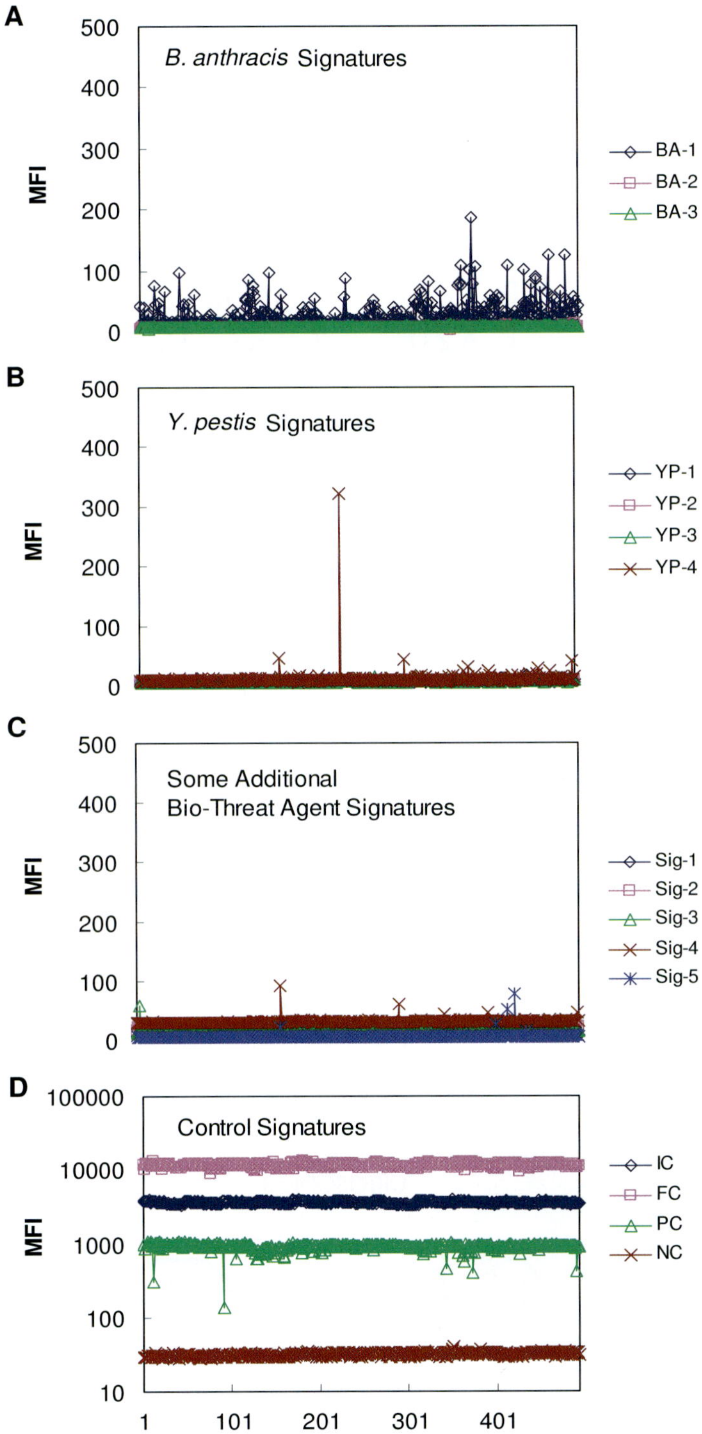
A
B. anthracis Signatures
MFI
500
400
300
200
100
0
BA-1
BA-2
BA-3
B
Y. pestis Signatures
MFI
500
400
300
200
100
0
YP-1
YP-2
YP-3
YP-4
C
Some Additional
Bio-Threat Agent Signatures
MFI
500
400
300
200
100
0
Sig-1
Sig-2
Sig-3
Sig-4
Sig-5
D
Control Signatures
MFI
100000
10000
1000
100
10
IC
FC
PC
NC
1
101
201
301
401

Figure 16.9

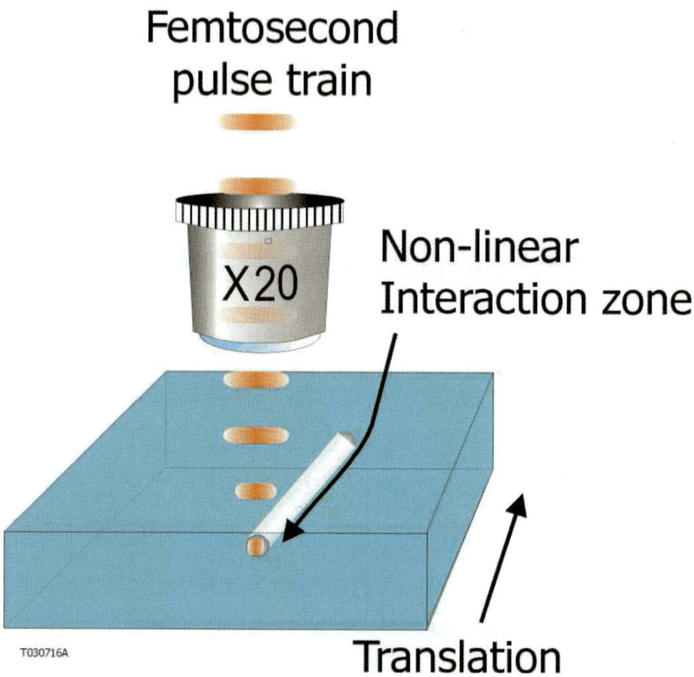

Figure 17.1

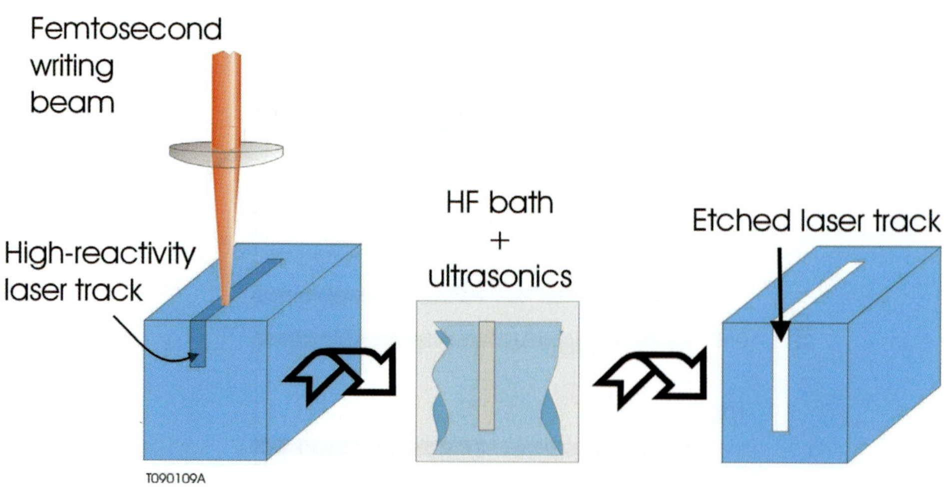

Figure 17.2

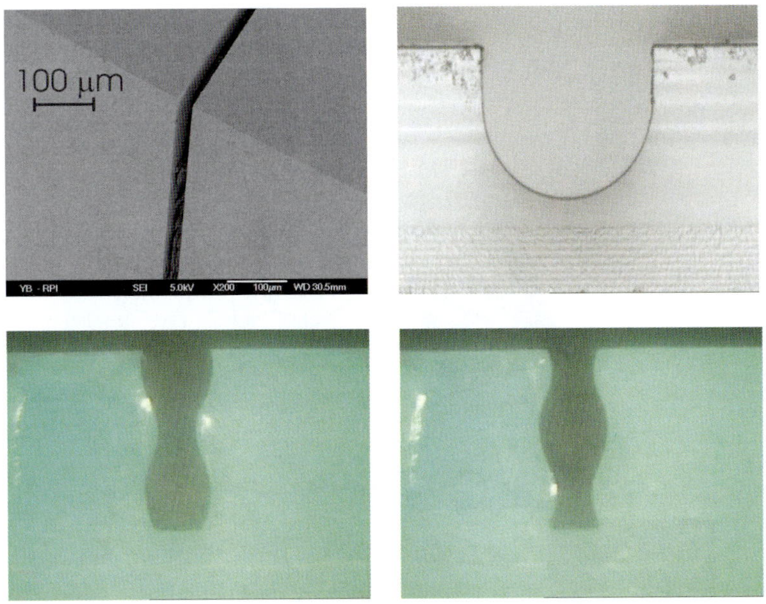

Figure 17.3

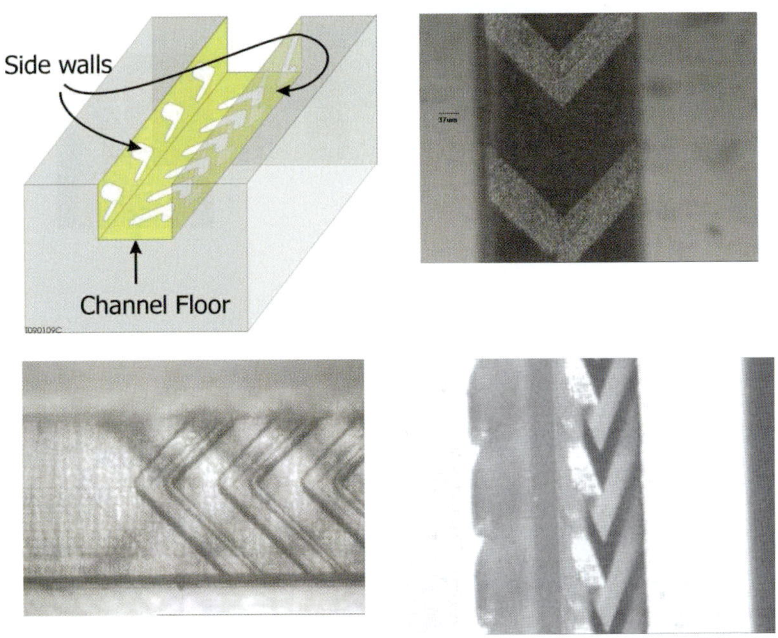

Figure 17.4

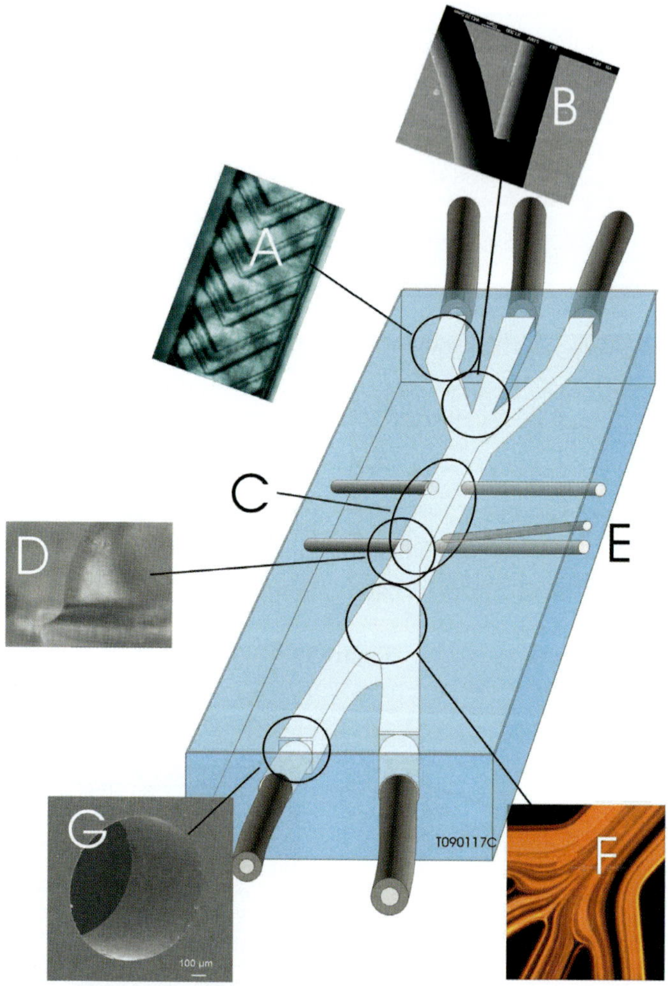

Figure 17.5

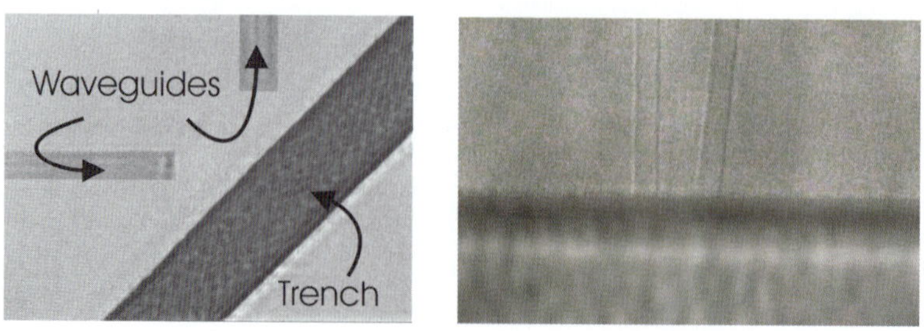

Figure 17.7

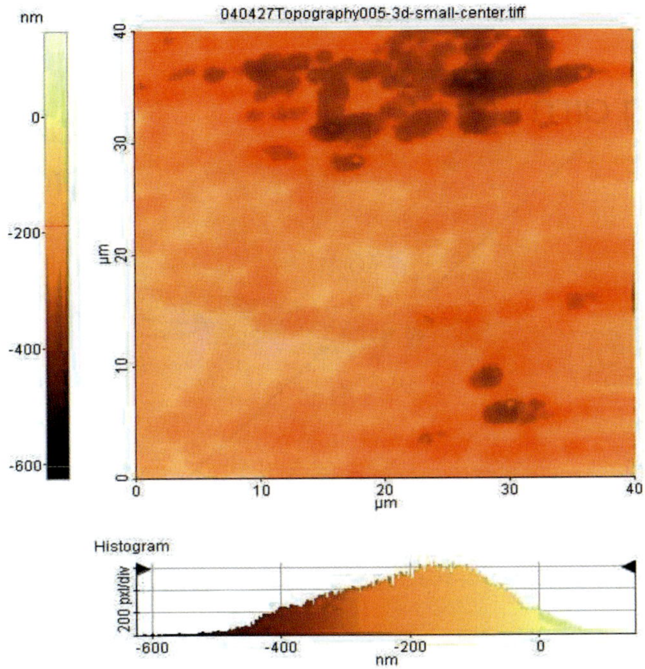

Statistics

Region	Min(nm)	Max(nm)	Mid(nm)	Mean(nm)	Range(nm)	RMS(nm)
Red	-624.124	146.883	-238.621	-195.233	771.006	126.902

Figure 17.8

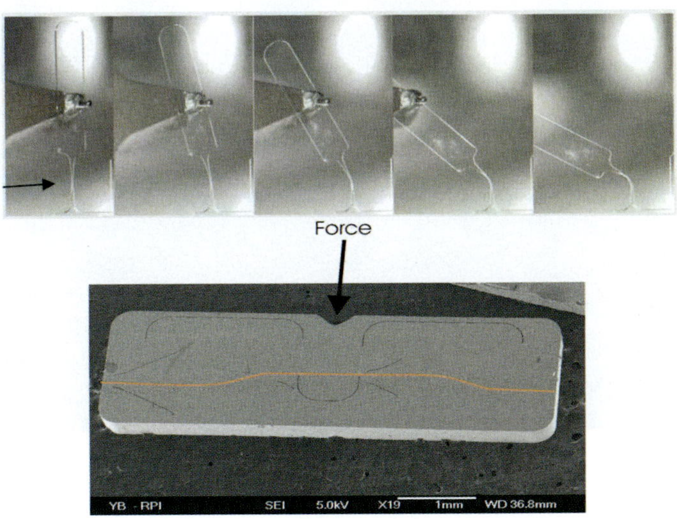

Figure 17.9

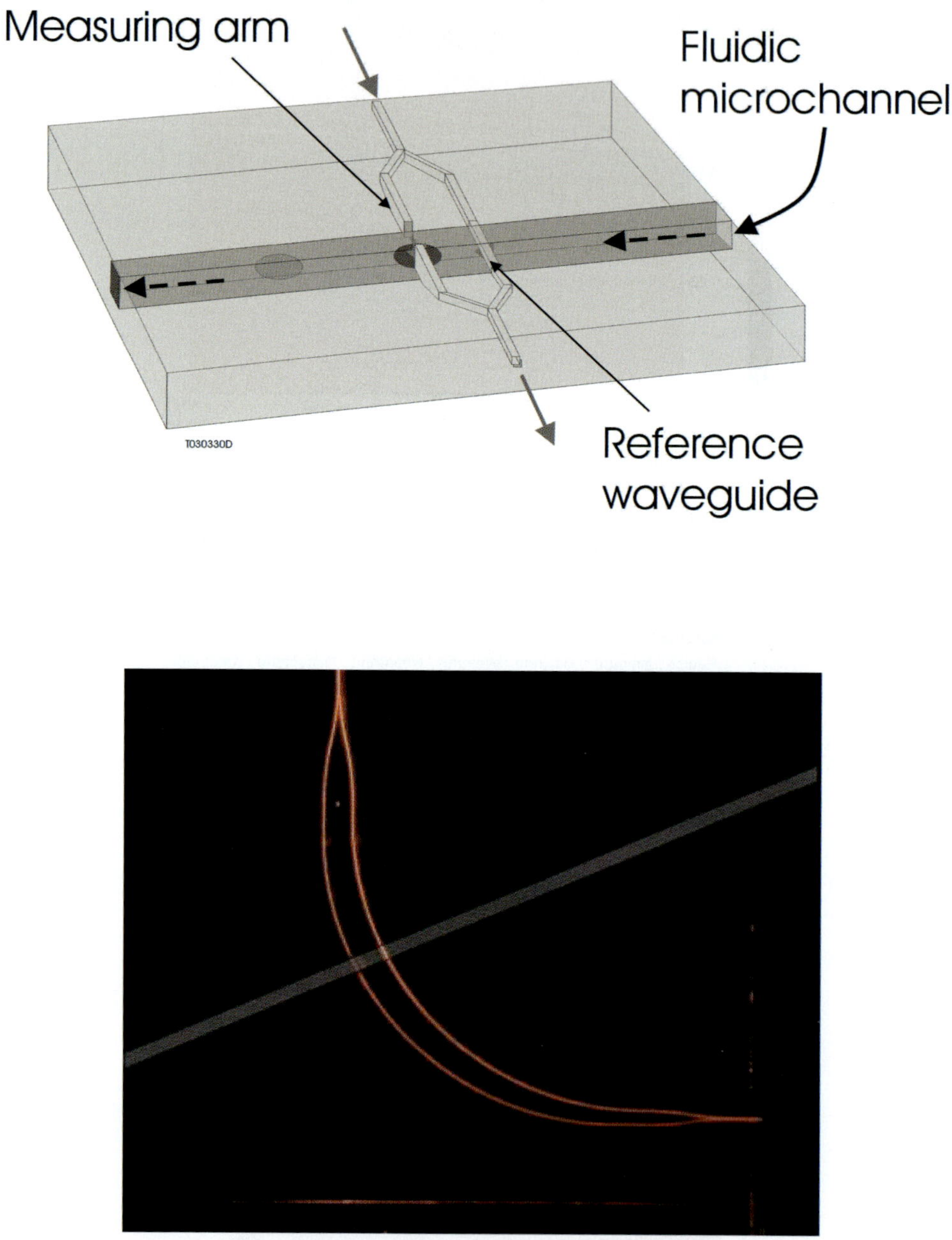
Measuring arm
Fluidic microchannel
Reference waveguide
T030330D

Figure 17.10

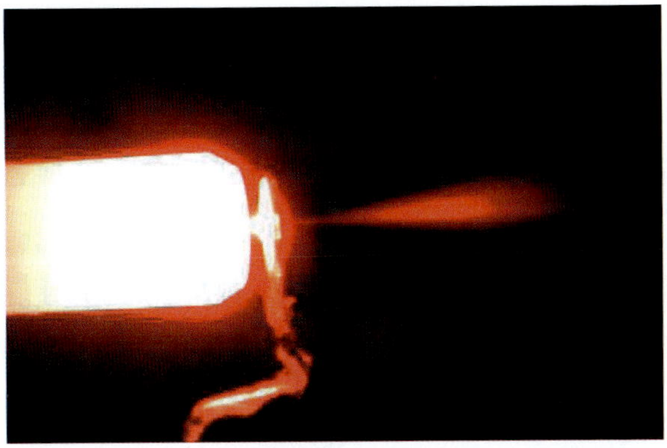

(a)

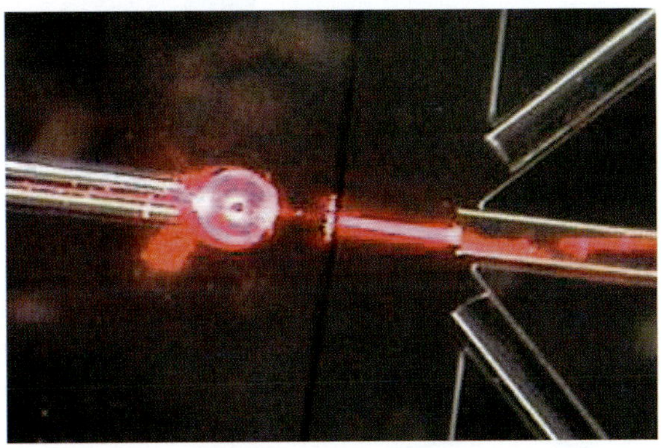

(b)

Figure 18.1

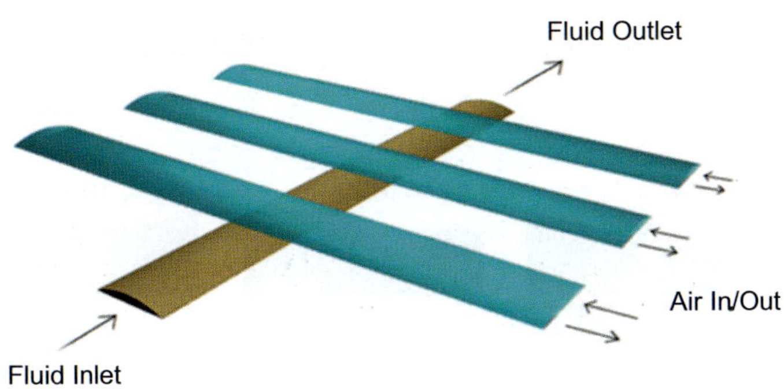

Figure 18.4

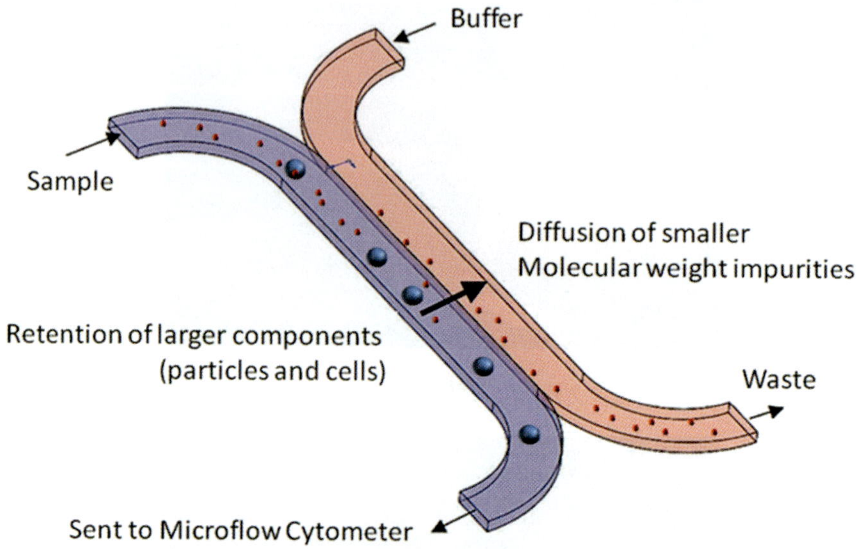

Figure 18.6

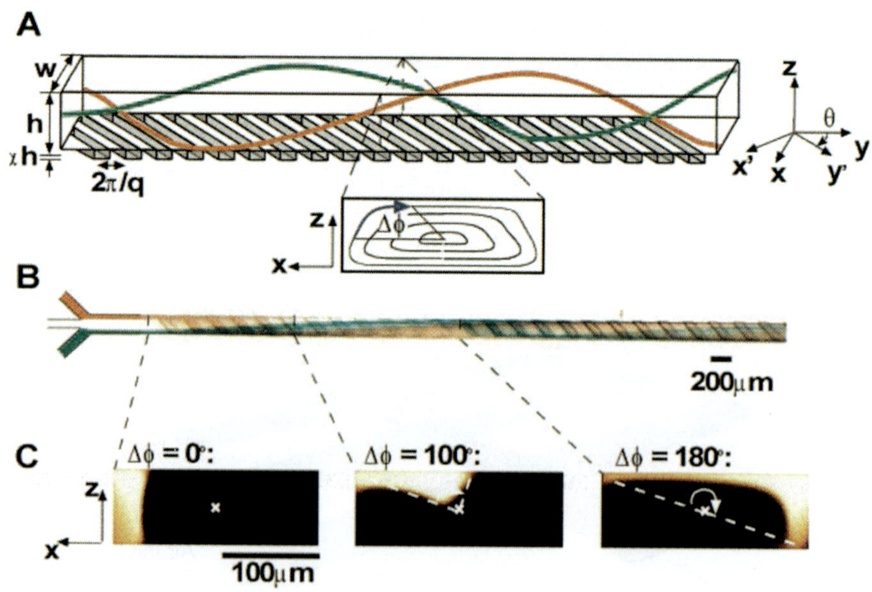

Figure 18.7

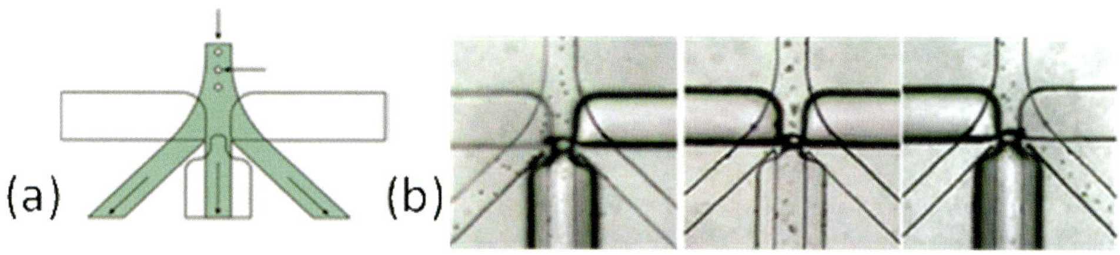

Figure 18.10

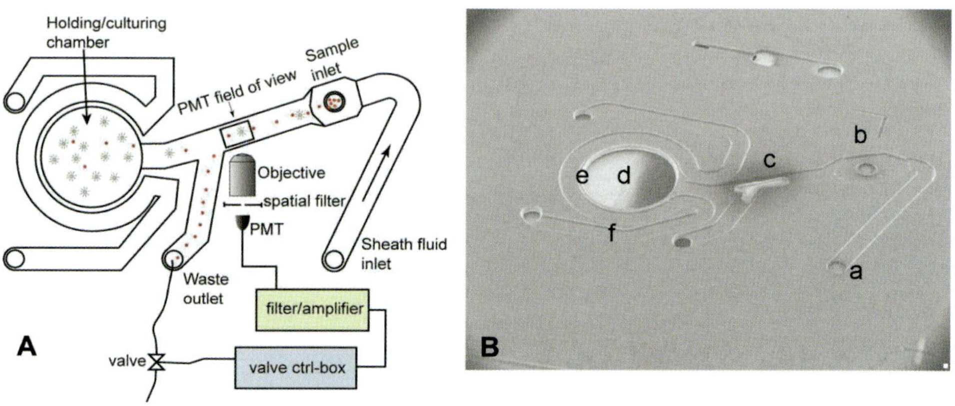

Figure 18.12

Index